D1070705

EVOLUTIONARY POLITICS

Glendon Schubert

Southern Illinois University Press / Carbondale and Edwardsville

Printed in the United States of America
Edited by Yvonne D. Mattson
Designed by David Ford
Production supervised by Natalia Nadraga
91 90 89 88 4 3 2 1

Library of Congress Cataloging-in-Publication Data

Schubert, Glendon A.
Evolutionary politics / Glendon Schubert.
p. cm.
Bibliography: p.
Includes index.
1. Political psychology. 2. Political sociology. I. Title.
JA74.5.S356 1988
320'.01'9—dc19 87-35305
ISNB 0-8093-1430-4 CIP
ISBN 0-8093-1431-2 pbk.

For my grandson,
Lee Glendon Kaminski,
in the hope that he will grow up
in peace

Contents

Figures and Tables ix
Acknowledgments xi
Introduction xiii

1. Biopolitical Science 3
2. Biopolitical Behavior 7

PART I POLITICAL ETHOLOGY 21

3. Human 25
4. Classical 32
5. Modern 41

PART II POLITICAL EVOLUTION 65

6. Hominid 70
7. Primate 83
8. Religious 103

PART III EVOLUTIONARY THEORY 123

9. Genetic 127
10. Cultural 156
11. Eugenic 169
12. Catastrophic 171

PART IV EVOLUTIONARY DEVELOPMENT 181

13. Periglacial 188
14. Epigenetic 199
15. Senescent 223

PART V POLITICAL THINKING 243

16. Animal 248
17. Human 258
18. Creative 289
19. Quantum 296
20. Life-Science Politics 324

References 333
Name Index 381
Subject Index 391

Figures

1. Academic ideologies 5

2. The social-science paradigm of political behavior 28

3. The biological paradigm of animal behavior 30

4. A life-science paradigm of political behavior 31

5. A hypothesis about the structural configuration of Findlay and Lumsden's table 1 291

Tables

1. Levels of Political Analysis, Disciplinary Sources of Core Political Concepts, and Related Political Concepts 8

2. Politically Relevant Core Concepts of Modern Biology 9

3. Eras of Decisional Periods 238

Acknowledgments

My greatest indebtedness, in regard to this book, is to a dozen of my professional colleagues for many research conferences, panel discussions, and dyadic conversations out of which came my ideas for pursuing further research on what resulted in the score of papers included here (as well as the other score *not* included here). Of these my longest and most enduring associations have been with Thomas Wiegele and Albert Somit, who organized and directed many of the conferences and institutional arrangements that made possible the organization of persons sharing a mutual interest in the interfaces of biology and politics: the Association for Politics and the Life Sciences, which now is an organized section of the American Political Science Association, as well as of the International Political Science Association.

Meredith Watts and Heiner Flohr each has (and on more than one occasion) involved me—and always to my intellectual benefit—in symposia that have turned me along paths that otherwise I might not have pursued, and for that I am grateful. More than a decade ago, Ernst Reese (the leading ethologist at the University of Hawaii at Manoa) facilitated my then budding interests in his subject in a variety of ways. Also at that time Gerard Baerends, Nobel Prize winner Niko Tinbergen's first doctoral student, graciously accepted me into the environment of his research laboratory at the University of Groningen, at a time when he was wrapping up a career of preeminent study of herring-gull behavior and communication; while I, notwithstanding the closeness in our ages, was a mere apprentice in classical ethology.

James C. Davies, Elliott White, and Steven Peterson all have pioneered as political scientists who have examined decision making from a neurobiological and/or a psychobiological point of view; and I have learned much and been stimulated to learn more from their writings as well as from my many discussions with them. During the most recent half dozen years I was truly surprised to discover—as I'm sure he was too—a substantial and increasing communality of intellectural interest and professional association alike with my younger son, James Neal Schubert, whose research in famine politics, aggression theory, and nonverbal communication in political group decision making increasingly drew him into the same vortex of biopolitical analysis in which I had already been engulfed for more than a decade. So now we are professional colleagues, too, presently occupied with planning collaborative research in our common subject.

Several academic institutions have provided direct support to the production of either the research on which this volume is based or of the book itself. The University of Hawaii at Manoa for more than a dozen years, and in particular during the eighties Dean Deane Neubauer of the College of Social Sciences, have provided fiscal support to enable my participation in what in aggregate have been dozens of research conferences and panels on life-science politics at professional association meetings throughout North America and Europe, and in Africa as well.

Southern Illinois University at Carbondale has, through its Department of Political Science, provided major assistance in the production, and through its University Press, the publication of the manuscript for this book. I thank John Foster, the department chairperson, and Dean John Jackson of the College of Liberal Arts for making the administrative arrangements; and Cathy Croquer (with the help of Aline Wilson) for undertaking the typing of several successive drafts of the manuscript. I also am grateful to Chong Phil Ra, my doctoral student at the University of Hawaii at Manoa, for his help in similar word processing of the extensive list of references.

I am indebted to the Netherlands Institute for Advanced Study in the Humanities and Social Sciences for having provided simultaneous fellowships during the academic year 1978–1979 to Albert Somit and myself, and for having sponsored the pioneering symposium "The Biology of Primate Sociopolitical Behavior," organized jointly by Somit and me and convened at the institute in Wassenaar in April 1979, and reported verbatim in Schubert and Somit (1982).

Introduction

This book is a result of my increasing involvement in the biopolitics movement for twenty years. It is based on a selection of approximately half of my forty-odd research papers written on many aspects of the biology of political behavior during the past dozen years and published originally in more than a dozen refereed periodicals and half a dozen symposia. Most of these were research articles appearing in either newsletters of interdisciplinary associations or specialized political science journals or in journals of behaviorally oriented disciplines other than political science. Only two of my biopolitical articles were published in mainstream (i.e., general) political science journals; and one of these is not an American journal. Each symposium focuses, of course, upon a specialized theme that is explored in depth by many contributors. So my biopolitical writing has appeared in diverse and widely scattered sources, very few of which are read by more than a very small minority of political scientists; and by an even smaller proportion of social scientists, or of biological or life scientists, or of other academic persons.

In addition to the symposia to which I have contributed, there has been barely a handful of other biopolitical symposia or research monographs. Elliott White and Joseph Losco (1986) have edited a symposium on biosocial organizational theory, and Roger Masters (Gruter and Masters, 1986) has co-edited one on theories of exclusion from human social groups. Works by Thomas L. Thorson (1970) and Ralph Pettmen (1981) are book-length essays on biopolitical philosophy; and Peter Corning's (1983) is an encyclopedic commentary on the evolutionary roots of organizational theory. Certainly there is yet nothing even remotely like a textbook on biopolitical behavior.

Given the unlikelihood that very many political, other social, or biological scientists will have read my original articles and chapters, plus the lack of any booklength work suitable for use as an introduction to the study of biopolitics, I have constructed the present book in an endeavor to make available a work that can serve as an introduction to biopolitics. With such aspirations, I have selected the papers included herein with a strong concern for their interrelationships, integration, and aggregate comprehensiveness in the development of a cohesive and consistent approach to the understanding of the biosocial and biopsychological substrates of political behavior. Such an approach must deal with change at the evolutionary, not merely historical, level. But the substance of that evolution has to be—at least, for humans—as much concerned with cultural as it is with genetic change; and hence with cultural even more than genetic influences on human behavior. That in turn requires that the method of analysis of political behavior be epigenetic and transactional, because of the dynamically reciprocal effects of specific and historical environments on any individual genome—and vice versa—in the determination of the boundaries within which behavioral choice roams psychobiologically throughout any human lifetime.

At a more earthy level, I should explain that in reproducing these earlier writings for republication, I have taken various liberties with them, some more and some less and no two in quite the same regard: (1) a uniform system of in-text citation has been substituted for all except a very few of the original footnotes, and those surviving ones are substantive rather than merely citational; (2) I have updated in some places by adding more recent references, but in others by inserting original material that was not included in the papers' initial publication; (3) a few chapters have been cut considerably in length; and (4) I have tried to avoid the republication of earlier typographical errors, although no doubt not without some unintended but also undetected production of new ones, for which I hasten to apologize in advance.

The stucture of the book is as follows. The opening two chapters are introductory, with the first discussing what is implied by a biological approach to the study of political science generally; and the second, to the study of political behavior in particular.

The next seventeen chapters are apportioned among the five major sections of the work. Each of these sections begins with an introductory essay to define its scope and put into context its subject chapters. The first of these focuses on an *ethological* approach to politics. Ethology is the naturalistic study of the behavior of all types of extant animals, including insects, birds, reptiles, fishes, mammals generally, other primates than humans, and—of course—humans too. (Because they both are long extinct, neither Pterodactyls nor Australopithecines can be studied by ethology; instead deductions and inferences about their behavior are made by paleobiologists and paleoanthropologists on the basis of skeletal and ecological evidence.) Many ethologists observe and report primarily on the behavior of animals, in many leading works such as the excellent text by Robert Hinde

(1970) and E. O. Wilson's (1975) magnificent monograph on the behavior of social animals. The three chapters that comprise this section discuss: (1) the ethology of human behavior; (2) the relationship between human ethology and that of other animals, and the transition from classical ethology (i.e., naturalistic observations in the field, primarily by European pioneers and their students) to the modern discipline (which combines laboratory and experimental research with field observations and is centered now at least as much in the United States as in Europe or elsewhere); and (3) how modern ethological theory and methods can and should be applied in the analysis of human political behavior.

The evolution of political behavior is the subject of Part II. Its major themes are: (1) our ambiguous heritage from the hominid species ancestral to *Homo sapiens sapiens* who, as symbolized by either Lucy or the preserved footsteps of the Laetoli family (Lewin, 1987: 167, 279), were probably much more carnivorous than are now contemporary simians, but genetically far more similar to modern humans than any true carnivores have ever been; (2) the extent to which comparisons between humans and other modern primates (such as chimpanzees, gorillas, and baboons) in their social group relationships can augment our understanding of how and when political organization and leadership evolved culturally (and see Goodall, 1986; Strum, 1987; Schubert and Masters, forthcoming; cf. Fossey, 1983)—the roots of political behavior, so to speak; and (3) beginning about twelve thousand years ago with the transition from gathering and hunting (as the universal human group adaptation), to pastoralism and agriculture via the processes of domesticating other animals to our needs and uses; and focusing then upon the close integration of political with religious leadership and functions when food surpluses made possible urbanism and complex (both vertically and horizontally) social organizational structures. Clearly politics as we understand and practice it today stems from this very recent archeological-to-historical period of no more than 125 centuries, and *not* from Lucy of 30,000 centuries or the Laetoli footprints of 37,500 centuries, ago.

The evolutionary theory of Part III constitutes the substantive as well as the structural epicenter of the book. It deals with the principal components of the twentieth-century neo-Darwinian modern synthesis: the integration of Medelian genetics with natural selection and its subsequent critique as the prime movers of biotic evolutionary change. That requires: (1) an exposition and analysis of the sociobiological paradigm, which has been widely touted as the cutting edge of ethological evolutionary theory since the mid-seventies; (2) an analysis of the consequent and repeated claims of E. O. Wilson (and his associates, and many social-science acolytes) that econometric-type models are sufficient (i.e., even in the absence of relevant empirical data) to justify the postulation of bean-bag genetics as the dominant paradigm for understanding human cultural evolution; (3) a concise and skeptical essay on the ultimate domestication of humans via biotechnical reengineering of the evolved species' genome; and (4) a review of the contemporary research literature of biotic extinction in relation to punctuated equilibrium and catastrophe theories of speciation in evolutionary change.

Evolutionary theory may be this book's epicenter; but its heart—its "true" center—reposes in Part IV, which discusses developmental change as growth spurts, plateaus, and decrements in the life spans of individual animals, from conception to death. The thrust of the emphasis in Part III is on how changes occur in the structure of systems of genes per se; the emphasis shifts in Part IV to the significance of genomic differences—of how, that is, a genome gets expressed in terms of animal behavior—which always involves the stimulation and constraints of a specific empirical environment (habitat), which includes of course conspecifics as well as other biota, to define the niche of the organism whose genome is the subject. This theme is itself developed in three major respects: (1) by examining how and why the exceptionally dynamic habitats associated with advancing or retreating continental ice sheets influenced human phylogenic development during the ice ages of the most recent hundred thousand years in particular; and why that type of environment is prerequisite to the optimal expression of the human species genome, in terms of both health and behavior; (2) by turning to the ontogeny of individual development, in terms of the epigenetic theories and professional career of Conrad Hal Waddington, an exceptionally articulate experimental biologist who created the most influential model of the transactional relationship between habitat and genome in the growth and behavior of individual organisms; and (3) by focusing on decremental growth, the senescence that begins (differentially, for at least some brain cells, as early as birth) in general at the achievement of sexual maturity and accelerates with the loss of reproductive competence. (At least this is true among humans, who are one of the few species of animals whose longevity, since the transition from gathering/hunting as the species adaptation, permits the survival of postreproductive individuals.) The chapter on senescence reports empirical research designed to test competing hypotheses: the conventional cultural one, which asserts that humans become more politically conservative with age only if they were socialized during a relatively conservative sociocultural era; and a biological hypothesis, which associates ideological conservatism with increasing physiological (including cognitive) incompetence.

Part V deals with thinking, and ultimately with the relationship between the characteristics of the basic mammalian and the general primate brain, and the social and cultural parameters characteristic of political decision making. The four chapters explore these relationships in terms of: (1) a critique of several of the best-known laboratory research projects designed to investigate language use and decision making by chimpanzees; (2) an extensive review of research in neurobiology and the structure of the human brain, in relationship to thinking, consciousness, and political behavior; (3) an examination of creativity as a characteristic of human thinking; and (4) an examination of political theory from the interacting perspectives of modern biological theory and twentieth-century physics theory, with an emphasis on the role of quantum theory as a component of psychobiological theory in the scientific explication of human thinking.

The book concludes with a chapter that makes no attempt to summarize the preceding contents, but that does attempt to point out some of the most important ways in which the use of evolutionary theory, ethological methods, and an epigenetic stance toward human development would revolutionize the study of politics by forcing its attention upon how change is the law of life, even if institutional stability remains the law of the land.

Evolutionary Politics

1
Biopolitical Science

The biological revolution came much sooner and more easily to psychology and anthropology than to sociology and political science, and for very basic reasons relating to the fundamental structures of these academic disciplines. Comparative and developmental psychology have been involved in natural science research procedures and in anatomical-physiological work for over half a century; anthropology also divides into sets of dissimilar subdisciplines, with a physical side integrated with the natural science disciplines of geology, botany, and zoology (inter alia), and indeed is itself considered to be a branch of biological science. There never had been such a natural science component in either sociology or political science; and there was none yet in 1975. Such sociological subfields as demography and human ecology and aging take a social-science approach; the field structure of political science is discussed in chapter 2, below. It is hardly surprising, therefore, to find even a dozen years ago a variety of basic textbooks in psychology (e.g., Eysenck, 1967; Kimble, 1973) or of monographs in anthropology (e.g., Callan, 1970, attempts to integrate ethology and social anthropology; Chapple, 1970, integrates cultural anthropology with behavioral biology; see also Greenwood and Stini, 1977); but neither is it surprising to find biologically oriented works such

---

This chapter is a revised version of "Politics as a Life Science: How and Why the Impact of Modern Biology Will Revolutionize the Study of Political Behavior," ch. 6 in Albert Somit (ed.), *Biology and Politics: Recent Explorations* (The Hague: Mouton, 1976), pp. 155–57, 163–65. Reprinted with permission of the publisher.

as these just beginning to appear in sociology,[1] as well as in economics (Boulding, 1972; Tulloch, 1971) and political science (see also Wiegele, 1982).[2]

Keith Caldwell (1964; but see also Roberts, 1938) heralded the advent of what has now become a new approach to the study of political science. It was Caldwell's view that "biopolitics" suggested "political efforts to reconcile biological facts and popular values . . . in the formulation of public policies," a synthesis made both timely and urgent by the "population explosion" and the "concurrent explosion of biological knowledge, an accelerating geometrical expansion of knowledge, the culmination of long years of accumulating inquiry in the various bio-sciences." It is, he said, "the contemporary convergence of these two explosions—of people and biology—that justifies, indeed necessitates, a focus on biopolitics." Caldwell went on to discuss primarily issues of ecology and of human physiology, in relation to governmental science policy and the difficulties of translating biological knowledge (or the lack thereof) into political action (see Caldwell, 1966). Biopolitical themes explored by other political scientists during the sixties covered a wide range of questions, including ethological theory (Pranger, 1967; Somit, 1968; and Adrian, 1969), comparative mortality in relation to political-psychogenic causes (Haas, 1969; Rogers and Messinger, 1967), the effects of malnutrition on political participation (Stauffer, 1969), neurological and endocrinological substrates of political behavior (Davies, 1969), and psychopharmacology (Somit, 1968). Already by the summer of 1970, when a score of papers were presented at panels in Munich organized by Albert Somit for the Eighth World Congress of the International Political Science Association, it was apparent that so many different political scientists in diverse countries were becoming involved in biopolitical research that a new dimension was being added to the scope (and competence) of the discipline (see Somit, 1972, for a review of biopolitical research up to that time).

The biopolitics movement was organized in 1980 as the Association for Politics and the Life Sciences (Corning, Losco, and Wiegele, 1981: 590), began to publish its own journal in 1982, and in 1986 became an official section of the American Political Science Association.

Traditional political science certainly continues to be an integral part of the publication, the teaching, and the thinking of the profession; and it remains the only emphasis in the discipline that journalists understand, and hence to which they pay any attention. By *traditional* I mean those political scientists who look

[1]For further discussion of some of the reasons why sociology has been slow in developing a subfield of social biology, see Means (1967; particularly his analysis of over a hundred sociology texts, turning up negative evidence of interest in biology, p. 202). (The journal for this subfield is, of course, a recently rechristened *Eugenics Quarterly*, to which sociologists occasionally contribute but in which they play a minor role.) However, by 1975 there was at least one introductory text in sociology that explicitly embraced an approach adapted from human evolutionary theory in biology (Lenski, 1970); and Lenski (1972) believes that "the rapprochement between the biological and social sciences is long overdue. Human societies *are* part of the biotic world, and by denying or minimizing this fact we impoverish both theory and research" (cf. also Ball, 1973).

[2]Thorson (1970) is not about biopolitical behavior except in his critique of Easton's political systemizing, which he castigates as an organismic theory that goes to great pains to include no organisms in its political system. Thorson dichotomizes biology and science, between which he claims a choice is necessary (p. 96); cf. Thorbeck (1965), which also is metaphysically inspired by Teilhard de Chardin.

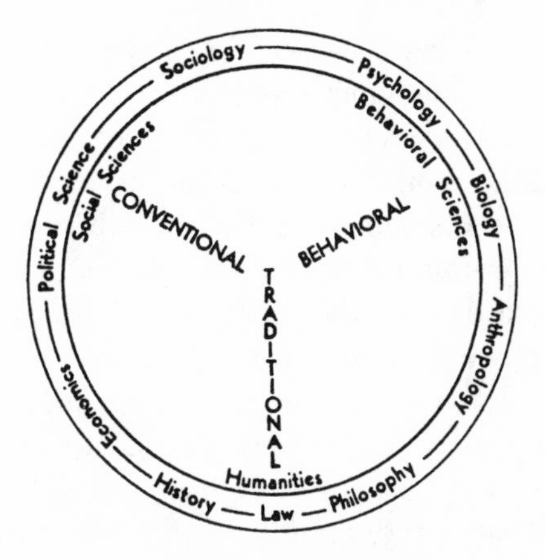

Figure 1 Academic ideologies.

to law, history, and philosophy for their models and inspiration, as indicated graphically by figure 1. The original figure from which figure 1 is adapted related specifically to a particular subfield of political science; but its implications are perfectly general, as I thought at the time when I first proposed it more than two decades ago (Schubert, 1967a: 7, and 1967b: 108). In this model the professional discipline of biology marks the core of political behavioralism, and biology is flanked by anthropology on the one side and psychology on the other; sociology, political science, and economics are positioned at successively greater distances from the biological core of behavioralism.

By now, however, it is possible for anyone teaching political behavior to include biopolitics as part of her or his course: the revised edition of Manheim's (1982) undergraduate text on political psychology includes an entire chapter on "The Body Politic: Physiological Bases of Political Behavior"; and Gianos' (1982) undergraduate text on political behavior concludes a chapter on "Politics as Biology" with a discussion of neurophysiology. Within the past decade no less than three directly relevant professional journals have been established—*The Behavioral and Brain Sciences*; the *Journal of Social and Biological Structures*; and *Politics and the Life Sciences*—and since the beginnings of all of these journals, political scientists have been and remain importantly involved in their editing, writing, and development. Furthermore, during the same period, the "Biology and Social Life" section of *Social Science Information*, the journal of the International Social Science Council, has been edited by political scientist Roger D. Masters. Books on biopolitics published since 1980 include: Blank (1981, 1984), Laponce (1981), Pettman (1981), Watts (1981, 1983), White (1981), Wiegele (1982), Schubert and Somit (1982), Corning (1983), Wiegele et al. (1985), and White and Losco (1986). Bibliographies of biopolitics are published annually in *Politics and the Life Sciences*; and for a specialized bibliography on psychobiological politics, see the appendix to Schubert (1983a).[3]

[3]All references to Schubert denote Glendon Schubert, unless specific reference is made to J. Schubert.

The implications of the biological theory, for the perpetuation of much of traditional political science, are much more revolutionary than those of political behavioralism as it has been understood heretofore. Our Scopes trial as a profession has yet to play out its drama, as the merest glance at where we begin our study of politics makes evident (see ch. 7, below). Biological theory implies the rejection of the presumption that our political theory as a species began 2500 years ago in Athens, or, alternatively, as described in "naturalistic" fables (whether optimistic like that of Rousseau or pessimistic like that of Hobbes), or according to the authoritative allocation of values in the even more popular fable of Genesis. The roots of political behavior (and cf. Hofer, 1981) go back not thousands but millions of years (Fried, 1967; Tiger and Fox, 1971; Pilbeam, 1972; Tiger, 1979b; and chs. 5 and 6, below); and political man did not spring (garbed in full civic regalia, and uttering a partisan war cry) from the forehead of Socrates—as our teaching of the wellsprings of political philosophy might lead innocents to infer. The implications of contemporary research in physical anthropology, archeology, paleontology, and related sciences are going to jack political philosophy off its classical assumptions—once political scientists become better educated in, and start facing up to the facts of, biological life, including their own life history as a species.

It seems altogether likely that the apparent things (e.g., Storing, 1962; Berns, 1963) about political behavior that could be learned through cognizance by one human of another already have been learned during the course of the past half million years. It is also entirely possible that the Greeks of Periclean Athens (and for a century thereafter) had a particularly felicitous way of stating such cognitions, which can provide us with at least a benchmark against which to measure what we may have learned since then as a species. But we would not wish to rely upon either Hippocrates' or Aristotle's knowledge of human anatomy or medicine. It is at the level of folk wisdom (disregarding questions of general knowledge base, cultural dissonance, language, etc.) that Socrates excelled; and one can appreciate the attractive simplicity of confining political analysis to this level of understanding: it is indeed basic. But there are other things about human behavior that could not have become observed until the development of appropriate technologies: photography in general is one example, and high-speed photography of human facial changes is a more particular example (Tomkins, 1962: 206; Leventhal and Sharp, 1965; Hass, 1972; Ekman and Friesen, 1975). Behavioral science has not hesitated to take advantage of what additionally could be learned about human behavior by adopting technologies that enhance the capacity to observe organisms in action. What remains to be learned will be primarily the previously *unobservable*, and this will consist mostly of events that take place *within* the human body or of human behavior in the context of environments that previously were inaccessible to observation (like the long political past that antedates the invention of writing, or that parallels the absence of documentary evidence about it).

2
Biopolitical Behavior

POLITICAL BEHAVIOR

If biology is to make a greater contribution to political science, then it becomes necessary to be able to specify what political science is about. As I have argued elsewhere (Schubert, 1976)—and as even superficial examination of the relevant evidence makes clear—there is no simple and easy answer to the latter question (but for a sophisticated one, see Laponce, 1974). Clearly it is no longer the study of government alone; nor has it for a long time been merely a branch of the humanities. Defining it as the study of politics begs the question, unless and until we are better able to break that tautology of circular definition. It is true that many (and probably most) departments of political science continue to display topical structures that exhibit little change from a now remote past when the functional specializations of the persons teaching the courses did correspond to political philosophy, international relations, comparative government, political parties, public administration, public law, state and local government, and methodology. But research categories are a far better guide than teaching rubrics to the locus of the cutting edge of inquiry for the production of new disciplinary knowledge; and political scientists no longer write books or articles to elucidate the classic structure of the profession, however much the latter may continue to

---

This chapter is a revision of "Biopolitical Behavioral Theory," *Political Science Reviewer*, 5: 406–10, 412–25 (1975), © Intercollegiate Studies Institute, Inc. 1975, reprinted with permission of the publisher; and "Politics as a Life Science: How and Why the Impact of Modern Biology Will Revolutionize the Study of Political Behavior," in Albert Somit (ed.), *Biology and Politics: Recent Explanations* (The Hague: Mouton, 1976), pp. 168–71, reprinted with permission of the publisher.

Table 1 Levels of Political Analysis, Disciplinary Sources of Core Political Concepts, and Related Political Concepts

Level	*Source*	*Core Concepts*	*Related Concepts*
Macro	Mathematics/ Biometrics	Methodology	Empirical or Systematic Theory
	Biology	Systems/Functions Structures	Change Policy making Stratification
	Physics/ Electrical Engineering	Communications/ Control	Leadership Manipulation/Persuasion Symbolization
	Anthropology	Culture	Democratization Integration/Fragmentation Transmission
Micro	Sociology	Socialization	Reinforcement Desocialization
		Role	Recruitment Elite/Mass
		Groups	Nuclear/Institutional/ Situational
	Psychology	Personality	Needs Motivation Physiology/Perception Cognition
		Attitudes	Ideology Alienation
	Economics	Conflict/Decision making	Participation (Activity/ Inactivity) Aggression/Violence

dominate the manner in which departments still advertise themselves to (necessarily) naïve outlanders.

Table 1 reports ten core concepts in terms of which political analysis tends to be organized. These core concepts include: methodology; systems, structures, and functions; communications and control; culture; socialization; role; groups; personality; attitudes; and conflict and decision making. My specification of concepts related to the ten core concepts certainly is intended to be exemplary rather than exhaustive; and it should be noted, moreover, that many of these other concepts relate to more than a single core concept (e.g., ideology relates to culture and socialization as well as to attitudes; and leadership relates to personality and to groups as well as to communications and control). The first four core concepts are classified as macro because they tend to direct attention to aggregated, collective correlationships at an institutional level of social organization; the remaining six are, relatively, micro because their focus is more upon relationships within or among individuals. Evidently these core concepts at the micro level are derived from the cognate disciplines with which political science has been most closely

Table 2 Politically Relevant Core Concepts of Modern Biology

Biopolitical Concepts	*Biological Concepts*
Evolution [Human Nature]	Biological Structures—From Molecules to Man (only the last two sections on "Anatomy and Evolutionary Biology of Primates" and "Applied Human Anatomy," of this chapter)
	Heredity and Evolution
Ethology [Behavior]	The Nervous System
	The Biology of Behavior
Ecology [Environment]	What Is Ecology?
	On Feeding Mankind
	Renewable Resources (management of the environment to control growth of human and other forms of life)
	Biology and Industrial Technology
	Environmental Health
Equality [Welfare]	Science and Medical Practice

allied—sociology, psychology, and economics—at least during the third quarter of this century. Of the four macrolevel core concepts, culture has been the outstanding intellectual artifact of the anthropologists; the cybernetic language, theory, and model underlying communications and control stem from mathematics, physics, and electrical engineering, while systems, functions, and structures, like methodology, come directly and primarily from biology. The systems model has been the overarching paradigm of biological science throughout this century; and although work in probability theory of classical mathematics was an earlier forebear, it was work in biometrics before the First World War that contributed directly to the developing fields of social, psychological, and educational statistics; therefore, biology was also the cognate disciplinary source for the methodological approach of contemporary political behavioralism. So it is not really very radical to argue for an extension of biological influence upon political inquiry as we now practice it: *both* our meta-theory (Miller, 1971; Easton, 1973) and our methodology already are primarily biological in their orientation; and what I am proposing is that our substantive theoretical concepts should become better articulated, deliberately and consistently, with *their* biological wellsprings.

MODERN BIOLOGY

Let us follow the lead of a consortium of eminent biologists charged with the task of surveying the life sciences' impact on public policy. Their report (Handler, 1970) fills a book of twenty chapters and almost a thousand pages, more than half of which discusses subjects of direct and immediate implication for the study of politics; these chapter titles are, as regrouped by me, the basis for table 2.

The first group of chapters above on human evolution specifies a subject that is essential to an appreciation of both the present physiological constraints upon, and the future possibilities for, political behavior; evolution explains how and why humans have become what they are. Thus evolution is the behavioral history of our species over a time span of four million years—which provides a broader perspective from which to appraise both present events and future probabilities than do the more ephemeral population groupings and epochs that necessarily are deemed to be of interest by historians of our written culture. Evolution deals also with the biological foundations of language, upon which depend virtually all of those aspects of political behavior which—however incomplete in scope—presently constitute the subject matter of political science (see Masters, 1970). Behavioral genetics bears directly upon the composition and possibilities for future growth and change in both human populations and other forms of life as well. It is therefore involved in all of the other major biopolitical concepts to which we now turn: ethology, ecology, and equality.

Ethology investigates how and why animals behave as they do in relation to their physical environment, other life forms, and each other. It does so by putting the subject of human political behavior within the context of comparative animal behavior; and as Roger Masters has demonstrated in a series of important papers (1973a, 1973b, 1974, 1976), it is not necessary to settle for the popular versions of the "naked ape" syndrome (Morris, 1969; and cf. Lorenz, 1966; Ardrey, 1970; and Pilbeam, 1972) that have titillated casual readers of what is displayed at supermarket paperback stands. Among the principal topics of animal behavior, within which human behavior is included as a particular type with its own special mechanisms, are learning, orientation, perception, exploration, locomotion, circadian rhythms (cf., for example, Marler and Hamilton, 1966, ch. 2; and Chapple, 1970, chs. 2 and 3), hormonal and neural systems and behavioral integration, motivation and appetitive behavior, development, reproduction, communication, conflict, and social behavior. The comparative and interdisciplinary sciences of animal behavior raise and investigate questions in regard to each of these concepts that do not and cannot arise for students of human behavior only, because of the closure imposed by presuming that the particular, species-specific evolutionary solutions with which humanists happen to be familiar are the only possible or most efficient ones; a comparative approach makes possible the investigation of relations that do not exist for social scientists. As Robert Hinde (1970: vii) has remarked, "the obsolescence of the old distinctions between the biological sciences is nowhere more evident than in the study of animal behavior. The problems it raises not only provide a meeting ground for psychologists, zoologists, physiologists, anatomists, geneticists, ecologists and many others—they demand their cooperation." Those many others must surely include students of social biology and the biology of politics, because the work in animal behavior bears directly upon most of the core concepts of political behavior, including communications and control, socialization, role, groups, personality, attitudes, and conflict and decision making.

The study of ecology shifts attention away from how and why humans and other animals behave and focuses instead upon the ways in which their actions affect both the living and the physical environment to which they are (or had become) adapted, interactively with the reciprocal positive feedback effects of those environments, in the form of the great unfree bads, upon the subject species. (Indeed, the major policy problems confronting the world today are precisely a consequence of imbalance among biological demands, cultural demands, and limitations in the responsive capacities of the natural environment; see Brown, 1974; Masters, 1978a; and Green, 1986.)

The five chapters that define ecology preempt over a fourth of Handler's book, discussing the interrelationships among natural ecosystems, the nutritional needs of the species, and the prospects for attempting to meet these by changes in environmental management. They deal also with the (again, interactive) effects of industrialization, population growth, and urbanization upon health and survival—not "systemic" survival (to invoke David Easton's wonderfully abstract and vacuous sense of the term) but rather *species* survival in the Darwinian sense. The latter clearly is an appropriately important subject for any science of political behavior to concern itself with; and one might observe that it escalates our political span of attention, beyond nuclear annihilation, to include a variety of alternative if equally deadly global scenarios. Even more familiar and less prepossessing topics, such as mass starvation and overpopulation—long recognized as subjects appropriate for political discussion and policy making—will be better informed if they are placed in the broader and complex context that the biological concept of ecosystemics requires.

Equality deals with the transactive limits between biological factors tending to favor variability and cultural norms favoring uniformity (White, 1975) in regard to several major facets of human morphology; I list these under their social science attributions (with corresponding animal behavior analogues in parentheses: health (survival), age (development), sex (reproduction), race (speciation), and intelligence (adaptation).

Handler's chapter on the application of biological knowledge through the practice of medicine discusses two interrelated attributes of human individuals and populations: health and maturation. Postindustrial societies support policies and practices which have attempted—though with differing success—to counteract by equalization the biological inequalities arising from age differences (at various stages of the life cycle) and health differences (which are biological at least to the extent that these stem from genetic influences). Evidently, public policy with regard to public health (and "private" health as well) and public welfare, at least as concerns children and the elderly, are among the major problems of politics in any society, but especially so for industrialized societies today (Fries and Crapo,

1981). Of no lesser significance are the culturally inspired egalitarian drives to overcome any biological differences in regard to sex, race, and intelligence—or at least so it must seem to observers of recent American and West European politics. Behavioral genetics and animal behavior are the fields of biology that impinge most directly upon the politics of sex, race, and intelligence.

These latter three parameters of biological differentiation arise among all sexually reproducing species of animals: the sex differentiations reflect the most popular (in recent eons) evolutionary adaptation to the problem of reproduction; the race differentiations reflect relatively much more minor adaptations, of particular but geographically isolated subpopulations to the differential impact of environments in different ecosystems; and differences in the intelligence of individual animals reflect unavoidable consequences of a breeding system that operates to sustain a heterogeneous gene pool. A homogeneous one proffers a poor survival strategy for most species, however much we may admire sharks and plankton for their superior adaptations to *their* chosen environments. It is true that, for those who may disagree with the preceding assertion, biological engineering offers, through cloning, a possible (and error-free) equalitarian solution for human problems of sexual chauvinism, racial discrimination, and stupidity (see ch. 14, below). But even cloning implies the direct invocation of biological knowledge and skills; so whichever way we go we shall need to come to terms with our biology if we are to act at all effectively politically.

THE BIOLOGY OF POLITICAL BEHAVIOR

A major limitation of political behavioral research is that it is premised on a paradigm of human behavior that is only skin deep, because the observations never attempt to get inside the soma except inferentially. Even what political scientists know of cognition is based upon externally observed, or at least auditioned, behavior. But what is always and everywhere most important to all human beings is what is going on *inside* of themselves at any particular moment—and we must specify "moment," because change is often so rapid. But what social science political behavioralism begins and rests its case with is what can be perceived to have occurred in some monitorable and measurable way *outside* of persons. Our task is to get beyond interactions and into transactions; and the only way to do that is by invoking theories, methods, and skills that permit us to include biological variables in our Riemannian geometries. Environment must be seen as a set of variables, not as the sole primal cause of all behavior; the history of politics should begin before our species can be differentiated as such; and our political behavior must contribute to, and become reconciled with, a theory of animal behavior that includes life forms whose morphological appearance is strikingly unanthropomorphic (see, e.g., Campbell, 1972).

The most extensive and probably the best work on political equalization of the biologically inequitable has been done on various aspects of health in relationship to political behavior. Harold Lasswell has explicitly included health as a variable in his set of key political values for a long time; and considerable work in biopolitics strongly supports his position. A series of studies by David Schwartz and his students explored the effect of health differences upon political attitudes and participation, of postural and speech differences as indices of states of arousal, and of the remarkably close correlation between the way people feel about their own bodies and their feelings about the body politic (Schwartz, 1970, 1973; Schwartz and Zill, 1971; Shubs, 1973). My colleague Stauffer (1969) has studied the effect of malnutrition upon political participation and attitudes, while another colleague (Haas, 1969) has analyzed the relationship between disease and foreign and domestic violence as causes of death in both industrialized and other countries. Wiegele (1971, 1973) and his associates (1973) have investigated how health differences might affect attitudes toward foreign-policy decision making; and indeed, a special note of urgency is lent by a serendipitous byproduct of their research. Notwithstanding the prodigious efforts of modern medicine to keep people alive, there is less evident success in keeping them healthy (to say nothing of *fit*). The proportion of American youth of college age—which from a biological point of view is the life-cycle time for optimal health and fitness—who are *un*fit, is alarmingly high and increasing; and none of them is likely to get remarkably healthier as he or she grows older. Almost a third are clinically obese; and it is notable that the experimental findings of Wiegele's paper were largely vitiated because of the constancy (in the event) of what had been hypothesized to be the independent variable: all fifty of the white, middle-class, minimally twenty-one-year-old student subjects rated themselves as being in good to excellent health; but when placed on a treadmill, almost half performed *very* poorly, another fourth poorly, less than a fifth were rated fair, and only the remaining 12 percent were placed in one of the upper four categories (ranging from average, through good and very good, to excellent). If this is a typical sample of the American citizenry who is going to be running (if that is the appropriate word) the country during the rest of this century, then there surely is a crying need for the higher priority in relating health to politics that some writers (e.g., Chase, 1971) have demanded.

A related field in which several political scientists contributed to what seemed, about two decades ago, a promising beginning is that of political psychopharmacology (Somit, 1968; Davies, 1969; and Stauffer, 1971). But there has been no follow-up, probably due more to the tightening up of academic policies and procedures relating to experimentation with human subjects than to the attenuation in political salience of the subject itself (and certainly not of the public health and personal security problems to which it appertains). The one experimental study that was attempted, which ran into the kinds of technical difficulties typically encountered by prototypes, is suggestive of the kinds of drug effects upon political behavior likely to be found in a variety of different arenas and situations (Jaros,

1972). There are major correlationships not only between drugs and personality, but evidently also (lest we forget!) between drugs and health—both pro and con; and between drugs and nutrition (because from the point of view of appetitive behavior, drugs are food with special effects—but then, all foods have their particular effects as ingestants for biological processing systems).

The political scientist who undertook the experimental study, Dean Jaros, apparently also introduced the concept of *desocialization* into our professional vocabulary. After surveying the literature on political socialization and remarking on its presumption that the process relates primarily to learning among children, Jaros observed (1972: 6) that the predictable psychological effect of biochemical changes at the physiological level will be loss of (or change in) learning, and therefore at least temporary *de*socialization. But the concept involves much more than merely drug effects; it implies the loss or modification of political cultural learning for whatever reasons, and two other variables that can be expected to have similar effects are (negative) health and aging (see ch. 15, below). Illness (including physiological loss and impairment) minimally results in some degree of political inactivity and alienation (Schwartz, 1973); typically it will also have other desocialization effects, some temporary and some (e.g., paraplegia) permanent. Aging, however, is a process whose genetic effects (once maturity has been achieved), subject of course to possible epigenetic effects in the opposite direction (e.g., those of aerobic exercise on human cardiovascular development, or stopping the smoking of tobacco), are strictly unidirectional. "Time's arrow" points only and compulsively in the direction of permanent political desocialization, and the relevant questions all relate to differentials in rates of decay in regard to different physiological structures and processes, and therefore also to political cognition and behavior. Only God—or if purists insist, DNA—can make a synapse; and even She does not repair them. The loss of memory, which begins much earlier but troubles most people by initial middle age, is only one of the most literal and obvious behavioral indices to the universally deleterious impact of physiological aging upon political socialization.

In a political society whose population "pyramid" rapidly is beginning to look like a parallelogram, we might expect to find greater concern for the policy problems, posed by both qualitative and quantitative changes in population growth (Clinton, Flash, and Godwin, 1972; Clinton and Godwin, 1972; Clinton, 1973; Fries and Crapo, 1981), reflected in political science research into the politics of the later period of the life cycle (J. Schubert, Wiegele, and Hines, 1985, 1986). There is an excellent beginning in work at the level of social science analysis (Cutler, 1973; Cutler and Bengston, 1974), but the major causative variables here clearly are operating at the biological level. We can anticipate a great deal more success in our attempts to develop solutions at the level of public policy, if the biogenic causes of social-psychological-economic-political desocialization effects of aging are better understood and consequently can be shifted out of the category of error variance where they now repose and into the same interaction matrix with our behavioral (intervening and dependent) variables.

Whatever our difficulties, as a profession, in coming to grips with health and age as *biologically* (and not merely socially) important influences upon political behavior, at least we are not inhibited by ideological commitments that preclude us from examining the relevant questions. It is by no means clear that we are equally open-minded in regard to the political significance of biological differences in regard to race, intelligence, and sex. One can find strong support in behavioral genetics for an acceleration of present trends in Western postindustrial societies in the direction of greater racial heterogeneity in breeding populations of humans. The general argument is that a more heterogeneous gene pool is more likely to produce the necessary critical number of individuals adaptable in whatever may be the appropriate ways in an unpredictable future, the better to assure species survival (see chs. 9 and 13, below). Given the apparently high level of racial autarchy that will predominate in the international (to say nothing of domestic) politics of the next generation, it might seem unfortunate if we fail to take advantage of biological knowledge that bears upon the problem. It is well established, for instance, that there are important relationships between race and health (Allison, 1971); and the failure to include, in the policy process, available biological knowledge can only produce results that frequently boomerang (Frankel, 1974).

An exceptionally literate study by Elliott White (1971) argues that effective political leaders must be, but must avoid seeming to be, of exceptional intelligence. Psychologist Herrnstein (1973; and cf. Young, 1961) advanced a more general statement of the thesis, namely, that social, economic, and political leaders (inter alia) constitute a natural meritocracy whose superior intelligence explains their status (cf. Deutsch and Edsall, 1972). Attributing intelligence variation to genetic differences hardly explains it, however (see ch. 18, below).

A more promising line of inquiry is into interaction between intelligence as a neurological (biophysical and biochemical) phenomenon and political behavior (Laponce, 1972, 1976, 1981, 1987); this approach explores laterality, the neurophysiological predisposition to favor and develop the use of one side of the body (eye, foot, hand, etc.) over the other. Our predominant dexterity results from the interaction of both culture and evolution, over a very long period of time; but its very universality makes the question how different societies deal with the problem of the "deviant" left-hander of interest to students of political behavior, among other reasons because of the expected correlation between left-handedness and left-wing political attitudes and affiliations, and also because the degree of left-handedness in a society may be a reliable index to the degree of practical democracy operating in the polity. In any case, inquiry into laterality raises important questions concerning other physiological correlates of political behavior. Laponce himself has pursued some of these matters in the leading article of a volume of the *American Political Science Review* (1975: 11–20), in which he puts laterality in the context of a broader set of spatial dimensions, as the basis for an investigation of the semantics of nonverbal political communication and symbolization; and as in his other papers, the questions that he raises demand that biological as well as cultural evidence be examined. The implications of Laponce's work potentially involve several other behavioral concepts in addition to such obvious ones as communications and control, conflict and decision making, socialization, roles, attitudes, and culture.

The other major biological attribute, which remains almost completely absent—except in a formal sense—in the work of political scientists, is sex. From a population genetics perspective, not only is a chicken the egg's way of making another egg but also human sexual structural differences retain a biological function to perform. Certainly this is true if the species is to postpone somewhat longer its ultimate fate of extinction due to its incapacity to change rapidly and drastically enough to adapt to environmental demands. But at least two questions seem to be in order. First, assuming that the gender stereotyping of a still only partially enlightened society entails antidemocratic consequences for personality development and therefore for political attitudes and role playing by both males and females, is there also a positive influence of androgyny upon personality development and hence upon political attitudes and role playing (Schubert, 1987)? Second, in the many societies in which traditional gender distinctions still persist, is it possible that on physiological grounds females are better qualified than males to play certain political roles (Schubert, 1983f; and Lovett-Doust and Lovett-Doust, 1985)? This latter would seem to be a particularly pertinent question to ask at a time when cultural norms, which in the past have discriminated against full and equal political participation by females, are undergoing rapid change (Schubert, 1985c). Laponce indicated, for example, that there is evidence suggesting that left-handers are more creative and more emotional as well as more independent in personality than are dextrals, for physiological as well as cultural reasons. Similar arguments are advanced in much of the feminist as well as in the human development literature (Schubert, 1988b) as a basis for distinguishing the characteristic personality of females from that of males (Barwick, 1971; Maccoby, 1966); and of course here again we must presume a complex transaction of biogenic and sociogenic causation. A relevant paper by a political scientist (Dearden, 1974: 28) reports the finding that human (like other primate) males are more aggressive than females in every respect save one—verbal aggression. This finding implies that Americans ought to man the hustings in full support of a women for President and do the same for her counterpart in the Soviet Union. Surely the hand that rocks the cradle is the better one to place at both ends of the hotline, or, if one prefers to vary the metaphor, on the fail-safe button. There are, of course, empirical obstacles to the implementation of such a policy even in polities less enlightened than those that already have passed into postindustrialization; but recognizing the need for such a revolution in policy is surely a step indispensable to confronting its entailed consequences.

HUMAN BEHAVIORAL ECOLOGY

There are two principal respects in which the environment makes possible both the satisfaction of basic human needs and the practice of politics. The first and most fundamental is ecological and involves energetics, the global distribution and other aspects of the species population, and the niches occupied by various subpopulations of humans in relation to other species with whom we share the biosphere (Green, 1986). Less basic (in long-run terms of *species* survival) but at the same time more obvious in its impingement upon political behavior and also more immediately critical for the survival of any particular individual is the social environment, which educates her or him into a particular culture. To a greater or lesser extent the social environment facilitates the manipulation of an individual's psychophysical systems in an attempt to influence that person's behavior (including political behavior).

Energetics is so painfully obvious as a critical dimension of public policy for all countries today that little argument seems necessary in support of the proposition that political science is likely to remain preoccupied, during any future that now seems imaginable, with problems of energy types, quality, production and distribution costs, dissipation (pollution), consumption priorities, and conservation. Indeed, I cannot imagine a viable politics (or political science) that will not *increasingly* be so preoccupied during any finite period of time that can possibly be of personal interest to anybody living today.

Evidently the distribution of human populations in relation to the availability of resources for the satisfaction of human needs, and in relation to different cultures, is likely to continue to be of concern to political scientists. Other aspects of population dynamics, such as density differentials and differences in both the absolute sizes and growth rates of human populations, seem at least equally viable for political analysis. Except for growth rate, these have long been the stock in trade of traditional studies of international politics; but it is likely that a more technical level of knowledge, reflecting the escalating growth of demography as a specialized field, will characterize future political analysis of population policy.

The biological niche of *Homo sapiens* is a much less obvious matter of concern for most social scientists, largely because they seem to feel, with almost incredible smugness, that they can take the niche of humanity for granted (see E. P. Odum, 1971: 214, 234–36, 510–16). Such humanistic conceit is credible only on a hypothesis of sheer and utter ignorance of the manner in which the biosphere has operated historically, works at present, and seems most probable of evolving in the future.

The success of humans in the competitive exclusion of other living species (of both flora and fauna), particularly during the most recent ten to twelve thousand years since we began in a serious way to scarify the natural land with our agriculture, has by no means necessarily been adaptive for our species except from a point of view with as short a range as that. Our increasing technological capacity and tendency to eliminate, as often unwittingly or accidentally as by design, other living species upon whom we previously had relied for sustenance, as well as ourselves, is an index of the extent to which our trophic niche, defined as the *functional* status of an organism in its community, is being redetermined as much by the indirect as by the direct effects of our predatory activities; and the fossil record is replete with evidence of extinct species whose predation was so successful that they themselves starved to death (see references cited in ch. 12, below). The issue goes far beyond the restoration of token vegetation to, and the elimination of domestic pets from, urban areas or even the apparent trends in the direction of human conspecific predation ranging from licensed hunting in season (under circumstances such that other hunters present more frequent targets of opportunity than the crops of ruminants or rodents available for harvesting). It extends also to the ubiquitous predations now characteristic of all large urban areas in the United States (where only humans can be and are hunted by each other, at least in part because all other prey has been exterminated).

The human species cannot destroy the biological community of which humans have been a part without their degradation of that biological community returning as feedback to threaten the human political community; and I'd like to give two examples (Clapham, 1973: 234–36) which are textbook material for undergraduate zoology courses:

> The most obvious example of an ecosystem that has been altered by man . . . [are] agricultural fields. Essentially, one species of organism is allowed to exist while other species are removed by tremendous expenditures of energy, herbicides, and pesticides, and the abiotic environment is controlled by extensive use of irrigation, fertilization, and tilling. . . . [T]he successful exploitation of natural ecosystems in order to increase the production of human food has led in most cases to a dramatic lowering of the fitness of the exploited ecosystems. Technology can allow these ecosystems to remain viable and productive, but only at considerable cost. There are clearly limits to which technology can overcome the tendency of ecosystems to revert to natural equilibrium, and the instability and uncertainty of the system's capacity to produce products useful to man increase greatly as these limits are approached.
>
> Current agricultural economics dictates that livestock should be fattened in feedlots rather than on the range. On the range, the urine and feces from the animals fertilize the land naturally to replace much of the nutrients removed by herbivores. In feedlots, however, the nutrients contained in the feed are removed from fields in their entirety, causing a reduction in soil fertility which must be made up with massive fertilization, and the excreta of the livestock are too concentrated to be utilized by plants in the feed lot area, and so are washed into nearby waterways, where they become pollution problems. This is a classic example in which the disruption of a normal biogeochemical cycle has led to the deficiency of materials in one ecosystem and a surplus in another, reducing the fitness of both.

In the past political science has assumed the complete beneficence of both the fattening of livestock for market and the raising of grain with which to do it. But that complacency has been unwarranted and has skewed our assumptions about the political economics of food policy, at both the national and international levels of public policy making.

The fluoridation of watersupplies—though still controversial in some outback regions such as Hawaii—is now generally countenanced like chlorination, although it must be admitted that that was done without prescience of such serendipitous discoveries as the detection of carcinogens as a byproduct of the chemical interaction between chlorination and pollutants in metropolitan drinking water supplies. By the end of the twentieth century environmental poisoning of rich and poor alike, because the human species has become a captive population, may well become the major factor in the reduction of human population densities in urban centers throughout the world.

PART I

Political Ethology

Part 1 examines ethology as an approach to the study of political behavior. Ethology is the naturalistic field study of animal behavior, and it has emphasized the use of direct observational methods to describe, and provide an empirical basis for the analysis of, communalities and individual variations from species-specific behavior. Such studies of the social behavior of mammals have been especially well publicized; particular attention has been given to social communication among diverse genera, including those of birds and cetaceans as well as carnivores, elephants, and other (than human) primates.

Chapter 3 begins with a discussion of human basic needs theory in political science generally and political psychology in particular. The second part of the chapter contrasts social science's exclusively culturally based decision-making paradigm for political behavior, with the ethologically based biological paradigm of animal behavior. These previously independent perspectives are joined in a composite life-science paradigm of human political behavior, which directs attention to the close interrelationships between the ways in which humans, as animals, necessarily behave like them, but also to the ways in which human psychophysiology and culture make human behavior unique.

Chapter 4 consists of a peer comment, written as a response to a target article by famous European ethologist Irenaus Eibl-Eibesfeldt on the nominal subject of *human* ethology. The comment asserts that the subject in fact presented in the target article is methodological, that of *classical* ethology, rather than substantive. Classical ethology is the naturalistic field study of animal behavior, as developed most conspicuously during the three decades between 1930 and 1960 in the work of Niko Tinbergen and Konrad Lorenz (Eibl's mentor), for which the two men were jointly awarded the Nobel Prize. Classical ethology entailed a characteristic epistemology, and it eschewed both the laboratory experimental methods and the physiological animal biology that interfaced with classical ethology to establish the basis for *modern* ethology. The comment criticizes the target article for proposing that ethologists of the eighties interested in human behavior should encumber their work with an outdated theory especially inapplicable to human behavior as well as forego the many other advantages proffered by an ethological approach rooted in the modern synthesis which combines both field observation of social and other behavioral externalities and the experimental laboratory analysis of the physiological internalities of animal behavior. The comment discusses also the historic differences in the development of Continental, British, and American studies of animal behavior, most of which have been eliminated or attenuated by the modern ethological synthesis. The chapter concludes that human ethologists should continue to look upon their own work as a species specialization in the application of modern ethology and that what Eibl describes in his target article is better characterized as ethnography than as ethology.

Chapter 5 undertakes a detailed overview of the field of modern ethology, especially in its explicit and implicit implications for the study of political behavior. The first section of the chapter presents a synopsis of modern evolutionary theory (i.e., the neo-Darwinian modern synthesis with Mendelian genetics), a subject that is developed in considerably greater depth in part 3 (chs. 9–12) of this book. Then the field of modern ethology is analyzed, in its relationship to classical ethology and in relation to behavioral ecology and sociobiology. The discussion of human ethology emphasizes communicative behavior, ethological ethnography, and the study of nonverbal political communication. The use of ethological methods for political analysis is related to political socialization, play and aggression, war, and political change. The chapter concludes with an analysis of the implications of modern ethology for the revolutionary recasting of political theory, especially on the basis of several contemporary evolutionary-theory alternatives to neo-Darwinian gradualism. Some of these that are increasingly gaining acceptance include catastrophe ("punctuated equilibria") theory, epigenetic theories of rapid phenotypic change, and accelerated genetic selection due to the constraints and opportunities imposed or made available by dynamically altered (catastrophic) environmental stresses on the behavior and development of all forms of life subject to such effects. Perhaps ironically, such theories of non-Darwinian "evolution" as applied to human political behavior tend to generate theories of political *revolution*.

3
Human

Peter Corning (1971: 339, quoting Bay, 1965: 40) has pointed out that a fundamental challenge to political scientists was levied in "Christian Bay's assertion that a satisfactory political theory must be derived from an adequate understanding of the 'basic human needs,' as well as from man's overt and often variable preferences." Bay was anticipated, in at least this respect, by James C. Davies in his *Human Nature in Politics* (1963), a book that borrows its title from Graham Wallas and its substantive approach from Abraham Maslow's theory of human needs (with credits going also to Freud and David Krech). Davies' book is a creative attempt to tease out of social psychology the "satisfactory political theory" that Bay has called for; but events have demonstrated that *social* psychology—the mainstream of what has been perceived to be the political behavioral approach—is not enough.

It is now a matter of history that it was Bay's radical critique of political behavioralism that sparked the protest movement within the profession that led to the organization of the Caucus for a New Political Science and the professional confrontations of the late sixties (see Van Dyke, 1971, reviewing Surkin and Wolfe, 1970). One operationalization of that approach, to an adequate understanding of basic human needs, is to study those needs at a level of causation more fundamental than what can be inferred from observations of either what human organisms do, or what they can articulate about their internal states of being. This does not mean that we need to disregard either of the latter types of social-science

---

This chapter is a revision of "Politics as a Life Science: How and Why the Impact of Modern Biology Will Revolutionize the Study of Political Behavior," in Albert Somit (ed.), *Biology and Politics: Recent Explorations* (The Hague: Mouton, 1976), pp. 165–68 and 176–80. Reprinted with the permission of the publisher.

behavioral information; quite the contrary. But it does mean that other types of information, relating to independent observations of the operations of the neural, hormonal, and motor systems (among other internal ones) of humans must be taken into consideration if we are to become more realistic in our studies of such phenomena as political action. At the very least we need to understand—and to deal with in our own theories of political behavior—how the satisfaction of basic human needs affects the possibility and the modes of acting politically (cf. Madsen, 1985).

From a biological point of view, the most basic need of any animal is to survive long enough to reproduce (see ch. 7, below). It is the gene pool, to use the language of population genetics, and not the individual phenotype on which natural selection operates to preserve; but that can be done only through the successful adaptation of some minimally large population of phenotypes that do reproduce successfully and rear their young to the stage of self-sufficiency. To survive that long, an individual human animal must continuously satisfy certain physical and chemical requirements which have long been more or less understood: air, body temperature maintenance, sleep, water, and food. Aristotle was quite aware of the need for air, but he could not have known that it is specifically oxygen that is needed, nor why and how it enters into metabolism. He was aware also of both sleep and dreams; but more recent observations of brain-wave fluctuations are suggestive of support for a structural typology of sleep and a theory of its function for decisional purposes that became possible only with the development of the technology of the electro-encephalogram. Relatively sophisticated technical knowledge concerning most of these needs (adequate nutrition, potable water, and oxygen energetics) is indispensable for an understanding of what many biologists, and at least some political scientists, consider to be the most crucial problems of public policy confronting political science today; these problems involve the ecologies of land, sea, and air.

A fifth type of physical/chemical need is that for sensory stimulation. It is through the senses (tactile, visual, auditory, olfactory, gustatory) that an animal maintains contact with its environment, and such contact on a continuing basis is essential to survival. The evolution of internal homeostatic limits for animals presumes that appropriate stimulation will be forthcoming; hence the "vacuum activity" to which ethologists have directed attention, referring to extreme lowering of threshold for a behavior; and hence the probability that humans, like other animals, require a certain minimal continuing stimulation of at least several of their senses, quite apart from the question of the use of those receptors to provide environmental information needed by the animal for other purposes.

A sixth type of physical/chemical need is that of infants to be held, touched, and otherwise stimulated. Such maternal/paternal/surrogate care appears to be necessary to the survival of infants up to about two years of age; and it appears also to be essential (although not directly to survival) to learning, growth, and security for older human children as well as for other primate juveniles.

As Davies points out, Maslow classified "sex" (evidently signifying sexual intercourse) as a physical need; but that seems to reflect his feeling of having to categorize it into either the biological *or* the social *or* the psychological class of alternatives in his (the Maslowian) typology; and his decision resulted in the blurring of one of the unique characteristics peculiar to the human species. Here is an instance where reference to zoology results in a much more humanistic judgment than one based upon the softer side of psychology. In animals other than primates it is not usually necessary (or possible) to distinguish between sexual interaction and sexual reproduction, because of their almost invariate coincidence; but in primates these two functions are much more complexly interrelated. This is especially true of humans (because of the loss of estrusity in human females). It is the more puzzling and ironic to note that Maslow began his professional career as a biologically trained primatologist and experimental psychologist.

Social scientists and biologists agree that sexual interaction is a human *social* need, important to good health (both mental and physical), but not otherwise or directly essential to the survival of the interacting individuals. Much more controversial is the question of whether and how to limit sexual reproduction, with the primary objective of attaining zero or negative population growth on a worldwide basis, and eventually (i.e., in some future century) working toward a reduction in the absolute global level of the species population. Because neither voluntary nor democratic decision making is considered feasible as a means of attaining global ZPG, the issue will be a difficult one for political scientists to help resolve.

Other social needs, confirmed by both sociological and primatological research, include group association with and orientation to conspecifics. This means that humans need to touch, see, talk with, and otherwise interact with other humans—for direct sensory gratification, for information about the environment in addition to what an individual can derive directly from her or his own sensory contacts, and for the security in relation to the environment that the individual finds in association with other humans. During infancy such needs are critical to the point of survival; subsequently they are necessary if humans are to become sufficiently skilled and involved in social transactions to be able to act politically, as well as for the health and happiness of most persons.

Maslow's typology specified certain psychological needs of self-esteem and self-actualization; but evidence to test an operationalized theory of such psychological needs is at least as likely to come from physiological psychology and biology as from social psychology (cf. Nash, 1970; and Davies, 1963: 45–60). It is probable that the satisfaction of such psychological needs, before a person is competent to act politically, is just as critical as the satisfaction of the social needs described above. But not only are the physical/chemical needs literally, proximately, and *directly* prerequisite to the survival of individual adult humans; neuro- and psychobiology's increasing understanding of, e.g., depression in general and schizophrenia in particular indicates that *all* human needs, whether classified as "basic,"

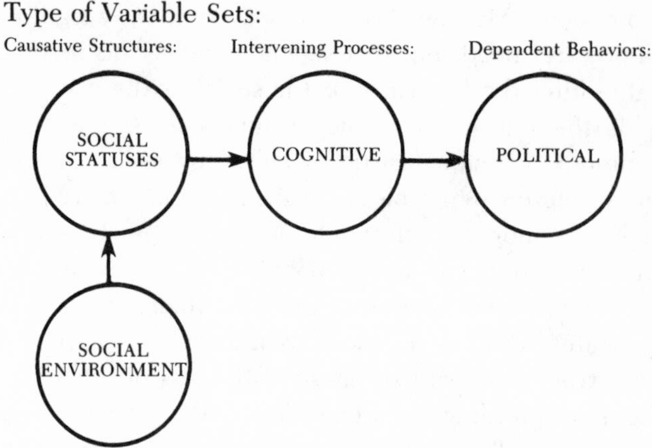

Figure 2 The social-science paradigm of political behavior.

social, or psychological, are transactionally active in some non-Euclidian space considerably more complex than a hierarchy, at the many levels of physical chemistry in the life processes (including the survival) of individual humans, as the first ten volumes (1978–1987) of a new journal, *The Behavioral and Brain Sciences*, exemplify.

MODELS OF BIOPOLITICAL BEHAVIOR

It is possible and useful to interrelate the conventional paradigm of political behavior with a radically different but nevertheless complementary paradigm of biological behavior. The conventional political behavioral paradigm is modeled in figure 2. Political actors are classified, for purposes of data observation and aggregation, according to a set of sociological indicators of their attributes that are conventionally deemed relevant for analysis (party, age, sex, socioeconomic status, etc.); these inputs of the system of political action have themselves been determined by the social environment. Political behaviors are classified in terms of a set of equally conventional action models (policy choice, vote, role performance, speech, etc.); these are the outputs of the system. Intervening between social attributes and political action are the cognitions of the individual actors (beliefs, attitudes, decisions, preferences, etc.); by making conscious choices among competing alternatives, political actors determine what their behavior will be (see Schubert, 1968a: 415, table 1).

Such a paradigm postulates a (social) *psychological* model of decision making; and this lies modally between the more strictly *logical* model of traditional political science and the *nonlogical* models that are suggested by clinical psychology on the one hand and comparative psychology and biology on the other hand (Schubert, 1968a: 417, table 3). The social indicators are relatively unsophisticated indices to learning and experience; and these indicators function as surrogates for the total life history and socialization of the individual. According to this psychological model and its underlying paradigm of political behavioralism, differences in political behavior are explained by differences in conscious choices as to how to behave, which in turn are explained by differences in the life experiences of the actors.

The biological paradigm of behavior is sketched in figure 3. It directs attention to the nonlogical influences—at least from the point of view of the logic of political behavior—upon all behavior and to the priorities of survival requirements, the satisfaction of which is preconditional to indulgence in political behavior. According to the biological paradigm, human needs have to be satisfied, like those of any other animal, through sensory and appetitive interaction with an environment that is both partly natural and partly social; and different kinds of needs find sustenance in differing parts of the environment. Such biological characteristics as age and sex are reciprocally engaged in interactions with both the natural and the social environment: how either sector of that environment will affect individuals differs according to their stage and type of development, and how they will respond to that environment is partially determined by such aspects of their development. Similarly, an animal's psychophysiological systems are directly affected by age, sex, health, and the other biological characteristics. Appetitive behavior involves searching of either the natural or the social environment (or both), and hence the feedback link indicated in figure 3; the satisfaction of physical-chemical needs necessarily entails feedback through psychophysiological systems to the organic causes of those needs. So the biological paradigm states that needs autonomically activate psychophysiological systems which initiate appetitive behavior, the consequence of which is to cause the animal to probe its environment, which in turn provides further stimulation for the animal.

Figure 2 puts humans more in the stance of gods than of animals and hence is incomplete and inadequate precisely to the extent that humans *are* animals. Figure 3 describes human behavior as animal behavior, and it is incomplete and inadequate precisely to the extent that humans are *cognitively* different from all other animals. Figure 4 attempts to put them together, so that humans can better be studied as and for what they are— not gods, but nevertheless the only animals that were, quite properly, characterized by Aristotle as "the *political* animal" (see Schubert, 1973).

Figure 4 depicts the psychophysiological system of political behavioralism as superior to, but continuously interacting with, the more fundamental system of biological behavior. The statement of a human need occurs initially not at any level of conscious thought, but rather involves autonomic invocation of psychophysiological systems, which in turn activate appropriate appetitive behaviors; those appetitive behaviors necessarily involve interactions between the person and the environment, and at some point of intensity either the behaviors or the needs motivating them become sufficiently amplified by the psychophysiological

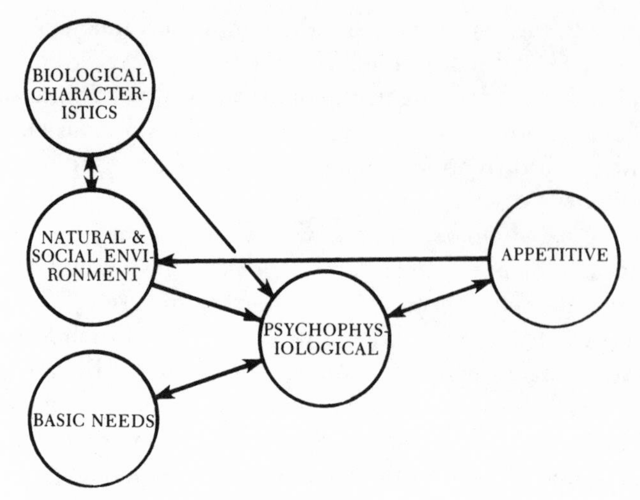

Figure 3 The biological paradigm of animal behavior.

systems to become perceived as *bio*feedback to cognitive systems of the human, who may then choose to impose various (perhaps cultural) restraints or reinforcers (as the case may be) upon the appetitive behaviors. In relation to basic needs, such restraints never can be more than a temporary reordering of preferences because the feasible time scale for postponement of basic physical-chemical satisfactions is strictly determined by the operating limits of interacting psychophysiological systems; these limits, if exceeded, result in the disablement or death of the organism and hence the elimination of both the necessity and the possibility for any kind of choice making. In addition to that limited and partial kind of cognitive intervention in appetitive behavior, cognition affects one's biosocial characteristics, because sex identification and attitudes toward age and aging, for example, are important feedback to the further development of these facets of a human organism (Schubert, 1987; and ch. 15, below). Moreover, cognitions about the environment can lead to choosing behaviors that will affect the environment as feedback to it.

The characteristics that are designated as "social" in figure 2 and as "biological" in figure 3 are redesignated as *biosocial* in figure 4 because those with which we are concerned here (age, sex, race, intelligence, and health) belong clearly in both realms of discourse: all have a clear and direct biological significance which is determinative of the social statuses that are derived from them. The environmental variables provide an indirect link between basic needs and biosocial characteristics; and consciously perceived *bio*feedback serves as one important link between autonomic, psychophysiological and conscious, cognitive systems, although direct two-way interaction between these two sets of systems (without biofeedback through self-conscious awareness) is much more important to behavior generally, and certainly also to political behavior. Biofeedback refers here

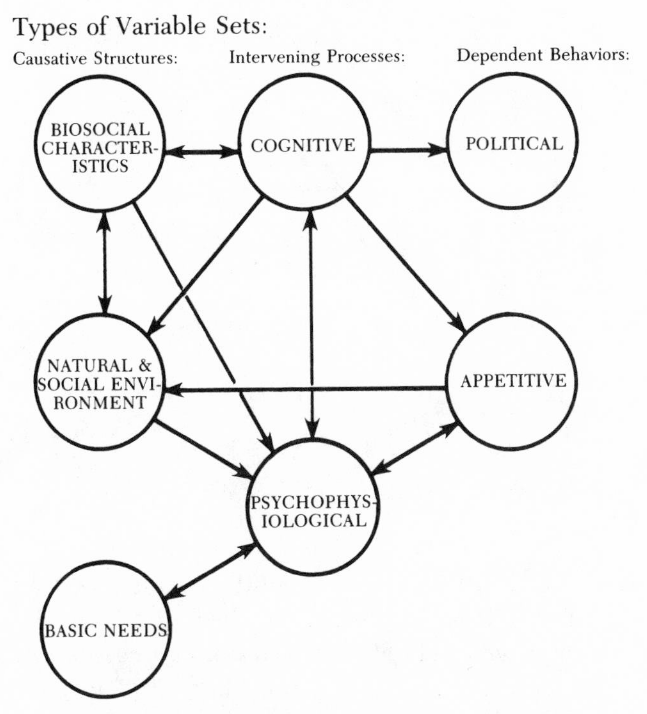

Figure 4 A life-science paradigm of political behavior.

primarily to physiological systemic effects upon cognitive systems, which are not necessarily perceived at all (let alone as such); but it also includes the reverse link postulated by humanistic psychology and psychobiology, with self-conscious cognition affecting physiological systems. No direct link is shown between political and appetitive behavior, which constitute discrete and alternative modes of action. (Of course it is possible for a dinner at the White House or a Washington cocktail party to function empirically and simultaneously as *both* appetitive and political behavior; but to suggest this is a semantic quibble and raises no question of analytical importance.)

What clearly needs to be done, in moving beyond political behavior to biopolitical behavior, is to design research that will explore the vertical as well as the horizontal relationships depicted in figure 4. That will necessitate consideration of the interfaces between the biological and the social sciences and will point political science in the direction of a study of political behavior which, for the first time, can deal with political life at the level it is lived and not as an exercise in puns, rhetoric, or intellectual dilettantism.

4
Classical

If Friedrich Engels had written an article entitled "Marxism—Concepts and Implications for the Social Sciences," which somehow (like Mendel's records of his experiments) came to light only after decades of cryptic storage and was then published in some journal of open peer commentary like *Current Anthropology*, the effect would surely be sensational. Historians of ideas would have a field day, together, of course, with radicals from various of the social sciences (for whom the event would take on the significance of the Second Coming). On the other

---

This chapter is a revision of my comment "Classical Ethology: Concepts and Implications for Human Ethology," *Behavioral and Brain Sciences*, 2: 44–46, on the target article by Irenaus Eibl-Eibesfeldt, "Human Ethology: Concepts and Implications for the Sciences of Man," *Behavioral and Brain Sciences*, 2: 1–26 (Mar. 1979). My comment also was published in March 1979. Barely four months later, the following news item appeared in the [Armed Services newspaper for U.S. soldiers stationed in Europe] *Stars and Stripes* (Jul. 26, 1979, p. 2, col. 1):

Engels Letter Found

> BERLIN (AP)—A previously unknown letter from Friederich Engels, Socialist leader and close friend of Karl Marx, has been found in an archive in Weimar, the East German news agency ADN reported.

My introductory paragraph was written in jest; but truth is obviously so much less probable, thereby irresistibly suggesting the serendipitous query: Was Engels, who, as we well know, wrote such other ethological classics as "The Part Played by Labour in the Transition from Ape to Man" (Engels, 1876), in fact also the author who wrote "Human Ethology: Concepts and Implications for the Sciences of Man"?

hand, it would have no effect whatsoever upon the current rulers of such countries as the U.S.S.R., the People's Republic of China, and Cuba; or their respective subject peoples. But social scientists who are neither historians nor ideologues would find more perplexing the task of commenting, from the perspectives of contemporary theory and methodology, upon Engels' prescriptions. A social scientist/human ethologist, confronted with the paper that E-E [Irenaus Eibl-Eibesfeldt] has written, finds himself in a somewhat similar position.

E-E is internationally known and respected for his prolific writing, both professional and popular, concerning nonverbal communication among primitive and sensorily deprived humans as well as the expression of emotions in contemporary human societies. He is the leading student and associate of Konrad Lorenz, the founding father of classical ethology; and he is the author of a leading undergraduate textbook on ethology, published in English by an American publisher. One turns to his article expecting to learn much from what he has to say about the theory and promise, for the life sciences, of ethological approaches to the study of human behavior. But the reader soon discovers that this article has little new to say; much that it does say seems appropriately directed to some other and less professional audience; and many of its figures have been published and discussed many times before (e.g., table 1 and figs. 9–11, 13: Hass, 1972; Eibl-Eibesfeldt, 1972, 1975b op. cit.)

The first part of my commentary will examine E-E's assertions and discuss the extent to which this article rehashes both ideas and data that have long since passed into the public domain, at least for an informed readership. The remaining three parts of my commentary will discuss: the relationship between ethology and experimental psychology; that between ethology and sociobiology; and in conclusion that between ethology and ethnology.

STONEWALLING STRAW MEN

Virtually all of the introductory sections of the article fall in the straw man category, together with much that follows; it is most unlikely, for example, that "whether the concepts of phylogeny, selection, and inheritance are applicable to human behavior . . . is still a matter of considerable controversy" for readers interested in the behavioral and brain sciences, which the editor of *Behavioral and Brain Sciences* has associated with the fields of "psychology, neuroscience, behavioral biology, or cognitive science." Perhaps E-E could provoke a debate at this level of distraction among readers of *Current Anthropology, American Sociological Review*, or *Psychology Today*; and there is an indication that he is confused about his intended readership, because the title of his article directs his remarks to the "Sciences of Man"—which the journal's first editorial (*BBS* 1[1], p. 1) and the editor's initial "Call for Papers" have explicitly identified with *CA* rather than *BBS*. Anyhow, E-E's article is replete with similar rhetoric, such as the claim that "one of [human ethology's] principal questions is whether biological heritage has determined human behavior to any significant degree"; rather, one ought to presume that the relevant questions for the behavioral and brain sciences, as distinguished from at least substantial parts of the sciences of man, are not whether, but rather how, why, and by means of what processes of developmental interaction. The same point is repeated in the summary (sec. 7.4), although some might question whether "genetic" determinism is synonymous with "biological." But even if the author were right about his audience, he seems to underestimate the extent of sophistication about evolutionary theory, even in such relatively retarded life sciences as political science (White, 1972; Masters, 1975; Somit, 1976), that has emerged in phase with the preceding decade and a half which E-E concedes to the "formal" study of human ethology.

It is surprising to encounter, in the discussion of environmentalistic ideology (as restated in the concluding summary, sec. 7.19), the argument that "strict cultural relativism may lead to ethnocentric political strategies." It is the conviction of many social scientists of man that it was the proponents of biological determinism in the form of racial eugenics who, well within the memory of many still-living survivors, became conspicuous for both the promotion and the execution of ethnocentric political strategies; of course, the same sort of thing can be said about American treatment of blacks, tribal genocide among blacks in half of the countries of Africa today, and so on throughout the world and, as far as we know, throughout human history in recent millennia (see Harris, 1977). To the extent that we understand such matters scientifically, it is in spite of—not because of—cultural relativism. The usual understanding is that cultural relativism has debunked ethnocentric political strategies, not fomented them.

There must be very few human ethologists today who would deem it necessary either to rebut Ashley Montagu or to defend—to say nothing of building directly upon—Lorenz (see Blurton-Jones, 1976; Travis et al., 1977). Montagu continues to maintain the beneficence of innate human nonaggression just as Ardrey continues to argue the opposite; but much of their debate is at the level of public rather than of scientific interest. A focus upon the critique of E. O. Wilson by his colleagues Lewontin and Herrnstein, or of Trivers by Sahlins, might not be any more enlightening (see Sahlins, 1976; Caplan, 1978; Gregory, et al., 1978); but at least it would better illustrate the cutting edge of debate concerning the interactive effects of phylogeny and environment in the course of human development. An example of contemporary discussion is provided by the "forum" of the current issue of the *Human Ethology Newsletter*, which includes three such articles, all on the sociobiology of human sexuality, by Daly and Wilson (1978), Travis (1978), and Barkow (1978). The announced topics for the next two issues are: methodology for studying human adaptation (January 1979); and the relevance of modern hunter-gatherers to the evolution of human behavior, particularly from the points of view of theory and methodology (March 1979). But in E-

E's article, sociobiology appears only as the last three paragraphs of his "discussion"; it is such an afterthought that it is neither anticipated by any of the preceding data, nor does it get summarized: sec. 7.20 is conspicuous by its absence, and so are considerations of or references to most of the contemporary protagonists and antagonists of sociobiology. The more one reads E-E's article, the more the conviction grows that its principal difficulty is timing: its publication comes at least a decade too late.

In order to examine that hypothesis on a less intuitive basis, I undertook to compare the bibliography of E-E's article with an alternative and putatively representative collection of contemporary research in human ethology. The then current issue of the *Human Ethology Newsletter* (no. 23, Oct. 1978: 2) refers to "Human Ethology Abstracts" of which two of three parts have been published so far. The first part was readily accessible, so I used that. E-E is listed in the same column of the same page of the *Newsletter* as a member of the executive board of the International Society for Human Ethology, as the organization publishing the newsletter recently had decided to redesignate itself.

The 134 references of E-E's article are apportioned, roughly by decades, as follows: 78 for 1968–78, 34 earlier during the sixties, 10 during the fifties, 4 during the thirties, and a couple of others before that. The year 1968 is the mean for citations during the past half century, so the works cited are about a decade old on the average, even though a majority of them are less than that. A virtually identical total of 133 of the annotated "Human Ethology Abstracts" (Travis et al., 1977) are works published during the decade 1968–77, 4 others fall in the preceding decade, and 19 are references to undated manuscripts. The average age of these references is five rather than ten years, so they are somewhat more contemporary as research literature.

The scope of human ethology is twice defined by E-E to be congruent with "biological research," to wit: "morphology, ecology, genetics, phylogenetics, developmental biology, sociobiology, and physiology." The abstracts classify items under rather different concepts: general human ethology (theory and methods, general reference, group formation and mating strategies, sex differences); aggression, altruism, and cooperation; children and infants (neonate behavior, attachment and separation, attention and cognition, general behavior, play and peer interaction); communication (vocalization and language, the face, gesture, and posture); and social spacing (personal space, social distance, environmental design). Even though "biology" conceptualizes human behavior rather differently than does "human ethology," there appears to be no obvious reason why both systems could not be used to classify similar sets of "references on human ethology." But that is clearly not what happened: of the 157 items listed by the abstracts, only 5 appear among the 134 references to E-E's article; and of those 5, 2 are to works authored either by E-E himself (1974) or by persons associated with his own research institute (Pitcairn and Schleidt, 1976). The other three works appearing in both bibliographies are: Wilson (1975), Blurton-Jones (1972), and McGrew (1972). This is hardly impressive evidence that the two bibliographies purport to deal with the same subject.[1] I shall now turn to the very good reasons why they do in fact deal with quite different subjects.

ETHOLOGY AND EXPERIMENTAL PSYCHOLOGY

One difficulty lies in the disciplinary chauvinism supporting the educational systems out of which classical ethology developed (see Presthus, 1977; H. T. Wilson, 1977) and which still tends to characterize the thinking of its remaining spokesmen. Classical ethology was, and to a large extent still is, a naturalistic activity pursued by persons trained as zoologists, who do their own experimental work and engage in relatively little collaboration with experimental psychologists. In the United States, to the contrary, such organizations as the Animal Behavior Society (ABS) and the American Society of Primatologists are much more equally divided between zoologists and comparative psychologists (among other supporting disciplinary components), and the representatives of both groups are accustomed to a symbiotic relationship very different from what is found in continental Western Europe. The United Kingdom is somewhere in between, as symbolized by the subtitle of Hinde's textbook (1970); and it has been more to the U.K. than to the U.S. that Continental ethology has turned. *Human* ethologists are organized neither on the Continent nor in the U.K.; but they have been for at least five years in the U.S., piggybacking the annual meetings of ABS.

Probably it is innocence of what American human ethology involves that explains the prominence E-E gives to such putative goals as "to bridge the gulf between ethologists and opposing [sic] groups of behavioral scientists." But this article frequently manifests a barely repressed fear of experimental psychologist strangers (such as Tobach and Eisenberg); as Pogo says, "We have met the enemy and he is us."

Rapprochement is not likely to be facilitated by the persistence of classical ethologists in the use of theory, methods, and concepts that are idiosyncratic and that have long since been given up or greatly modified (always in the direction of indeterminacy and flexibility) by an intervening generation of students of animal behavior. The Lorenzian hydraulic model of animal motivation discussed by E-E, for example, is treated as a relic in modern textbooks on animal behavior (Hinde 1970: 201–2; Beer, 1973: 34; Alcock, 1975: 189–91; Mortenson, 1975: 41–42, describes it as the "flush-toilet model" and portrays a figure of it, and Dawkins, 1976: 8, includes it in his assertion of "the general, deserved destruction of sim-

[1]The second part of the "Human Ethology Abstracts" (Travis, 1977) is an annotated bibliography of 222 items, almost entirely from the seventies, of which precisely 1 (Ashley Montagu's *Man and Aggression*, 1973 ed.) also appears among E-E's references, cited in an earlier edition (op. cit., 1968).

plistic energy models"). Concepts of fixed action patterns, innate releasing mechanisms, drives, and releasers are too rooted in Newtonian mechanics (see Landau, 1965) to be useful in guiding contemporary research (Hinde, 1970: 21, 121–213; Beer, 1973: 37–38: "The hierarchy theory . . . was the culmination of the development of theory at the heart of classical ethology. . . . Later thought and study questioned and tested the assumptions and implications of the theory, and in so doing carried ethology from its classical period into a period in which theoretical unity gave way to differentiation in theoretical interests and in research"; Mortenson, 1975: 41–43). At least this is true in the study of mammals; and it is worth remembering that the animals with which both Lorenz and Tinbergen, as well as most of their students, worked primarily were either fish or birds, whereas the thrust of interest in recent animal research in the U.S. has been in mammals, for which the early ethological research apparatus was not designed. Even at the level of stickleback courtship discussed by E-E, it is by no means evident that it makes the slightest difference whether it is stated that the male's zigzag course through the water ("courting dance") *releases* or *stimulates* presenting behavior on the part of the female conspecific—except that "stimulates" articulates with a very much larger corpus of research on human behavior, while "releases" does not.

Imprinting, perhaps the best-known concept of classical ethology, is now understood to be very much more complex and subject to many more qualifications than could have been realized at the time Lorenz reported his early observations (see Rajecki, 1978). It is one thing to find "imprinted" goslings waddling after Lorenz, or even jackdaws trying to feed him warm minced worms; it is quite another to hypothesize that a prenatal human imprints on the mother's heartbeat. Everything that we know about the phylogeny of humans ought to impel us to anticipate a great deal more flexibility in perception and response alike than has evolved in the behavior repertoires of either geese or jackdaws. The latter, incidentally, are animals whose social behavior I have investigated at first hand (see Roell, 1978). Many of these differences in human flexibility are discussed in a paper by Blurton-Jones (1976) which is also on the subject of the implications of ethology for social science, but in which Lorenz is neither mentioned nor cited.

E-E's emphasis upon animal "drives" is the wrong way to establish ties for or with human ethology; it would be much easier to link up with more of the life sciences if the concept of "needs" were substituted. It then immediately becomes necessary to reconsider E-E's remark that "the format application of ethological methods to the study of man began about fifteen years ago." That statement may hold for Lorenzian ethology; but there are certainly many instances of earlier beginnings in human ethology. Tomkins (1962) was reporting at international congresses his theory of human facial nonverbal communication a decade earlier; Bowlby (1958) was discussing ethological approaches to attachment behavior; and various amateur primatologists (Marais, 1934) and humanist precursors (Day, 1920) were much earlier in discussing animal models of human behavior. Most important was Maslow (1954) who, trained as a primatologist, had formulated his theory of human needs as early as 1943. His influence has spread beyond psychology to include many social biologists (see Corning, 1977; Davies, 1977).

HUMAN ETHOGRAMS

E-E's assertion that "there exists a universal 'grammar' of human social behavior" suggests a dubious way to conceptualize the basis for a theory of human (and probably also most other forms of mammalian) social behavior. How dubious is exemplified by the basic primate political biogrammar proposed by Tiger and Fox (1971: 32) as a model for human political behavior. Grammars constitute systematic analyses of the structure and functioning of communication systems, especially natural languages; and however important communication is to social interaction, and vice versa, the two are by no means the same thing (see Burke, 1945). Dawkins (1976: 43–47, esp. 46) has provided an elegant discussion of grammatical models, with an analysis of some of the problems involved in biogrammars. He concludes that "the problem is this. In the case of human language the criterion for grammatical correctness is the judgment of a native speaker of the language. In the case of animal behavior 'correct' and 'incorrect' have no such meaning."

Sociobiology represents a type of grammar, at least in the guise of the models of population and behavior genetics applied to humans, and as understood by its most vociferous opponents (The Ann Arbor Science for the People Editorial Collective, 1977). The Hamilton-Trivers-Dawkins grammar of hard-core sociobiology specifies how rational actors will behave if they are to maximize their individual genetic contributions to the survival basis of their species. But sociobiology also has a more general and omnivorous form, as synthesized by Wilson (1975), who is said to speak openly and often about the "cannibalization" of other disciplines, including the social sciences, that are to be subsumed under the aegis and imprimatur of a dominant (if not rampant) supersociobiology. Such excesses indeed invite retorts such as E-E's that "the new field of sociobiology has certainly justified its own existence as a part of ethology." But aside from tit for tat, E-E's comment is simply wrong and misleading. Evolutionary biologists, population geneticists, behavioral geneticists, and others who have contributed—mostly during the past fifteen years—to the postulation of the hypotheses associated with hard-core sociobiology certainly have not been a "part of ethology" in the past; and E-E is the first to suggest that their work should be so classified in the future. It seems more likely that ethologists, including human ethologists (Daly and Wilson, 1978; Travis, 1978; Barkow, 1978), may well be involved for some time to come, together with experimental psychologists, primatologists, and many other life scientists, in attempts to test empirically the hypotheses of sociobiology.

By chance (see Aubert, 1959), one of the many ongoing real-world experiments in alternative life styles has recently provided some negative evidence bearing upon certain of E-E's sociobiological speculations, with particular regard to Trivers' (1971) most basic hard-core propositions about altruism. Suicide may indeed be the ultimate form of altruism; and E-E is probably right that in hunter-gatherer bands the populations typically consisted of "fairly closely related individuals, so that investment [in suicide] should pay off [in terms of genetic 'inclusive fitness'] for any group member and not just for immediate kin." But in the largest-scale and best-documented report of mass suicide in recent human history, the population members, although otherwise conforming to the size scale and at least some of the other criteria for neolithic horticulturists, were all unrelated genetically except at the level of nuclear family groups, of which many were mother-infant combinations. Jonestown indeed testifies to what E-E terms "man's astonishing ability to identify with his larger group (band, village, people) to the extent of self-sacrifice." Unfortunately for Trivers' theory, the motives of the deceased had to be strictly cultural; they died for social and ideological reasons, not for their genes. From a biological point of view, they were an aggregation of poorly adapted and relatively unfit persons who, from a genetic point of view, took the action best designed to assure the minimization of their inclusive fitness, both individually and collectively (Schubert, 1985a).

HUMAN ETHNOGRAMS

It may well be that "ethologists prefer nonparticipant observation techniques in the natural setting," but where in the world are they going to be able to indulge such preferences? A mirror-lens camera may be perceived, by primitives unsophisticated about cameras, to be less intrusive than cameras that are aimed at them (Hass, 1972: ch. 7, "A Voyage of Self-Exploration," describes how the photos were taken); but a primatologist who attaches herself to a band of nonhuman primates is engaged in participant observation just as surely as Max Gluckman (1955) was a participant in the council of *indunas* when he sat in on their deliberations, while exploring legal ethnography during the colonial era in what is now Zambia.

Still, there is no doubt of the value, at least to ethnologists, of E-E's recurring global expeditions during the past couple of decades to undertake filming of NVC (nonverbal communicative) behavior sequences, of the surviving remnants of various scattered hunter-gatherer bands that have been "sidetracked into a cultural blind alley" (Morris, 1967: 10). They are evolutionary losers only in the genetic sense of inclusive fitness; from the perspectives of group and kin selection, they have usually been very well adapted, which leads most ethnographers to object vehemently to their characterization by Morris. But the question still remains as to how much more value such film records have than, say, the surviving prints of Fitzpatrick travelogues, as contributions to the theory and method of human ethology.

It seems apparent that E-E visited his primitive subjects with various hypotheses about nonverbal communication (NVC) in mind; or, as he might prefer to put it, with mental templates of the fixed action patterns (FAPs) that he hoped to release for purposes of his photography, the task of which was to record on film what he perceived to be acceptable examples of the occurrence of such FAP-NVC behaviors in the subject populations. But the same behaviors may have been associated with different social contexts; other social situations may have produced the same behaviors, or the same social situations may not have produced the target behavior on other occasions. The problem is that even if he did undertake to film (as Fitzpatrick might well, however inadvertently, have done) the undoubtedly long stretches of facial and gestural communication when nothing relevant was being signaled, such data are neither analyzed nor discussed in his reports of the behavior. Instead of evidence pro and con, we are offered examples of good behavior—or, what amounts to the same thing, good examples of the behavior. Hence the nominal objects of his observation, "unstaged social interaction" as he calls it, must be qualified by the phrase "highly selective" for purposes of recording, or at least those of publication. The quest for data not inconsistent with an hypothesis may produce data that exemplify the theory; but they cannot help in the more mundane but scientifically crucial task of attempting to refute the hypothesis if possible.

Are the resulting film images better classified as ethograms or as ethnograms? Certainly the task to which E-E aspires (sec. 7.2), of trying "to understand the evolution and functional aspects of cultural patterns, in the perspective of their contribution to overall fitness," is one that is ordinarily undertaken by social anthropologists, and hence, in the ordinary understanding of the words, constitutes the practice of ethnography rather than ethology.

5
Modern

By the end of the seventies, political scientists had published little that could fairly be described as ethological analyses of political behavior—if by that we mean studies designed, carried out, and interpreted in the ways that zoologically trained students of animal behavior do their field research. Such studies with humans as the subject animals are rare (Masters, 1976; Schubert, 1982a); and such studies by political scientists of other than human animals are nonexistent. Dominance hierarchies of human three-year olds (Barner-Barry, 1977) and maternal aggression in mice (Corning, et al., 1977), are both *pre* or *proto*-political, although in a different sense; and neither proffers an example of political behavior per se. One reason for the dearth of research in political ethology must be the lack of awareness, among political scientists, concerning the types of questions about politics that are appropriate for ethological inquiry and the methodology for such ethological research, given such a suitable focus of investigation (Schubert, 1980a, 1981a).

Why is ethology potentially important to political science? Because its acceptance and development inevitably will bring about a revolution in the theory and practice of the discipline (Schubert, 1976, 1975a). The result will be a political science based upon biological rather than theological reconstructions of human history (Schubert, 1986a), including the political history of our species; and by conforming to—for the first time—and dealing with the transactions between

---

This chapter is a revision of "Political Ethology," *Micropolitics*, 2: 51–86 (1982). Reprinted with permission of Crane, Russak, and Company, Inc.

human biology and human culture (White, 1972), politics necessarily will become concerned with both a broader and deeper concept of humanity and of the political behavior based upon it.

All major subfields of political science will be affected: political theory, for example, will be concerned with the behavior of a species whose history is measured in evolutionary rather than merely historical time. Public administration will have to confront a basic challenge to conventional organizational theory (Caldwell, 1980), in terms of an analysis of the natural units for group size in humans (Tiger and Fox, 1971). International politics will have to eschew the simplistic neo-Freudian theory of aggression upon which present analyses are based and deal with human political aggression in terms of physiological as well as sociopsychological variables (Pettman, 1975). Political socialization needs to enlarge its vision and concerns, giving up its continuing preoccupation with whatever knowledge of law and institutional labels can be imparted by exposure to what purports to be formal education in primary and secondary schools and in its place substituting a concern for political learning that extends ontogenetically from at least the cradle to the grave (White, 1980b). Such changes in all four of the subfields mentioned above—political theory, public administration, international politics, and political socialization—are discussed in more detail below. Comparative politics, which now accepts a lesson from social anthropology (that cross-cultural analysis is indispensable), will learn from physical anthropology that cross-*species* analysis is prerequisite to an adequate understanding of what is common to our species—now that cultural relativism has taught us how different people can be in their beliefs and behavior (Greenwood and Stini, 1977). Studies of political decision making, which have become increasingly analytic during the past generation, will be transformed by hologic theories of brain function that have emerged from neurophysiology during the past decade (Pribram and McGuinness, 1975; Pribram, 1979).

The structure of this chapter is as follows. First we shall consider the scope of evolutionary theory as the governing paradigm of ethology and its implications for the appraisal of competing (feral canid versus contemporary nonhuman primate) models of prototypical human social structure, and also for the evaluation of models of cultural evolution. Next we shall take up contemporary ethology, distinguishing it from the related fields of classical ethology, animal behavior research (including cetacean and canid), primatology, behavioral ecology, and sociobiology. Then we shall turn to ethological approaches to human behavior; and following that we shall consider several examples of appropriate political targets of ethological analysis, for possible future research. These include: a return to first principles in the form of descriptive empiricism, the function of play in political socialization and aggression, the biological roots of political aggression (including war), political statics versus revolutionary theories of political change, and the relationship between the species genotype and cultural aspirations in political ideology and public policy.

EVOLUTIONARY THEORY

Acceptance of the modern synthetic theory of evolution will require political scientists to look upon Homo sapiens as biologists do, as one among a host of life forms that are temporarily and more or less well adapted (Lewontin, 1978) to present geospheric and biospheric conditions (Dunbar, 1960; Ray, 1978). Many of those life forms are in direct or indirect competition with each other for access to the peculiar combination of circumstances for needs satisfaction that defines the unique niche of each species so that its reproduction and continuation may be possible. For a typical simple animal species, even including chordata as complex as top-predator sharks, adaptation has persisted for hundreds of millions of years; for extant mammals the corresponding time is much shorter; and for our own present species, the past includes no more than forty thousand years (Washburn, 1978), while the future seems likely to be very much shorter than that. Even for the hominoids that contemporary research in physical anthropology nominates as probable ancestral types of the genus *Homo*, the relevant era spans perhaps four and a half million years.

For any species chosen at random, the highest probability at any time—and the one certainty, given enough time—is extinction. There are special reasons why the prediction applies with particular urgency in the case of humans at present. Those reasons, which empirically are only too familiar to us all, concern (1) the short-range effects of a form of cultural evolution that is of direct major importance *only* for our own species, although its indirect effects increasingly apply indiscriminately to virtually all other life forms of the biosphere; and (2) the longer-range effects of our specific evolution, as positive feedback upon the geosphere's capacity to sustain a biosphere in which human adaptation—even in accelerated generic rather than specific form—may be possible. It is the differential calculus, the rate of disintegrative change, that is the cause for concern in these matters; and that is due almost entirely to human cultural evolution. This is why evolutionary theory is indispensable as the prerequisite to greater realism in political analysis: the still-prevailing, nonevolutionary science of politics is a theory of political statics. And as long as it remains that, it is doomed to remain a form of history, dealing with dead or dying events while political life moves on.

Hunting Bands

We can begin a more realistic approach by taking seriously the hunting-band model, as a highly probable descriptive statement about human and protohuman social organization and about our generic evolutionarily stable strategy of adaptation for several millions of years up to about twelve thousand years ago. Since then—which is to say, since the onslaught of really rapid cultural evolution (Pettersson, 1978)—increasing numbers of humans, both absolutely and relatively, have shifted their adaptation first to pastoral-agricultural and then to industrial niches; meanwhile, the hunting-band niche continued to be occupied by a few hundred thousand humans, on all of the continents, into the middle decades of the present century. Now individuals survive, but the niche is gone, ploughed under as part of the burden visited upon those hunter-gatherers by more "civilized" (i.e., differently adapted) neighbors who usurped or despoiled the enclaves within which the hunters had been confined. Yet the values of our industrial culture remain so committed to the rationales of imperialism that we spend more TV money and consequently shed more tears commiserating with threatened marine mammals than we do with doomed hunter-gatherers.

Remnant hunting-band populations have preempted the attention of ethnographers during the past two centuries (Harris, 1968) and more recently have attracted that of some ethologists (Eibl-Eibesfeldt, 1979). But we do not know that they are the descendants of populations that had persisted without deviation in a neolithic adaptation. Medawar (1976), 504) has pointed out "that human beings could revert to the Stone Age in one generation if the cultural nexus between one generation and the next were to be wholly severed," in view of "the *reversibility* of exogenetic evolution"—and, for that matter, of genetic evolution as well (Birdsell, 1972, 165). The exogenetic reversibility signifies that, after having made the shift to pastoral or agricultural adaptation, many human populations may have been forced by more numerous or more efficient or luckier competitors into environments so marginal (like the Ik; cf. Turnbull, 1973) that reversion to a hunting-band adaptation was the only alternative to extinction.

Now that we are becoming aware that the dangers of "peaceful" uses of atomic energy may in the long run exceed the short-run dangers of atomic warfare, in terms of degradation of the biosphere, we may be tempted to look upon the hunting-band niche as neither a rhetorical oversimplification nor a romantic fantasy, but rather as a realistic alternative—and certainly the option to be preferred among those likely to be available in the decision set. But the same processes that can be expected to make earth uninhabitable for either agricultural or industrial man certainly will eliminate existing food chains along with their constituent elements, so that whatever plant and animal life survives is likely to be available in patchy dispersion as well as in mutated form. That means that the real question will be whether the hunting bands of the future will be able to find enough to either hunt or gather, thereby postponing the extinction of the species yet a while longer. Any human survivors of nuclear holocaust will themselves undoubtedly be diverse mutants of the present species (cf. Hildebrand, 1968: 250); but that could prove to be their saving grace, in a biosphere that is as unrecognizable from the present as the present is compared to the world of twelve thousand years ago.

In the meantime, bioanthropologists tell us that in biological terms, humans continue to remain optimally adapted to the hunting-band mode of living, even though the niche *is* gone. Assuming that this is true of our physiology, emotions, and both our physical and mental health, then the consequent discord between what we are prepared to do (by genetic evolution) and what we are expected to do (by cultural evolution) ought to produce characteristically deleterious effects upon human social organization and behavior. Barkow (1977: 139), for example, has suggested that

> human behavior is ordered by social norms and these norms vary greatly from society to society. This variation generates a question central to a human ethology: under what biological constraints do normative systems develop? It seems obvious that some systems put more stress on the individual than do others. . . . Presumably, the amount of stress is determined by the degree of departure of the norms from those holding during the latter stages of human evolution. Indeed, were it possible to somehow hold extraneous variables constant, it might well be possible to utilize the cross-cultural measurement of modal psychological stress to gain insight into the nature of early man's culture(s). *Ceteris paribus*, the less modal stress the closer the culture to the normative environment in which we evolved.

And Tinbergen (1976: 520) points out that "ethology is well placed to play a part in re-shaping both environment and our own society because, after a phase in which it stressed mainly the genetic aspect of behavior programming, it is now moving to a position where it begins to map the interplay between our genetic blueprint and phenotypic flexibility, and to spot the pressures which overstretch even our exceptional adjustability" (cf. Lee, 1978; Underwood, 1975, 1979).

Tiger and Fox (1971) have proposed a political biogrammar, with rules of behavior appropriate to the hunting-band mode of living. Antony Jay (1971) provides an analysis of economic competition, based on the model of the entrepreneurial group as hunting band; and a cognate analysis of political competition would prove most instructive as the basis for a better understanding of political organization and behavior. Such an inquiry should also lead to policy changes to reduce the gap between what the human animal is capable of doing well and the tricks that it customarily is asked to perform. This ought to be pursued by ethological methods of research.

Homology?

A question of increasing interest to ethologists and anthropologists alike, although of little apparent concern to political scientists (but cf. Willhoite, 1976: 1112, n. 10), is: which provides the more useful model, for purposes of comparative study of human social organization, the other primates or the carnivora (and especially, the canidae)? Reflecting no doubt the influence of physical anthropologists and their earlier battles with theologians (cf. Day, 1920), the only hypothesis discussed until recently concerned the extent to which human society resembles or deviates from the social behaviors typical of one or another of the great apes, thereby treating the question as one of homology (cf. Masters, 1973a). But the homological approach, however tautologically important in physiological, genetic, and immunological approaches to the taxonomy of the animal kingdom, provides no certain basis for predicting behavior, as Darwin's studies of the finches of the Galapagos showed so indisputably (Wallace and Srb, 1961: ch. 7); and it is especially unsuitable for predicting social behavior. As Tinbergen remarked, "we should seek analogies no less often than homologies—meaning by analogies 'similarities convergently evolved in less closely related species in adaptation to similar niches'" (as quoted by Medawar, 1976: 505). And Blurton Jones (1975: 83) has queried why humans are not "compared with all other mammals so as to discern the likely adaptations to [the] hunting [of] animals larger than oneself (social hunting in wolves, wild dogs, spotted hyenas, and lions) in an omnivorous ape moving into savannah countryside? This is a more strictly zoological comparison [than the comparison with other primates], stressing the ecological niche as well as the geographical habitat, and trying to take into account the phylogenetic relationships. It is a more cumbersome approach and one that leads to no immediate answers, but it is the approach that I and many other zoologists would support." Since the midseventies a rising chorus of support has become apparent, including the book *Wolf and Man: Evolution in Parallel* (by Hall and Sharp, 1978; and cf. King, 1980).

Crook (1975: 103) has conveniently summarized Schaller's findings about the effects of the formation of hunting groups in carnivores, adding conclusions of his own concerning their significance:

> 1. Group hunting results in greater success than individual hunting, and groups take larger prey animals.
> 2. Group size is flexible in relation to the character of the prey hunted.
> 3. Group hunting often involves cooperative circling, shortcuts, and relaying procedures which reduce the energy expenditure of the group. These tactical manoeuvres thus develop prior to language.
> 4. Division of labour occurs in some species, some individuals hunt, some track, some kill, some carry, while others care for the young. Yet others may maintain vigilance. These functional roles are also interchangeable.
> 5. In dogs and wolves greeting ceremonies prior to hunts seem to strengthen bonds and synchronize behavior.
> 6. A tendency for intraspecific aggression to out- groups tends to produce separate group ranging or group territorialism, as in the hyaena.
> 7. Food storage in trees, in water, or under a stone may occur within the range.
>
> If we superimpose these characteristics upon a chimpanzee-derived picture of possible hominid hunting techniques, we get a vision of a process remarkably resembling that of some primitive peoples. . . . The origins of inter-group aggression and warfare are clearly deducible from such an account through considering the effects of food shortage on the relationship between groups in neighboring territories.

Thompson (1978: 951, 982–83; and cf. King, 1976) has tested mathematically several hypotheses about the evolution of territoriality and society in top carnivores. He notes that, as Alexander (1974) has pointed out, evolution has had to overcome the disadvantage that sociality facilitates the spread of parasites; and Thompson concludes that "in the nonhuman primates, sociality is explained almost entirely by anti-predation . . . and differential foraging success—plus some intergroup dominance." Although "in man, sociality is further developed than in any of the nonhuman primates. . .this evolution does not seem to represent an intensification of any of the nonhuman primate patterns." Furthermore,

> the carnivores, like the primates, are more social on the whole than the majority of mammals. . . . Man has become more social than any other mammal. . . . [Apparently] human evolution has converged with carnivore evolution in many ways. This convergence might arise from the fact *that man* (unlike the nonhuman primates but like the social carnivores) shares food, possesses weapons capable of inflicting fatal injuries on his conspecifics, *is not himself subject to predation*, and often expresses territoriality toward the limiting resource food. There is no nonhuman primate which manifests all these traits, and there are only three species of carnivores (plus man) which do. . . . Explanations in human evolution have tended to ignore the possibility that natural selection may act directly on behavior. (emphasis added)

Geist (1978: 116–23) has generalized and emphasized the latter point, that natural selection acts directly on behavior. Michael Chance (Chance and Larsen, 1976: 328) had made earlier the same and manifestly incorrect italicized statement (e.g., that "human beings in human societies are not subject to predation"). Certainly humans have been subject to predation by felids, canids, and occasionally by raptors, sharks, and bears (among other predators) throughout recorded history; and it is almost certain that our species was even more vulnerable to predation during the hunting-band adaptation era of more than twenty thousand years ago, when gigantic species of felids (inter alia) were in competition with us; not to mention of course the conspecific predation at which we still excel. And surely conspecific predation affects many species, generally even more strongly than does predation by top carnivores. Probably for at least a million years, man has remained "the most dangerous game" of all.

Cultural Evolution

Cultural evolution has been frequently discussed (Masters, 1975; Corning, 1973; Cloak, 1975; Lancaster, 1975; Underwood, 1975; and Hill, 1978). Mainardi (1980: 242–43), a leading ethological student of animal cultures, states that

> a comparison between the mechanisms of biological evolution and those of cultural evolution . . . has shown . . . how certain factors play the same role in both types of evolution: natural selection (the maintenance and diffusion of both cultural and genetic mutations depend on their survival value), chance, and migration. One should recognize, however, that genetic mutation differs from cultural mutation in that genetic mutation is always random while cultural mutation may be purposeful. But beyond that, the most important and significant difference between the two may lie in the mode of transmission: constrained and slow in chromosomal evolution but malleable and swift . . . in cultural evolution. A genetic mutation occurring in an individual must await the next generation before it can start to spread, and then only to the offspring of some of them. Each further step requires waiting another generation, and then it can occur only along the lines of kinship. . . . [But] non-

> genetic evolution of behavior can be an efficacious means for populations to adapt rapidly to changing situations or to conquer new and different environments.

Mainardi's paper discusses a broad array of instances of social learning and the transmission of traditions of behavior among nonhuman animals, including such popularly cited examples as termite fishing in some chimpanzee populations, and the consequences for one troop of Japanese macaques of the creative discovery of first potato and then grain washing, by "the brilliant Imo" (as she has been saluted by Harvey Wheeler, 1978: 315).

Humans have no monopoly of either social learning or cultural transmission (Bonner, 1980); but they surely have such a preponderant share that this is one topic concerning which social scientists have much more to teach ethologists than vice versa. But if they seek to express their understanding of culture metaphorically in terms of evolutionary theory, then it behooves social scientists to learn considerably more about genetic evolution before they presume to articulate theories of cultural evolution.

But even if they do, there are problems in working out the analogy. Genes exist *only* in vivo; while culture exists only partly in the brain (or the mind's eye) of the beholder: much of culture obtains in vitro much of the time. Nothing is being said here about the intrinsic meaning of unperceived objects; but rather that many cultural elements are physical objects, and sometimes they persist (like the Dead Sea scrolls) for relatively extended periods of time, independently of the perception of any humans. It is doubtless true that culture, like genes, must be transmitted in vivo; but unlike genes, culture is not necessarily stored or preserved in vivo. And Blurton Jones (1976: 440) has indicated other difficulties.

> Cultural inheritance is not a complete replacement for the theory of natural selection. It is merely a means of transmission of behavior. Some essential features of evolution by natural selection are not paralleled. The source of specificity, a substitute for differential mortality, must be specified. If cultural inheritance is to be adaptive, then only certain individuals must get their traits transmitted to subsequent generations and they must be traits that are in some ways adaptive: something must happen to individuals that do possess/acquire these traits. Cultural inheritance as formulated at present is merely a means of transmission.

What, indeed, *do* we mean by the "adaptation" or the "selection" of a culture, or of any of its constituent elements? Surely *not* that those persons who maximize their inclusive *genetic* contributions are best adapted to their culture (but see Durham, 1976, contra; and also Morgan, 1979). We can reduce the problem to potsherds and arrowheads, and it may make preeminent sense to do so when these are the only evidence available, just as extrapolations to social organization as well as to physiological structure can be based upon a few sets of footprints preserved in volcanic ash for four million years.

A spate of recent efforts to present theories of cultural evolution includes those of Alexander (1979a, 1979b), Darlington (1978), Durham (1979), and Lockard (1980). Masters (1970) proposed a linguistic model, with phonemes acting as units of culture, analogous to the complex protein molecular compounds (genes) that are conceptualized to act as the units of biological inheritance (Underwood, 1979). Hill (1978) has proposed a "concept pool," with ideas in competition with each other for survival. Other theories are discussed critically in a modern text in bioanthropology (Greenwood and Stini, 1977: ch. 20).

A potentially more fruitful approach to understanding how culture evolves builds upon the work of Wright (1977: 471–73; 1978: 452–53; Eshel, 1972; Durham, 1976), whose theory of group selection finds increasing support, for humans, as paleoanthropologists extend our knowledge of human social structure and behavior during the tens of thousands of years before the era of history begins. Also of critical importance is epigenesis (Waddington, 1957, 1960, 1975), a theory that explains how the environment, which during the ontogeny of social animals includes social groups and which for humans also involves the group's culture, affects selection. Waddington (1975: 218) has defined epigenesis as the causal study of development, and he argues (1960: 89, 94) that

> natural selection pressures impinge not on the hereditary factors themselves, but on the organisms as they develop from fertilized eggs to reproductive adults. It is only by a piece of shorthand, convenient for mathematical treatments, that indices of selective value are commonly attached to individual genes. In reality we need to bring into the picture not only the genetic system by which hereditary information is passed on from one generation to the next, but also the "epigenetic system" by which the information contained in the fertilized egg is translated into the functioning structure of the reproducing individual. As soon as one begins to think about the development of the individuals in an evolving population, one realizes that each organism during its lifetime will respond in some manner to the environmental stresses to which it is submitted, and in a population there is almost certain to be some genetic variation in the intensity and character of these responses.

He adds that

> the genetic system in a population may determine a phenotypic appearance which previously could only be obtained under the combined influence of the initial genotype and a specific environmental stress. . . . Moreover, not only do animals exhibit behavior which can be considered as the exercise of choice between alternative environments but in many cases they perform actions which modify the environment as it is originally offered to them; for instance, by building nests, burrows, etc. Thus, the animal by its behavior contributes in a most important way to determining the nature and intensity of the selective pressures which will be exerted on it. Natural selection is very far from being as external a force as the conventional picture might lead one at first sight to believe.

Geist (1978, 116–23; Schubert, chs. 12 and 13, below) accepts Waddington and adds that behavior is the cutting edge for change in first physiology and then morphology, with genetic change coming only after the limits of plasticity for behavior, physiology, and morphology all have been exhausted. But for humans, culture is determinative of many kinds of behavior; and it is transactive with physiology and morphology in influencing much of the rest of behavior. Cultural evolution ought to be studied, therefore, as both a cause and a result of the behavior of specific human groups in maintaining their respective adaptations to particular habitats.

MODERN ETHOLOGY

Ethologists tend to be methodological individualists whose overall emphasis is at least as strongly upon individual as upon social behavior. Furthermore, although many ethologists also tend to be ecologists, they are distinct from microbiologists, geneticists, veterinarians, anthropologists, neurobiologists, biochemists, biophysicists, and comparative psychologists—all of whom are among the scientists concerned with the study of animal behavior from a biological viewpoint. Summarizing a classic essay by Tinbergen, Blurton Jones (1975: 70–79, 88) has explained that ethologists distinguish at least four different types of causes underlying animal behavior. The most proximate answer to the question "Why does the animal do that?" is physiological motivation. The second level is that of development and involves embryology, learning, and related disciplines. Third is the function of the behavior, which relates ethology to ecology and population genetics in appraising the survival value of the behavior. Fourth is the phylogeny of the behavior, using methods analogous to those of comparative anatomy to investigate the behavior's evolutionary history. Moreover, Blurton Jones continues, ethology assumes certain characteristic attitudes toward research. These include: describing as the result of painstaking observation what an animal does as a prerequisite to analysis of the behavior; learning the natural history of the animal and its habitat prior to attempts to experiment with it; and recognizing that theory is concerned with the explanation of *observed* behavior. This leads to empiricism, operationalism, and quantification and applies equally to social as well as individual behavior. "The basic belief," says Blurton Jones, "is that the animal, or the data, can generate hypotheses and that these will be better hypotheses than those derived from elsewhere." He then expresses skepticism that the findings of "experiments actually account for anything in real life," or that nontautological hypotheses about human behavior can be deduced from either ordinary experience or cultural images of it. He expresses great confidence, however, in the ability of inspired observation to denote unexpected associations.

Charlesworth (1978: 248–50) has suggested that there are two distinctive features of ethology that together distinguish it from other behavioral and social sciences. The first is acceptance of the modern synthetic theory of evolution as the governing paradigm to guide research into present animal behaviors and their developmental history. The second is the goal of explaining behavior as it occurs in natural habitats. Consequently, naturalistic observation and description predominate as the methodology during the early stages of investigation, with laboratory research following later. Acceptance of the evolutionary paradigm demands comparison of present behaviors with what can be reconstructed about the habitats and behavior of an animal's ancestors, as well as comparison with its most closely related contemporary species. Also involved is ontogenetic information about the animal's development and learning, its neurophysiology, and both its population and behavioral genetics. Ethologists assume, according to Charlesworth, that information from these disparate sources should dovetail into a consistent and coherent synthesis to explain the animal's behavior. In particular, the evolutionary concepts of selection pressure, adaptation, survival, and reproductive success provide useful conceptual tools for guiding empirical research in animal behavior and for analyzing the cultural phenomena that are the indicia of the conceptual change studied in the history of science. A second benefit of evolutionary theory is that it keeps research focused upon animal-environment interactions, which according to Charlesworth have been neglected in empirical research by psychologists.

Classical Ethology

Modern ethology should be distinguished from classical ethology (Schubert, 1979, ch. 4, above), which began half a century ago in the naturalistic observations of Continental zoologists, especially Lorenz and Tinbergen, and primarily of birds and fish. Classicists describe manifest patterns of species-specific behavior, related to such primal functions as feeding, mating, reproducing, parenting, and defending against predation. Most behavior of the typically nonmammalian animals observed is attributed to direct genetic controls, as in the case of "fixed action patterns," such as a herring gull nestling pecking vigorously at the red spot on the beak of a parent (or an artificial replica of such an object) in order to induce ("release" in the vernacular of ethology [Heyner, 1978]) regurgitation, and hence feeding; similar genetic controls are observed in the "imprinting" that initially was thought to be irreversible and only minimally dependent upon learning, but determined by attainment of a species-typical growth stage plus the statistical presumption that, in the habitat defined by the niche, an appropriate "releaser" (stimulus) is a highly probable event.

There are at least three principal ways in which modern ethology differs from classical ethology. Modern ethology is committed to the substitution of detailed physiological knowledge, at all levels of biotic organization (ranging from genetic to the social structure of populations), in place of the black box that had surrogated for endocrinology, neurology, cytology, and even embryology, in classical ethological research. That commitment has entailed a rapprochement with psychologists generally and comparative psychologists in particular; it has also resulted in a much greater emphasis upon laboratory research, as a complement to, and in counterpoint with, field experimentation. The third major difference concerns the predominant class of animals to be investigated. Here mammals have tended to displace birds and fish, at least as the center of attention. This change reflects in part the resurgence of primatology and the flourishing of field studies of top carnivora during the sixties—at the precise moment in history when awareness became widespread of the universal and inexorable destruction of the niches of these animals, which, in one case homologously and in the other analogously, are closest in their behavior to humans.

The rapprochement with psychology resulted not only in changes in the paradigm and the workways of ethologists. It became reflected also in the organization of professional societies, so that by 1966 Robert Hinde could describe the new field of animal behavior as a *synthesis* of ethology and comparative psychology.

Current Emphases

Recent interest has focused on marine mammals, canids, and of course primates. The cetaceans illustrate the characteristic orientations of the two major constituent professional groups, with ethologists typically studying subjects such as reproductive behavior of feral populations of sea lions on off-shore archipelagos, while comparative psychologists are more likely to keep a couple of dolphins in tanks where they can test the communicative and learning capacities of the animals, as most dramatically illustrated, perhaps, by the experiments in linguistics and cognition of psychologist John C. Lilly. Public awareness has been reinforced, of course, by the well-publicized activities of environmentalist organizations such as Greenpeace and by entrepreneurs who exhibit various cetacea in circus-type performances. Environmentalist groups also have supported the proliferation of captive groups of wolves in a score of pack reservations scattered around the country, with considerable research focusing upon social behavior, especially sexual behavior and dominance (Klinghammer, 1978). At the same time, ethological research has flourished on feral populations of wolves, jackals, hunting dogs, coyotes, foxes, and other wild canids (Fox, 1975). Equally important has been the eruption of primatology during the past two decades (see Willhoite, 1976), with the establishment and expansion of primate research centers and facilities throughout the country; this comes in addition to the increased use of nonhuman primates as well as humans as laboratory animals in biomedical experiments in medical research centers and also in addition to the continuing exploitation of nonhuman primates in zoo exhibitions (Hediger, 1964, 1968, contra).

The behaviors of marine mammals and canids are studied as specializations, within the field of animal behavior, that are of particular relevance to human ethology. Primatology, however, represents a distinctive focal point and synthesis, with a differing disciplinary base and certainly with broader consequences for human biology than are implied by either dolphins or dogs (to say nothing of rats and pigeons). Physical anthropologists have taken a leading role in field studies of the quasi-feral populations of gorillas, chimpanzees, orangutans, baboons, macaques, langurs, lemurs, and American monkey species still amenable to such investigation. Those primates are only "quasi-" feral because of the extent to which human penetration (including the effects of primatological participant observers) of the habitats of nonhuman primates has altered their niches (Rainier and Bourne, 1977), so much so that primatology now is confronted with a problem analogous to that found in ethnography, that is, how much and in what ways has our modern species already altered, in this sense artificially, the behavior of the vestigial remnants of primitive human populations? Considerable publication has been directed to the work, mainly by psychologists, on the likes of Sarah, Washoe, Lana, and their cohorts, in attempts to teach human language to chimpanzees and gorillas (Sebeok and Umiker-Sebeok, 1980; Terrace, 1979). Two other facets in particular, of recent primatological research, are relevant here: one is the strong concentration on comparative psychobiology undertaking research with the same design applicable to both nonhuman and human neonates or juveniles. Initially such studies were done seriatim as between the subject species; now they are often carried out in parallel (Chevalier-Skolnikoff and Poirier, 1977). The other facet is the reciprocal flow of knowledge and techniques, developed in human medicine, for use in the care and treatment of nonhuman primate laboratory animals.

Behavioral Ecology

Ethologists and behavioral ecologists alike study how animals behave in response to their environments; and both are concerned with adaptation and selection under evolutionary theory (Krebs and Davies, 1978; Pyke, 1978). Behavioral ecology has roots in population biology and has been influenced strongly by the rise of sociobiology and by the even more elegant mathematical statement and testing of evolutionary hypotheses developed by the late Robert MacArthur of Princeton, particularly in the form of optimality theory and evolutionary stable strategy.

Sociobiology

Sociobiology (Barlow and Silverberg, 1980), also stemming from population biology, emerged in the sixties as a rebuttal of group selection theory and of the hypothesis of innate biological mechanisms for the control of upper population size limits, proposed by Scottish ecologist Wynne-Edwards (1962). But sociobiological theories of kin selection, inclusive fitness, parental investment, and reciprocal altruism have little to contribute to the better understanding of political behavior. According to Harvey Wheeler (1978: 307), future developments in neurobiology may furnish a sound foundation for human sociobiology but until then we must depend upon a phenomenological analysis of consciousness and symbol formation which together can help explain the development of human social organization and also provide the basis for a taxonomy of the evolutionary trajectories of symbolic idioms, and ultimately of a politics of symbol formation." Social scientists ought to entertain strong reservations (see also ch. 9, below) about the scientific propriety of indulging in such simplistic metaphors as that of a human species "gene pool," plus the anthropomorphisms with which sociobiological thinking is rife. Hull (1978: 687) has remarked that "various biologists may have been attracted to this highly individualistic view of evolution for deep-seated psychological reasons. Others may be repulsed by it for similar reasons." Barnes (1968: 100–101) has discussed how difficult, and perhaps impossible, it is for scientists to think about their objects of inquiry other than subjectively; and we can accept his caveat that simplistic metaphors may initially and for a while do more good than harm. But modern genetics is no infant industry in concept construction; and the time has come when we should do better than to use such analogies as "the culture pool."

Genetic evolutionists are concerned with a process that in humans is extremely slow, cumulative, and largely uncontrollable for ethical, social, and political reasons. From a measurement point of view, the genes embodied in the cells of any population of animals, extant or extinct, and most certainly including humans, are much better thought of as an open than as a closed system. In their pleiotropic and polygenic effects some such genes in part are expanding while others are contracting in response to chance biochemical transactions; in response to environmental effects upon the organic systems of the phenotypes of the animal; and (particularly in the case of humans, but by no means limited to them) in response to the impact of cultural evolution upon their behavior, physiology, morphology, and, ultimately, genotypes (Waddington, 1957, 1960, 1975; Geist, 1978; also ch. 12, below), as well as upon their environment (including other animals).

HUMAN ETHOLOGY

Here we shall consider the relationships between animal signals and human nonverbal communication, ethology and ethnography, and nonverbal communication and politics.

Signals

The ethological approach to the study of language behavior does have direct and important implications for the study of political behavior. Crook (1975: 105) undoubtedly is correct that "man's linguistic ability to refer to objects and their relationships using a verbal code of nouns and verbs has had psychological effects that raise Man to a distinct grade of psycho-social complexity." Much of human consciousness is mediated by language, and the critical matrix of human culture and cultural evolution also consists of language. Social scientists and physiological psychologists are better qualified than ethologists to study most aspects of language (cf. Lyons, 1972); but there are aspects of the evolution and function of language that are better analyzed by ethologists (Andrew, 1963, 1972; Dawkins and Krebs, 1978). Language is by no means all of human vocalization, however; nor is it even, in many social situations, the most important aspect of human communication. It is here, in regard to nonverbal communication (cf. Hinde, 1974: 139–46, on nonverbal communication and human language) that ethology has much to contribute to political analysis.

Studies of human nonverbal communication constituted one of the major emphases and developments in human ethology during the seventies. Operationalized categories are proposed and discussed by Ewan Grant (1969) for facial expressions and by Brannigan and Humphries (1972) as well as by Blurton Jones (1972: ch. 4) for facial and other modalities of nonverbal expression. Hinde (1974: ch. 10) provides a general introduction to NVC in humans; and so does social psychologist Michael Argyle, but with greater emphasis upon social interaction (1972). Crook (1975: 104; and cf. van Hooff, 1972) has emphasized the importance of homologous comparisons with other primate species because "many aspects of human non-verbal communication by facial expression and gesture resemble those of non-human primates. They appear universal to our species and are therefore likely to depend upon innate bases. The same appears true of the components of certain communication rituals, the greeting interaction for example. . . . Yet these same elements in the communication repertoire mingle with culturally induced ways of expression to produce complex codes of manners that carry the communication of affect in a given society."

Contemporary research by human ethologists links NVC with social behavior as in the study of human postural signals (Lockard et al., 1978), or of the predictive value of facial signals to the outcomes of encounters between young children (Zivin, 1977), or of long-distance transmission of facial affect signals (Hager and Ekman, 1979). Symposia exemplify current emphases in the analysis of NVC (e.g., Key, 1980, 1982); much of the relevant research of the past decade, including all aspects of human ethology discussed here, appears in Travis et al. (1977), Travis (1977), or subsequent compilations in the same series.

Ethological Ethnography

Blurton Jones has heralded the rapprochement between ethology and cultural anthropology, and he exemplified this by his collaboration for more than a decade with the Harvard Kalahari Research Group (cf. Lee and DeVore, 1976) in comparative studies of attachment behavior (Blurton Jones, 1972: ch. 12; and Blurton Jones and Sibly: 1978) among the Kung San peoples of southwestern Africa. Blurton Jones and Konner (1973: 695, 734) have remarked:

> We feel it may be significant that we find more vigorous activity by the Bushman girls than by the London girls; no doubt this helps them in their active adult life no less than the London middle-class tradition of rearing sessile women helps them in their housebound adult life. . . . The cross-cultural differences in the girls, having no possible relationship to the known endocrine differences in the populations, and plenty of possible relationships to the observed differences in infant care and opportunity for child-child interaction, could be taken as justification for those who argue that our culture makes girls unnaturally well-behaved and other than they might be.

Blurton Jones' relationship with the anthropological project involves possible integration of ethological with ethnological approaches, in the design and carrying out of future research; and this points in a direction that political science might well strive to emulate with ethologists. A somewhat related enterprise has been underway for a longer time by Eibl-Eibesfeldt (1974, 1979) and his collaborator Hass (1972), in their quest to record photographically good examples of fixed-action-pattern facial expressions that remain invariant across a variety of human cultures, especially primitive ones (see also ch. 4, above).

Nonverbal Political Communication

There are several recent attempts to look at political campaign gestures from an ethological perspective (Masters, 1976, 1978b; and cf. Beck, 1975; and Morris et al., 1979), but the surface has merely been scratched in terms of what might readily be done now, with available methodology and technology (Lehner, 1979; Schubert, 1980a, 1981a), in studying the nonverbal content of political messages and communication processes. One example of a line of inquiry already well established, in political science as well as in ethology, is voice stress analysis (Wiegele, 1977, 1978, 1979a) to reveal systematically the affective component of political speech, which can make possible correlational and other interactional analyses of both the manifest and the latent dimensions of oral political communication. Another major potential line of inquiry concerns the counterpoint between nonverbal communication and speech in the bargaining behavior of real political decision-making groups in the field, as exemplified by an observational study (Schubert, 1982a; subsequently also by the studies of village-council decision making by J. Schubert, 1983b, 1984, 1986) of the Federal Tribunal, which functions as the supreme court of Switzerland. Most of its panels are five-judge, bilingual, and multisubcultural panels, required by law to discuss cases and make decisions in courtrooms to which the public is readily admitted. Over a hundred and fifty cases/decisions were observed and audited, and both gestures and other modes of nonverbal communication and the substantive content of verbal argument (variously in either the German or the French languages, mostly according to the language socialization of the speaker), were recorded and subsequently analyzed in relation to decisional outcomes. Independently of cultural cues (which also were important), nonverbal communication proved to be an important indicator, for the way in which these Swiss judges in fact voted in their decisions.

ETHOLOGY IN POLITICAL ANALYSIS

Ethologically oriented biopolitical research has been recently and well summarized by Wiegele (1979b; Peterson, 1979; and Falger, 1978). Here we discuss several further ways in which ethology can contribute to more effective political analysis.

Methodology

To become practitioners of ethology, political scientists will have to return to their own roots and regain the lost art of descriptive empiricism. To be able to describe political behavior, they are going to have to observe it; and to do that with maximal control over their own cognitive and perceptual preconceptions and natural limitations, they had better follow the lead of contemporary ethologists and utilize appropriate technical props such as videotape equipment for the field (Lehner, 1979: 196–97) and specialized video and audio scanners and analyzers for the laboratory, portable polygraphic equipment (Tanenhaus, 1977) for recording individual physiological responses in the field, and portable electronic data collectors (Lehner, 1979: 162–68) for coding nonverbal communication and postural behavior of interacting small groups. With very few exceptions, such as the House of Representatives of the United States Congress, political decision-making groups will be found not to exceed the upper limit of size of hunting bands (about a hundred persons); and since that in itself is hardly likely to be coincidental, the social structure of most political behavior most of the time will be found preadapted to research inquiries and engineering applications premised on the overarching paradigm of human and biotic evolution.

Socialization

A symposium edited by Suzanne Chevalier-Skolnikoff and Frank Poirier (1977) provides an excellent introduction to cross-species analysis of primate infant/juvenile development and includes pioneering chapters (by Sue Taylor Parker, Kathleen R. Gibson, and Chevalier-Skolnikoff) on comparative cognitive development, testing Piagetian theory. But this book also, however incongruously, continues to project the sponge model of learning, in Poirier's definition (on pp. 1–2) in the introductory chapters:

> The socialization process simultaneously refers to the external stimuli received by an organism, the individual nature of the process, and to the end product or consequence of socialization. . . . Each individual is the outcome, the result, of a given socialization process and we must look at the variables influencing this output both vertically (through time) and horizontally (in terms of social interactions) [*sic*: which in context seems to mean exposure to contact influences rather than to a recursive relationship]. The results, or consequences, of socialization not only depend upon the original genetic material of the individual . . . and the degree to which climate, nurturance, and other factors permit realization of that poten-

> tial; they are also influenced by the behavior of the adults and peers with whom the individual has been in regular contact.

Most of the literature on political socialization is modeled after this sort of learning theory from experimental psychology and tends to view children as relatively passive laboratory animals, whose responses to sets of standardized stimuli (e.g., the home, the church, the school) supply a basis for the measurement of the success of their domestication in the prevailing civic culture; or in the lingo of an earlier political science, in the extent of their civil education (cf. Peterson and Somit, 1982; and Strum and Latour, 1988).

Blurton Jones (1975: 76–77) has discussed the implications of attachment research (cf. also Alloway et al., 1977) for social science concepts of socialization, which of course apply also to political socialization.

> The traditional view of socialization in psychology seems to rest on a false conclusion from the anthropological data. . . . The fact that social behavior and child-rearing practices vary from culture to culture implies that variations in social behavior are associated with variations in child-rearing, but it does not imply that the existence of social behavior depends on the kinds of features of child-rearing that are seen to vary between cultures. . . . In evolutionary perspective one can argue that much of the meaning of "socialization" derives from a mismatch between the preadaptations of babies and the "Western" views on what a mother should do. . . . There is according to Trivers' theory of parental investment a situation in natural selection that makes for a conflict of interest between the generations, with parents opposing aggression by one offspring towards the other. This makes all the more clear that one has to test whether socialization pressures are actually effective. Anthropologists assume that they are, parents find them amazingly inefficient, now biologists show that there is the possibility of a selective disadvantage for the offspring in succumbing to *some* of the socialization pressures. Some research workers keenly involved with ethology have criticized the whole concept of socialization. . .stress[ing] the active role of the infant in evoking the interactions and relationships that constitute its social environment. Indeed, there is now a fair amount of evidence from developmental psychology that the infant has a considerable influence on its caretakers.

Adolescents were not the only cohort to become radicalized, within the context of a diverse array of at least Western national cultures, transactionally with the political events of the sixties (cf. Beck, 1976; Peterson, 1973); and this suggests that political socialization is an *interactive* process, for humans of any age. Political socialization ought therefore to be viewed, like other cultural change, as a dynamic and reversible process and *not*, as so frequently has been the case in the past, as though it were a form of imprinting. So here is another respect in which the use of political ethology can bring political science into closer touch with reality.

Play and Aggression

Many political psychologists have presumed that, whatever may be true in the case of charity, tyranny begins at home. But one of the striking research findings of experimental primatology even a generation ago (Harlow and Harlow, 1962) concerned the extent to which play among infant and juvenile monkeys is critical to their development as normal adults. With the subsequent explosion of field primatology and establishment of primate research centers and field stations, there has been during the seventies a considerable expansion of research on the function of play and its role in socialization and in ontogeny of nonhuman primates, accompanied by a parallel although less striking development of the study of play in canids (Bekoff, 1974) and other mammals as well (Bekoff, 1977).

Play usually is discussed in terms of its positive contributions to socialization, the acquisition of the communication matrix, social integration and cooperation, and the learning of adult social roles (Baldwin and Baldwin, 1977b; Poirier et al., 1977; Bruner et al., 1976). The Baldwins (1977a: 384) point out that exploration and play can lead to both adaptive and maladaptive consequences: infant and juvenile primates (depending on the species and habitat) are particularly subject to predation, and "in addition, more explorative and playful animals may be exposed to higher risks of having dangerous falls, becoming poisoned, getting lost, or experiencing other harmful consequences. As every human parent knows, exploring and playing children are capable of exposing themselves to countless dangerous situations." Much of the recent research literature is discussed in the same thoughtful paper by the Baldwins (1977a), who provide an excellent description (pp. 365–66) of how horseplay turns into aggressive play, and with what consequences. They remark that "in the field it is usually difficult to distinguish when approach-avoidance play grades into play fights or aggressive play. . . . When rough play escalates to the painful level, the animals learn to discriminate between aversive, aggressive play (which they avoid) and controlled play (which they still seek out). Eventually, the level of play activity tends to decline because the aggressive play is often painful (and hence punishing) and the novelty of controlled rough play is exhausted."

A major work on aggression and play is a field study of rhesus monkeys by Symons (1978), who has suggested (1977) that evidence from both the literature generally and his own research fails to support the popular hypothesis that one major function of play is the learning of dominance rank; Symons argues that even if such learning were to occur during play, there is good reason for believing that it is not a *function* of play. But if dominance rank is not a function of play, it is unlikely to be a function of more supervised school activities either, which (if true) would certainly have implications for much recent and ongoing human ethological research on dominance hierarchies in children (e.g., Omark and Edelman, 1975; Omark et al., 1976, 1980). Captive play groups of human children are among the most common subjects of ethological study, but there is no reason why such study cannot be made also of unsupervised, naturally coalescing play groups, for comparative purposes (cf. Omark et al. 1976); and political scientists interested in ethological analysis of political socialization will find Symons' work (and Blurton

Jones, 1967) a useful introduction to the study of aggression and play—which ought to have an important bearing upon adolescent and adult aggression, political and otherwise (Watts, 1976).

Predatory Aggression

For many adolescent humans who find themselves in suitably stressing habitats, the jump may be a short one from techniques and psychic rewards of aggression learned in juvenile play to the more destructive violence manifested in territorial and status competition among youth gangs, and in predatory aggression by such youths upon other youths, younger children, the aged, and the infirm (van Dijk, 1977). Thompson (1978: 969–70) has reported that "the crucial ratio of attackers to defenders required before fatal aggression can be attempted can be accurately estimated at about three to one. . .[and] only in [canid or felid] species groups [that include three or more adult carnivores] has intraspecific killing been found to be a major mortality factor." If the same ratio holds for humans, it bodes some interesting implications for both research on human aggression and public policy. In any case, ethological research on carnivore predation has broad implications for political research in human aggression in human groups up to the size of hunting bands (Schaller and Lowther, 1969; Curio, 1976).

Blurton Jones (1975: 89) has reported that among the Zhun/twa Bushmen, all attacks by animals were "on teenagers and old people, except for people specifically hunting predators, a pattern strikingly similar to the pattern of deaths from predation in baboons. However, actual mortality from predators is *extremely* low in the Zhun/twa." Of course, the rate was undoubtedly lower, by the 1960s when Blurton Jones consulted his informants, than even a century ago when wild animals capable of predating upon humans were somewhat more abundant in southwestern Africa. On the other hand, aggregates ranging from hundreds of thousands to millions of humans are killed each year as the result of conspecific predatory aggression, including but by no means limited to international warfare and not counting the exponentially larger number of human injuries due to such causes.

War

International warfare remains a major subject for political analysis, notwithstanding three millennia of recorded studies from a multiplicity of perspectives. An evolutionary one indicates that humans and their hominid forebears probably have been at war for several million years. Long before the disappearance some thirty thousand years ago of the other recent human species, the Neanderthals—doubtless due to a combination of warfare and predatory aggression and outbreeding on the part of the *sapiens*—the only kind of war possible has been intraspecific. The hunting-band hypothesis provides little basis for optimism concerning the prospects for peace, and perhaps least of all through the further centralization and integration of all humans by means of a single "one-world" super-state. Any such government could act at all only by suppressing even what diversity now remains, with and under the weight of a crushing authoritarian bureaucracy. What is more, even if it could be established, it would confront a global array of civil wars and revolutions, on a scale that would make the past three decades look like the Pax Romana. As Roger Masters (1979a: 284; Masters, 1979b: 37, and Caldwell, 1980: 8) has so well said, "there is something unnatural about [totalitarian] social organizations in a primate like *Homo sapiens*. After living in small cooperative bands for over 3.5 million years, the admixture of aggression, fear, and attachment to which our species is functionally adapted is twisted and tortured by mass totalitarianism." Masters was speaking of the level of *national* totalitarianism; but his remarks necessarily apply with even much greater force to the prospect of *international* totalitarianism.

Political Change

Scott (1978: 19–20) was speaking of nations or their subdivisions when he remarked that "the solution of violent revolution is an attempt to speed up [the] process [of fundamental social change] by bringing about the early death of those individuals who have their behavior organized in what is considered to be an undesirable fashion." His observation represents an appropriate biological and evolutionary statement about how populations of organisms respond to cataclysmic change in their habitats, in an endeavor to adapt to that change in requirements. We need not assume, with Wynne-Edwards (1962), that old leaders are sacrificed by any process such as self-sacrifice—although that would be a better probability for humans than for any other species. All we need assume is that the individuals who best typify the ability and behaviors adapted to the previous conditions of the habitat will be among those *least* well adapted to the new environment; therefore they will be among the first to be eliminated, by natural selection. Individuals better adapted to the new habitat (i.e., those, in the case of humans, who can lead their population to an accommodation with the new requirements) will replace the maladapted former leaders. But Scott's entailed caveat, that his "own bias, *based on a lifetime of scientific research* is in the direction of nonviolent means of bringing about desirable social change" (emphasis added), is Pollyannaesque and runs directly counter to what should have been the lesson of that lifetime of scientific research (which happens to have been concerned with the social behavior of canids, mostly domesticated). Certainly in the case of the hypothesized one-world totalitarian state, with its international civil service, it would be a challenge to political scientists to devise a method of nonviolent orderly systematic political change that might avert the periodic necessity of having the evolutionary alternative of revolution asserted. It should be easy to demonstrate that neither the British nor the American Constitution, as the model for such a one-world super-state, could be relied upon to avoid either the establishment of a totalitarian political and administrative system, or recourse to revolution as the basis for régime changes.

Our traditional approach, in political science, to the study of political philosophy and political theory alike demarks those fields as exemplars in the practice of political statics. Socrates, Rousseau and Marx are understood to be addressing a common—and constant—species, humans who remain at root the same, whatever ephemeral changes may be brought about by cultural influences—which are, after all, entirely relativistic. If only we were sharks (cf. Day, 1920), how pertinent such a perspective would be! In fact, however, natural selection continues to work upon us, if in diverse ways according to our respective subpopulational genetic inheritances, in recent and present habitats; and it is extremely likely that such change will continue to escalate in phase with the accelerating pace (Pettersson, 1978) of contemporary cultural change—which does, for the epigenetic reasons discussed earlier, affect our species genome. It is understandable that even distinguished ethologists, to say nothing of distinguished political scientists, should look upon revolutionary political change as undesirable; but this reflects in part a linguistic hang-up in terms of which "revolutionary" and "evolutionary" often are used, certainly by political scientists, as though they are antonyms. In fact, evolution signifies change in organisms (including their behavior) due to uncontrollable and heretofore virtually unpredictable environmental changes; and revolution (like "catastrophe") is an index to one of the many possible rates of such change. Hence political revolutions demonstrate—they most assuredly do not contradict—that changes are taking place in political evolution. Our traditional static approach to political theory presumes that change is abnormal, undesirable, and a sign that something is wrong; but an evolutionary approach will presume that change is a function of "environmental stress" (Waddington, 1960: 95) and that the task of political science is to understand what is happening in the sociopolitical as well as the natural environment that stresses humans in particular political demes, and how and why political behavior is changing in a variety of responses to such perceptions of environmental stress. This can best be done against the background of earlier political evolution among humans, beginning tens of thousands of years before the era of recorded history. That will make possible the model that we now lack (but cf. Tiger and Fox, 1971, and Chagnon and Irons, 1979) of how political life began and changed during the long era of hunting-band, small-group political organization of our species (see also chs. 6 and 7, below). Then we can seek to understand how and why the transition to agriculture and then urbanism, and from tribes to their aggregation in nations, took place, at times ranging from twelve to six thousand years ago in different regions of the earth (Carneiro, 1970; also ch. 11, below). The systematic body of theory resulting from such studies would put political scientists in a position to analyze the more recent transitions of the past two hundred odd years, into the industrial and nuclear eras, and the political changes that accompanied them (cf. Green, 1986).

CONCLUSION

Evolutionary theory will lead to a new understanding of the human political past and to analyses of how political structure and behavior can be brought into closer accord with the limits of human biological needs and capacities. Humans have been trying to force their biology to conform to the dynamically escalating requirements of their cultural innovations for thousands of years; but now there is evidence everywhere in the world that the species is rapidly being pushed beyond the limits of endurance of many physiological systems, whose complex transactions we have only recently begun to comprehend.

The theory of political evolution will neither change the present nor predict the future for us. But it will make possible a more realistic delineation of present options; and our choices among these will certainly have some impact upon our future environments, our future species genome, and therefore upon the future options open to our future selves.

PART II

Political Evolution

These three chapters explore the question: what would our understanding of the origins and development of political behavior be like, if we were to presume not merely the relevance, but the critical importance, of evolutionary theory? This impels us to push back before written history; before the invention of writing itself; and indeed before humans had become farmers as well as herders; even before humans began to settle in relatively permanent villages (to say nothing of cities and city-states). This implies a truly radical break with political science's past exclusive reliance upon the theories (and consequent data) of the humanities and the social sciences, as the bases for reconstructing the conventional image of the roots of politics.

In addition to whatever we may know and can learn about political beginnings, from history, sociology, cultural anthropology, and social psychology, what could we learn from physical anthropology, experimental physiological psychology, zoology, primatology, geology, and archeology? This second section of the book attempts a partial answer, with exemplification of both procedures and findings and of the new questions that they suggest.

Chapter 6 undertakes to investigate the two major competing hypotheses in social biology concerning the genesis of sociopolitical organization and protopolitical behaviors among the hominids who were the predecessors of modern humans. One hypothesis proposes a homological explanation: that because humans *are* primates, the study of how *other* (than human) modern primates behave socially will indicate how similar social behaviors (to those observed among modern simians) probably evolved among our hominid forebears.

The alternative hypothesis proposes an analogical explanation: that nineteenth- and twentieth-century anthropological observations of surviving "primitive" peoples indicate that the social organization and behavior of such humans much more closely resemble the counterpart ethological descriptions of social organization and behavior among feral populations of modern canid species, such as wolves and African (or Australian) hunting dogs.

Chapter 6 explicates and operationalizes these two evolutionary models, infers explicit findings from them, surveys the relevant empirical evidence, and states conclusions about the kinds of changes these findings imply for political science research in several major fields. These implications are stated as questions for and about political theory, political socialization, political communications, political equality, and public policy.

Chapter 7 focuses on several alternative explanations (and associated theories) of simian behavior, proffered by contemporary primatology. Analyses are made of four case studies, selected for their popularity and prominence in primatological research literature, as theories dealing with major questions of primate "political" behavior. These include: attention structure; male competition; male infanticide; and differential functional competences among both males and females (in lieu of hierarchy, agonism, and dominance), as theories of group leadership. The comparative analysis of these theories and their supporting case studies are the basis for a reappraisal of what ought to be the functional requisites for a more general theory of primate politics, including: the avoidance of "primatizing" human behavior and/or anthropomorphizing simian behavior; the distinction between political and other forms of social behavior; and a definition of the qualitative (in terms of consanguinity) and quantitative (in terms of population size) criteria that make possible an operationalization of the necessary distinction, to indicate when extended-family social organization and behavior become one of several similar components of a more diverse and complex form of organization and behavior that *should* be recognized as political.

Religion is important to politics because of religion's modal function: to inculcate authority acceptance for the incumbent Caesar as well as for the prevailing Deity. Therefore, religious behavior is symbiotically enmeshed with political behavior, not only historically (see ch. 8, "Integrating Biology with Culture," paragraph 2, below) but also spatially, in most human societies, most of the time, everywhere.

Chapter 8 begins with a review of the ongoing public policy debate over "scientific creationism" (i.e., religious fundamentalism) in opposition to "secular humanism" (i.e., nonhumanist, orthodox neo-Darwinian evolutionary science). This suggests the reappraisal of two alternative theories of how humans came to be. On the one hand, the Judaic/Christian mythology enshrined in the Old Testament indicates that Yahweh created Adam and (then) Eve in the Garden of Eden, approximately six thousand years ago, having created the rest of the earth for their use and needs fulfillment slightly (i.e., approximately an allegorical week) earlier. On the other hand, twentieth-century astronomical/geological/biological evolutionary science reports that the earth's solar system became (for physical reasons) established, some 4.5 billion years ago (and about 13–14 billion years after the universe began—for reasons that remain somewhat more speculative), with unicellular life eventually (for chemical reasons) becoming organized from even simpler forms approximately 2.5 billion years later. Humans are thought to have evolved as their present distinctive species during the past one hundred thousand years, from hominid species that had evolved (during the preceding five million years) from ancestral hominids who diverged from dryopithecine primates up to twenty million years BP (before the present). The original higher primate ancestors are dated at thirty million years BP; and the earliest primates evolved from other rodentlike mammals somewhat less than seventy million years ago. (*All* of these traceable life forms immediately ancestral to humans lived during only the most recent 5 percent of the 1.5 *billion*-year epoch of life on earth.)

Each hypothesis is attributed to be the product of human culture at a particular historic stage of its development: the first, of a particular primitive tribe of the many in the region (of the Eastern Mediterranean) descended from Natufian aboriginals of circa twelve thousand years BP, whose tribal creation myth shares (and borrows) much of its content from those of other (and more dominant) peoples settled elsewhere within the general region (including Mesopotamia). The second hypothesis reflects the much less ethnocentric—indeed, the interdisciplinary and globally pancultural—*Weltanschauung* of literate, highly educated, technologically oriented, modern, twentieth-century research scientists (i.e., the modern culture of the post-industrial intelligentsia). But it is explicitly postulated that *both* hypotheses were products of the best information and thinking available to skilled human specialists in the knowledge possible, respectively, to the first century B.C. and the twentieth century A.D.

Chapter 8 focuses on the major shift in niche (biologically) when humans were forced by overpopulation to begin to give up the gathering/hunting way of life, in favor of the more stable food supplies but more restricted life styles of pastoralism and agriculture, in "the transition" that commenced approximately twelve to fourteen thousand years ago in the Fertile Crescent of the Near East. The analysis discusses the domestication of both plants and animals, the psychic significance as well as economic utility of animal slaves to humans, and the general theory of creation mythology. Appraisal of the empirical evidence indicates that both the Genesis and the Darwinian myths were reasonable products of their respective cultures, and indeed that their content has much more in common than usually is realized and that the so-called cultural myth (Genesis) and the so-called biological one (evolution) complement rather than contradict each other.

Chapter 8 then discusses a set of alternative models that purport to explain religious behavior, including primate dominance (God is the ruling monkey); hemispheric dominance (God emerges from the reversal of left-hemisphere cerebral dominance in humans); and the biochemically generated biology of hope (God is endorphins). It concludes with a reaffirmation of the symbiotic importance of other animals in human psychobiology and with a plea for intellectual tolerance in understanding that in human cultures, religion has its scientific function and propensities—just as modern science inescapably functions as a form of religion for its true believers.

6
Hominid

Political science has for the most part accepted a concept of political evolution that equates it with written political history, which typically is presumed to begin (at least, for teaching purposes) approximately with *The Republic* and the *Politica*. (When the late Leo Strauss, during his tenure as a fellow at the Center for the Advanced Study of the Behavioral Sciences, was approaching the end of his life work, he was engaged in the enterprise of pushing back the relevant bounds to include their sources in the work of Aristophanes—not that of *Australopithecus*.) Thus we teach such utopian fantasies of human political origins as Rousseau presents, as well as the Hobbesian dystopia that preceded it. Political scientists are of course aware of the work of cultural anthropologists in field studies of surviving primitive peoples; but unlike such anthropologists themselves (Fried, 1967; Carneiro, 1970; Greenwood and Stini, 1977), we have not taken seriously the proposition that we might learn about politics by studying political behavior among such indigenous populations (cf. Durham, 1976b; Bernstein and Smith, 1979; Chagnon and Irons, 1979). Our presumption instead has been—at least until quite recently—that *our* teaching politics to *them* (e.g., "political development") was part of our burden as civilized, Western, colonizing heirs to the great political traditions of Athens, Rome, the Renaissance, English common law, and the American Constitution. As political scientists we are aware also of the pop-ethology literature of the sixties (Ardrey, 1961; Lorenz, 1966; Morris, 1967), but less

---

This chapter is a revision of "Evolutionary Politics," *Western Political Quarterly*, 36: 175–93 (1983), which was presented as a paper at the symposium "Ethological Approaches to the Study of Politics" at the Annual Meeting of the American Association for the Advancement of Science, in Washington, DC, Jan. 6, 1982. Reprinted with permission of the publisher.

so of the work in modern political ethology that has displaced it, based on the burgeoning field and experimental research during the past generation in the developing disciplines of animal behavior and primatology (e.g., Tiger and Fox, 1971; Reynolds, 1976; Geist, 1978; Lockard, 1980; see also ch. 5, above). Greater awareness of the latter work would have suggested to us important insights into our own political organization and behavior. What might we discover by studying the sociopolitical life of prehistorical humans or their predecessor hominids? Or the social behavior of nonhuman primates? Or the behavior of other predatory animals who, like us, are adapted to a top-of-a-food-chain niche in the biosphere?

The attempt to answer any such questions will push us out of the social sciences into the biosocial and life sciences with disciplinary ties to the humanities (Marshack, 1972; Shepard, 1978; E. Fisher, 1979) as well as to human evolution, prehistory, and biology; it will also lead us to studies of behavior (including cognitive behavior; see Crook, 1980) in other species than our own and among animals other than primates (Sebeok, 1977a). This chapter aspires to demonstrate the possible relevance of those broader questions by exploring in narrower focus merely one of their many potential applications to the study of political behavior. Do humans act like apes because of their common evolutionary descent? Or do humans act like carnivores because of the ecological niche to which we as hominids have been largely confined for the last five million years or so? Or both? Or neither?

Assuming that some reasonably justifiable answers can be constructed to these narrower queries, we might then wish to consider what those findings imply for a redefinition of the scope, theory, methods, and content of political science. The bulk of this chapter is concerned with the primate/carnivore inquiry, concerning which other scholars have done considerable research. This is followed by a summary of the principal findings. What this means for political science can in part be illustrated by research already begun in biopolitics; but in other respects it necessarily remains speculative, involving as it does the imaginative construction of possible (and partial) futures for the profession.

EVOLUTIONARY MODELS OF POLITICAL BEHAVIOR

The presumption here is that political behavior is older than culture, and indeed that it foreshadows—probably by millions of years—the emergence of even spoken language among humans (cf. Masters, 1970). Consequently, the wellsprings of politics must be sought in comparative cognitive psychology (Peters, 1978a, 1979; Shepard, 1978; Sebeok and Umiker-Sebeok, 1980; Griffin, 1981, 1982, 1984; also ch. 16, below) with the aid of evolutionary theory and the sciences of physical anthropology, primatology, and ethology.

This chapter focuses upon competing models of evolutionary politics. The primate model suggests that if monkeys, apes, and humans all exhibit similarities in social organization and behavior, this is because the genotype for the order of primates is the product of their common prosimian ancestry possibly twenty million years ago: hence the communality in their respective social behaviors and organization that can be attributed to genotypic differentials among them resulting in and from speciation. The genetic relationship between humans and carnivores, notwithstanding its very considerable importance, is at the much broader level of the class of mammals and goes back at least seventy million years before a common ancestor is hypothesized. So it is assumed that similarities in social behavior and organization, as between humans and carnivores, represent an *analogy* due to convergent evolution, which can be attributed to similar adaptations to similar ecological demands and constraints. So what do the primate and carnivore models tell us; and how are we to choose between them?

Primates

It is now consensual that chimpanzees, on the basis of both genetic and other biochemical evidence (e.g., Yunis et al., 1980; Yunis and Prakash, 1982), are the living species most closely related to contemporary humans. According to a leading Dutch comparative psychologist who has specialized in ethological studies of the agonistic behavior of feral chimpanzees, chimps like humans are conspicuous for their "hunting and predatory behavior" (Kortlandt, 1972: 85–90); chimpanzees are also highly xenophobic, and they engage in "armed fighting" (which Kortlandt, 1972: 81–82, exemplifies with his detailed discussion of half a dozen specific subcategories and which Goodall et al., 1979, show involves offensive as well as defensive sorties). Itani (n.d.: 10) also speaks of an encounter between two groups of chimpanzees that "had an atmosphere of a skirmish in a war"; and in comparing chimpanzees to gorillas, he remarks that "the antagonistic interaction of a group versus an individual, or a group versus another group, with the intent to kill, is peculiar to" chimpanzees (cf. Nishida, 1980).

Savanna-dwelling chimpanzees, as distinguished from those living in forests, establish for their respective groups territorial ranges which, notwithstanding the chimpanzee handicap of vastly reduced bipedalism relative to humans, are remarkable for their similarity in size to the home ranges of contemporary hunting and gathering peoples (Suzuki, 1975: 273). Savanna chimpanzees, as compared to forest groups, also form much more hierarchical, dominance-oriented aggregates of mixed ages and sexes, led and protected by the adult males (Suzuki, 1975: 275). But the latter pattern is an archetype rather than a mode for chimpanzees, reflecting their adaptive response to the severe pressures of the savanna habitat. These include, for example, the dispersed and less certain availability of food, the increase in danger from carnivore (and hominid!) predators (Walker, 1984: 138–149), and their own decreased mobility. Many observers of their behavior in forest and woodland have emphasized the horizontal rather than the vertical structure of chimpanzee society (as summarized by G. King, 1980: 104) and what King refers to as "the fragmented state [of] a social unit. . .divided into subgroups that move independently of one another" while "integration of the larger unit is maintained because subgroups associate with each other and because individuals readily shift from one subgroup to another." Campbell (1979: 300–301) agrees that "the chimpanzee community emerges as a stable social unit with the capacity to vary between compact and fragmented states." The structural plasticity extends also to roles, as Yakimov (1975: 305) has remarked. "In chimpanzees there are special changes in the social relations among group members during hunting. When catching a prey a nondominant male can be in the role of a dominant male."

For more than a decade a continuing study of a captive group of chimpanzees in the Netherlands has focused on the importance of coalitional behavior by males in the establishment and maintenance of, and change in, the primary (i.e., the male) dominance structure of the group (van Hooff, 1973a, 1973b; de Waal, 1978). An independent female dominance structure is much more stable (see also McGuire 1982; Hrdy 1981). Dutch physiological psychologists (Noe et al. 1980; de Waal and van Hooff 1981; de Waal 1982) and American political scientists alike have discussed the dominance behavior of the Arnhem consortium chimpanzees in terms of cost-benefit analysis (Peterson and Somit, 1980), and explicitly as political behavior homologous to that readily observable among contemporary humans (Wilhoite, 1976: 1118, 1980; Schubert and Somit, 1982).

An expanding and extensive research literature deals with equivalent aspects of the social behavior of other simians. For example, Loy (1975: 156) distinguishes three types of dominance interactions among macaques: fights, unprovoked submissive gestures, and displacements. Itani (n.d.: 2) remarks the extent to which primates generally demonstrate what he calls "intolerance among males," which is the underlying cause of structural devices that reduce conflict between and among them. Thus many primate groups patrol their territorial boundaries and attack or defend against intraspecific intruders; but the claimed effect of the territories is to avoid conflict, or at least to reduce the level of it that otherwise would obtain, by reserving exclusive, or at least preferential, access for the residential group to the food within the defined territory, within which conflict is also reduced by hierarchy or fragmentation. Either fragmentation (viz., decentralization) or hierarchy lessens competition for amenities such as food and also for sexual and other social intercourse, both of which are also reinforced by pair bonding. For example, van den Berghe (1974: 273) states that territory and hierarchy function as *alternatives* for most simians: "The territorial species are typically minimally hierarchical (even to the extent of sex equality and minimal sexual dimorphism); conversely, the hierarchical species tend not to be territorial." He also notes a correlation between territoriality and arboreality, so that hierarchical species tend to be terrestrial.

With the exception of hostility toward species that prey upon them, simians manifest little interspecific aggression (Thompson, 1975: 118). However, some research reports have focused attention on such exceptions as chimpanzees killing and eating infant baboons and other monkeys and small animals (Teleki, 1975), and baboons preying upon various species of antelope infants, hares, and occasionally, birds (Harding, 1975: 247; Strum, 1981); but the ratio of meat (including that resulting from the *intra*specific killing discussed below) to the flora ingested by either chimpanzees or baboons is very limited, no more than 1 or 2 percent even including the invertebrates that constitute the staple source of animal protein for simians. Of course physiologically that small proportion of ingestion, of animal protein, is indispensable as the source of vitamin B-12 (inter alia) for simians; and many observers have consensually remarked how arousing it is for simians to participate in and to observe the killing and eating of a prey animal. But there are species differences: baboons never share meat voluntarily, so aggression levels within a troop rise with the competition among its members to gain access to the meat (Harding, 1975: 250); chimpanzees do share in response to solicitation so that some observers have described meat eating by chimpanzees to be a social, almost a ritualized, experience for the participating animals. King (1980: 104) likewise claims that "primate group confrontations are highly ritualized, injuries are rare and hardly ever serious, and fatalities are virtually unknown, even in reputedly aggressive species such as the common baboon."

Moreover, Harding (1975: 253) observes that "in both baboons and chimpanzees, the killing of small animals appears to be an activity carried on only by adults and almost exclusively by males." Strum, however, reports (1981: 270–71) that among her baboons, certain "females were present at kills and obtained meat with a frequency that fell well within the range for adult males" and that for these females, "meat-eating took precedence over other behaviors, social or sexual"; indeed such a female often "brought her consort to the kill." Nevertheless, "male predatory behavior may be more flexible and more subject to change than female predatory behavior" (Strum, 1981: 294).

> The. . .baboon data illustrate that [the] behaviors [of cooperation, strategy, division of labor, and certain cognitive abilities], which would be critical to a primate predator attempting to capture large game, have precursors among the nonhuman primates. . . . [Some, for example, Tiger] have speculated that the origin of human sexual division of labor is to be found in hunting for prey that included large animals. But the [baboon] data suggest that even among collector/predators a difference in male and female predatory behavior exists that may reflect basic differences in the way the male and female primates are integrated into the group, and perhaps differences in the reproductive strategies for the two sexes as well. . . . Sexual differences may have originated before the shift from collector/predator to gatherer/hunter took place, while males but not females engag[ed] in certain aspects of predatory behavior. (Strum, 1981: 298; and cf. Symons, 1979; H. Fisher, 1982; Watts, 1982)

Strum (1981: 291–92) has also noted a reciprocal relationship between predation and scavenging, as alternative search behaviors for acquiring meat (and therefore protein): "Opportunities to eat dead animals were ignored in previous years, whereas no such opportunity was missed in 1976–77; . . .as more chances to predatory behavior have been ignored, scavenging has increased." Thus "the evolution of behavior cannot be seen in orthogenetic terms, for reversals in behavior development may be the rule, rather than the exception in dynamic and complex social systems" (Strum, 1981: 293).

Particularly during the past decade, increasingly frequent and numerous reports of simian intraspecific aggression have appeared, especially in the forms of infanticide and cannibalism, which often go together (Bygott, 1972; Goodall, 1977; Nishida, 1980; Itani, n.d.; Kawanaka, 1981). Of special interest for present purposes is the infanticide among Hanuman langur monkeys, especially as reported and discussed by Hrdy (1974, 1977a, 1977b), who speaks in terms of the usurpation of power and a subsequent change in regime in langur troops and models her explicitly political analysis on Shakespeare's *Titus Andronicus*. Her scenario portrays an invading male who drives out the resident adult male and then attacks and kills the nursing infants of troop females, thereby inducing them to recycle more rapidly so that he can impregnate them sooner, all to the benefit of his own direct (genetic) fitness. Surely this is an outstanding example of biopolitics among simians; but the quality of the underlying data supporting the claims of infanticide, as well as of Hrdy's interpretation of causality in relation to competing theories and explanations, has been criticized by several commentators (Curtin and Dolhinow, 1978; Boggess, 1979; Vogel, 1979; Eibl-Eibesfeldt, 1980; Schubert, 1982b).

An alternative and in some respects an opposite concept to that of dominance hierarchies (Bernstein, 1981) is that of "attention structure" (Chance and Larsen, 1976), which suggests that the charismatic leadership of attractive animals provides a better and more general explanation of primate social structure. Several political scientists (Beck, 1975; Masters, 1978b, 1982b; Barner-Barry, 1978, 1979), along with other social scientists (Cavallaro, 1978; Abramovitch and Strayer, 1978; Abramovitch, 1980), were attracted to Chance's ideas during the seventies, the political scientists at least in part because it seemed initially to them to proffer a more democratic and less authoritarian biological model than dominance does for both evolutionary and contemporary study of human political structures. But it soon became apparent on both logical and empirical grounds, and to zoologists, developmental psychologists, and political scientists alike (Seyfarth, 1977; Corning, 1980; Vaughn and Waters, 1980, 1981; Schubert, 1983d), that Chance's approach adds nothing to the predictive power of contemporary nonlinear and multidimensional coalitional-theory approaches to the study of social hierarchy. It has even less to offer to the understanding of hominid evolution.

Carnivores

The ranges of carnivores tend to be very much larger, and characteristically so by an exponential factor, than those of primates, including baboons and terrestrial chimpanzees—but not as compared to either early hominids or contemporary (in the sense of early twentieth century) preagricultural hunting peoples (Peters and Mech, 1975: 295, Thompson, 1978). Although primates demonstrate detailed knowledge and memory of the relatively fixed and predictable features of their arboreal ranges, such features become much less fixed and predictable for primates in terrestrial habitats; and Peters has asserted that much more complex and sophisticated cognitive mapping for spatial relationships becomes a critical neurological function to have evolved among predators of large animals moving through extensive areas (Peters and Mech, 1975; 295–96; Peters, 1978a, 1978b, 1979). Such a cortical development is one that hominids shared, because of convergent evolution, with wolves, hunting dogs, and lions. Moreover, "cooperation, strategy, and cognitive mapping. .. have several significant properties [as] prime candidates for behavior precursors or contributors to the evolution of at least some aspects of intelligence. . .[and thereby] the early hominids would find their niche, for the role of the intellectual hunter was the only exploitive role in the large-animal economy that had not yet been filled" on the Pliocene savanna (Peters and Mech, 1975: 296; for a discussion of the implications of psychobiology for politics, see Peterson 1982a, 1982b; and chs. 17 and 19, below; for the implications of culture for psychobiology, see Gould, 1980a: 29–31).

Peters and Mech have discussed also the social structure typical of feral wolves, using as their example packs in the Superior National Forest of Minnesota, where territorial ranges vary from 125 to 310 square kilometers and are both stable and exclusive under normal conditions. Each pack is a nuclear family, with a single dominant pair (between whom, however, the male is the leader); younger animals of both sexes but graded ages reflect successive litters. Except for the alpha pair, age dominance is just as ephemeral among wolves as it can be observed to be among contemporary (and to have been among historical) humans.

> Thus a pack can be viewed as a group of related, interacting individuals with various social ranks that keep them compatible. However, as younger members mature, they may not tolerate being subordinate to the other members. Or, if they are high ranking, they may not accept domination by the alpha animals. In either case, the resulting disruption of the social order may lead to the departure of the individual from the pack. . .[so some] young wolves are forced to leave the pack and become loners. . .[to] wander far from the pack territory and become nomadic in an area as much as 20 times the size of a territory. (Peters, 1978a: 134–35)

Banishment or isolation—at least, at the behest of the wolves—is not possible for those in captivity, regarding which additional findings have been reported by observers of their social behavior. Lockwood (1979: 229, 232, 235) found that "agonistic encounters in wolves usually take the form of highly ritualized displays involving more than two individuals"; and in his quantified analysis, "agonistic rank-orders fail to reach the criterion for linearity in all cases"; nevertheless, "the high intercorrelation of a number of variables traditionally associated with social dominance suggests that the concept does have validity." Moran and Fentress (1979: 269–70) found that the conceptualization of longer-term relationships between individual wolves in terms of "roles" and "social hierarchies" implies "organization which is too unitary and static, in addition to their frequent confounding of description with interpretations of causation and/or function. As such, these constructs may make shift in the relationships between particular animals that (1) have limited ramifications throughout the social group, and (2) not only shift, but are also reversible over varying intervals of time."

Curio (1976: 19) has pointed out that, quite unlike their facial expressions when killing food competitors (ranging from conspecifics to humans), among lions "the face remains entirely unmoved when capturing and killing normal ungulate prey." More generally, "facial signals, body postures, and so forth associated with aggression are entirely absent in social carnivores in all stages of the hunt up until use phase, when we see them directed towards conspecifics" (Peters and Mech, 1975: 304); or, as Thompson (1975: 118) puts it, "among carnivores there is, in addition to predator-prey aggression, a general animosity toward other carnivores," who are, of course, food competitors. Even so, according to Alcock (1979: 443), social carnivores "engage in group aggression to defend or expand group territorial borders only when game is exceptionally abundant within certain defensible areas."

Hominids

Comparison of preagricultural humans with terrestrial primates and social carnivores indicates, as we might expect, many communalities. Like hunting bands and wolf packs "the typical primate troop contains at least a core of related individuals" (King, 1980: 106n). But of greater interest, for present purposes, are the respects in which primates and social carnivores resemble humans while differing from each other. As Alcock (1979: 445) has summarized, chimpanzees resemble humans in regard to degree of intelligence, prolonged maternal care and family formation, and, to a lesser extent, tool making and using, and bipedal locomotion; social carnivores, on the other hand, resemble humans in intense aggression toward conspecifics and other competitors, group territoriality, cooperation in hunting and food sharing, highly structured social organization, prolonged pair bonding between mates, and a sexual division of labor beyond infant rearing. In a similar comparison, Thompson (1975: 123) states that humans resemble carnivores rather than simians in regard to sharing and storing food, reliance upon more than one adult to feed an infant, cannibalism, the killing of more prey than can be consumed, and interspecies intolerance; humans are more like simians in that both rely on males for group defense, a function assumed by both sexes among social carnivores. On a carnivore sophistication scale constructed by Thompson (1975: 122), wolves, foxes, tigers, and leopards all score 1.00; humans score the same as lions at .86; while chimpanzees get only .43 and no other primate is rated higher than .14, with most scoring .00.

King (1980: 102) has proposed that the primate homologies and carnivore analogies ought not be viewed as components of alternative and competing models, but that instead they should be viewed as complementary data that, taken together, "illuminate different aspects of a simple hominid behavior pattern." This points in the direction of the conclusion that Bernard Campbell reached, independently and at about the same time. Following Teleki (1975), Campbell (1979: 293–96) asserts that

> the most significant developments in the earlier stage of hominid evolution were not so much predation, but scavenging, big game hunting, and collecting. The appearance of these patterns of behavior is the point at which the primate analogy breaks down and has to be replaced by a strictly hominid pattern of subsistence and social organization. This organization is associated with new technology and carnivore hunting behavior, characterized by cooperation and food sharing. . . . It appears that soon after 1 million years B.P. we can be fairly certain that cooperative big- game hunting was established in a number of places. . .[but] that food gathering is a more novel and original behavior in hominid history than predation or small-game hunting. . . . Thus the successful exploitation of big game and the efficient collection of small portable items of food took hominid subsistence and behavior into a new realm which is not paralleled either by living primates or living carnivores.

His conclusion (Campbell, 1979: 301) on the question raised by King is that "phylogenetically related forms are most valuable as models at the earliest phases of protohominid evolution [*Ramapithecus-Australopithecus*], but become less relevant with the appearance of big-game hunting and food collection [*Homo habilis-Homo erectus*]."

SPECIFIC CONCLUSIONS FROM THE MODELS

Primates

The primate model supports the following findings, which by extension and homology may have major implications for contemporary human behavior.

1. Only chimpanzees, the primate species most closely related genetically to humans, engage in group conflict similar to human tribal (or gang) warfare (cf. van Dijk, 1977).
2. Depending upon both environmental opportunity (change) and group culture (social transmission of information; see Bonner, 1980; Mainardi, 1980), scavenging/gathering behaviors can substitute for interspecific predation—and vice versa.
3. Savannah-based chimpanzees adapt to the higher environmental stress encountered there than in forest habitats by establishing greater hierarchy in social organization.
4. Territoriality is widespread among primates in suitable habitats (i.e., forest rather than savannah); its effect is to reduce conflict both within and between groups; and it also has the effect of reducing the hierarchical structure of group social organization.

5. Group hunting similar to that found among primitive humans can still be observed in some savannah-based primates (e.g., baboons), although it is probably due to convergent evolution just as it is among social carnivores.

Carnivores

1. Equivalent evolutionarily stable strategies (Maynard Smith, 1974), involving group hunting over much larger ranges than nonhuman primates patrol, induced convergent development in the brains of social carnivores and humans.
2. The structure of human families (van den Berghe, 1979) and tribes is much more similar to that of wolf packs than to that of chimpanzee (or other nonhuman primate) troops.
3. Social carnivores display (like humans) much higher emotional arousal in aggressive behavior directed toward conspecifics (i.e., food—and in the case of humans, other-resource—competitors) than in aggression directed toward *inter*specific prey.

Hominids

1. As in the case of primates (cf. item 2, primates, above), scavenging and collecting were the main sources of food most of the time (except for hominids in the exceptional and climatically extreme environments discussed by Geist, 1978, as progenitive of "K-selection").
2. The hominid adaptation, combining the exploitation of big game with efficiently collected small food items, was unique—unlike what occurs among either primates or social carnivores.

Humans

For humans today, some conclusions from and implications of these models are:

1. For evolutionary reasons, humans in some aspects of sociopolitical organization and behavior resemble other primates more than carnivores; but in other respects humans are more like the social carnivores than like other primates.
2. In combination with their hominid ancestors, humans have been omnivorous primates for at least a million years (and perhaps for five million)—but more carnivorously so than any simian primates; and less so than any social carnivores.
3. During most of the recent ten thousand years (until the nineteenth century), humans as farmers and/or herders occupied an ecological niche more like that typical of other primates than of social carnivores.
4. Since the advent of industrialization, and because of an increasing population growth and increased competition for shrinking natural resources (van den

Berghe, 1974), humans have moved increasingly into what is more typically a social carnivore than a primate niche (cf. Jay, 1971).

5. If public policy requires a reduction in interspecific aggression, then the increasing importance of the model of animal predatory behavior implies that public policy should support the design of social environments to *inhibit*, rather than to encourage, opportunities for predation (e.g., see Newman, 1972).

Implications for Political Science

A greater understanding of evolutionary politics will indicate for political scientists many changes in their discipline. Our objective in suggesting a few examples is not to be exhaustive but rather to specify some work that already has been carried out (Albin, 1981), or hypotheses that might be investigated, in contrast to the approaches employed in contemporary, historical, *non*evolutionary political science.

Political Theory

If political scientists were to understand Periclean Athens as having occurred at a quarter to midnight, on a twenty-four-hour clock timing human evolution during the past million years (cf. Pettersson, 1978), how would political theory build upon evolutionary biology, physical anthropology, primatology, and ethology to construct a scientifically based theory of the biosocial origins of political organization and behavior—not as a substitute for the Socratic dialogues, but rather to put the latter into a more realistic humanistic perspective? (See Jaynes, 1976; and chs. 2 and 8.)

Political Socialization

Instead of asking eight-year-olds whether they know the name of the Chief Justice of the United States, what would political scientists learn about political learning if they studied how children organize themselves in positions of leadership and followership and behave toward each other in their own self-governance of such social groups? (See Barner-Barry, 1977, 1981, 1982; Jones, 1983; Wingerson, 1982.)

Instead of assuming that the political attitudes that adults teach children tend to persist throughout the lifetime of those so educated, with very little basic change, how would political scientists deal differently with research in political learning if their hypothesis was that children (like other animals) learn more from each other than from adults and that the content of their learning necessarily changes as their brains and bodies and basic needs develop and change throughout their life experiences? (See Chevalier-Skolnikoff and Poirier, 1977; Symons, 1978; Peterson and Somit, 1982.)

What would such an orientation toward political learning imply for political gerontocracy (e.g., the Kremlin or the White House) and political desocialization, and hence for age-related changes and differences in political behavior among both political elites and political masses? (See J. Schubert, Wiegele, and Hines, 1985; and ch. 15, below.)

Political Communication

What would political scientists learn about the transmission and reception of political information, including the emotional states of those transmitting and receiving it, if instead of continuing to restrict their study of political communication to language and verbal content, the relevant content were defined more broadly to include nonverbal communication (both facial and gestural) and audiospectrographic analysis of the human voice in articulation of political speech? (See Masters, 1982b; Wiegele, 1980; Schubert, 1982a; Sebeok, 1977b.)

Political Equality

If for evolutionary reasons (Symons 1978)—and independently of reinforcing cultural reasons—human males are, to a highly significant statistical degree, more competitive and aggressive than human females, what would be the political consequences of not merely equalizing but reversing the proportion of females presently in roles of political leadership? (See Schubert, 1983f: 123–24; Hall, 1985.)

In what other specific ways than sex (e.g., age, health, intelligence, race) does culture serve to *accentuate* (rather than to constrain) genetically based differences among humans, and with what consequences for political equality? (See ch. 17, pp. 488–99, 506–09, below.)

Public Policy

If for the best of evolutionary reasons and as a species characteristic interspecific aggression in human males is highly correlated with endocrinal stimulation that is also positively correlated with social and political ambition (cf. Davies, 1980; McGuire, 1982; Schubert, 1982b), how is a general reduction in the level of such aggression to be accomplished (e.g., see van den Berghe, 1974; Reynolds, 1976)?

If such a change involves either mass surgery or mass pharmaceutical treatment, or manipulations of the species genetic code (or some combination of these three techniques), what are the implications of such therapies for the future evolution of the species? (See Blank, 1981, 1982; Handberg and Maddox, 1981; ch. 11, below.)

CONCLUSION

It is not presumed that any of the above questions are easy ones; nor is it claimed that these are necessarily the best or the right questions, or the ones that should be asked or answered first. But it is submitted that our articulation of them may give some sense of the radical changes in both the knowledge base and the policy orientation of the discipline that are likely to be entailed, if and when the time ever comes when political scientists substitute evolutionary politics for the social-scientific creationism with which the profession, on the whole, continues to remain content to rest its case.

7
Primate

During the past decade, there has been a convergence in the work of primatologists and political scientists as their common interest in the epigenetic roots of social behavior has gradually become recognized (Schubert and Somit, 1982). While a few political scientists have explicitly sought an evolutionary alternative to the contemporary research designs that prevail in their discipline, primatologists have independently invented concepts of human political behavior in order to interpret the complex and dynamic structures of roles that they observe in small groups of contemporary pongids. Thus there has been an asymmetry, in this convergence, that deserves more critical attention than it has received.

Until a decade and a half ago (Corning, 1970), political scientists had ignored the modern evolutionary theory that resulted early in this century from the synthesis of the Darwinian and Mendelian theories (see Williams, 1966). Their persisting fascination with classical political philosophy was limited to "the great books," even though Aristotle (Schubert, 1973: 240, n. 1) and Rousseau (Masters, 1967, 1978c), to say nothing of Darwin himself (1965 [1872]), should have led political scientists to the evolutionary approach to political theory that has now clearly emerged (Masters, 1977, 1982c; Corning, 1977, 1983; ch. 6, above).

This chapter is a revised version of my article, "Primate Politics," *Social Science Information*, 25: 647–80 (1986). Used by kind permission of Sage Publications Ltd., London. The article was a revised version of a paper presented at a symposium of the International Primatological Society meeting in Nairobi, Kenya, on July 24, 1984. The symposium, "Political Behavior as a Primate Social Strategy," was jointly organized by a political scientist (myself) and by primatologist Shirley Strum, as an interdisciplinary forum.

For primatologists, the problem was not to discover evolutionary theory, but rather to consider the relevance of formal theories of human political behavior to questions of pongid social behavior. The major resurgence of modern primatology was catalyzed by the "nine-months 'Primate Project' . . . at the Center for Advanced Study in the [Human] Behavioral Sciences, Stanford, California, during 1962–1963" (DeVore, 1965: viii). At first, the emphasis was upon the putatively homological use of field studies of contemporary baboons (Lee and DeVore, 1968; Loy, 1975) and chimpanzees (Goodall, 1977) as models—or metaphors (Landau, 1961)—for understanding the social life of early hominids. The primatological quest for political theory began later, as a younger generation of scholars noted that the earlier approach was substantially analogical (Hrdy, 1981).

FOUR CASES OF "PRIMATE POLITICS"

This chapter will present and discuss four of the most important examples of theory about primate politics from the perspectives of both primatology and political science. These examples can be conceptualized each as constituting a benchmark on a continuum extending from monkeys that are born, raised, and experimented on in captivity, subject to the unceasing constraint of indoor cages and behavioral manipulation by their human controllers, to relatively freely ranging simians in a savanna-type ecology. These examples will be described here in terms of specific works by individual primatologists, although the discussion will include related work by other primatologists and by political scientists.

First we shall consider Michael Chance's theory of "attention structure," which was initially proposed primarily on the basis of his secondary analysis of other researchers' reports of their data. That idea became further refined over the course of the next decade (by Chance himself and various of his students and other associates) and was mostly based on observations of the Basel Zoo colony of long-tailed macaques and/or the much smaller satellite group of monkeys and their offspring imported from the Basel colony to the Uffculme [Psychiatric] Clinic laboratory in Birmingham, England.

The recent publication of Frans de Waal's (1982) book on *Chimpanzee Politics* is an obvious choice for discussion, especially in view of the circumstance that both Albert Somit and I together visited de Waal at the Arnhem Zoo in the Netherlands at a time when his doctoral research was still in process. The Arnhem chimpanzees slept in indoor cages at night but were usually let loose in an enclosure of about an acre during daytime hours (from mid April to late November); the social behavior that was observed for the purposes of de Waal's study took place under these conditions.

Next comes Sarah Blaffer Hrdy's model of regime change among hanuman langurs at Mount Abu in western central India; the "ground" monkeys with which she worked cohabited symbiotically with the humans, whose presence and artifacts they tended to exploit and, of course, be affected by.

Last we shall consider Shirley Strum's use of ecological and evolutionary theory in presenting her observations of the olive baboons that she described as "The Pumphouse Gang" (Strum, 1975), on Kekopey Ranch in Kenya's East African Rift country beyond the eastern shores of Lake Victoria.

Hedonic Macaques

The three succeeding sections of this chapter all deal with primatological theories of politics based on more or less extensive—more in the instances of Strum and de Waal, less in the case of Hrdy—observations of the behavior of groups of nonhuman primates. In each of those instances, primatologists developed hypotheses about what they had conceptualized to be political behaviors of the animals, in all instances relying heavily upon evolutionary and ecological theories that had been proposed by zoologists or comparative psychologists. None of these primatologists had undertaken to explore the professional political science literature to examine what formal political theory might propose regarding the human political behavior that the primatologists, in every instance explicitly, analogized in describing nonhuman primate data (Schubert and Somit, 1982: ix).

Although Michael Chance's "theory" of attention structure failed to reflect any study of political theory, the third component of his model—which he designated as "advertence/abvertence"—was derived not from observations of monkeys but rather was based on observations of the behavior of human patients in psychiatric wards of English hospitals. As a pharmacologically trained lecturer in ethology whose subdepartment was housed, both literally and figuratively, in the Department of Psychiatry of the Medical School of the University of Birmingham, it is understandable that Chance was in a position to make the observations on which his hypothesis was based. And one can agree that the practice of anthropomorphism in studying nonhuman primates ought to be undertaken as explicitly as possible (which is surely what Chance and de Waal do), rather than unconsciously and sub rosa as is often the case among both field and experimental primatologists. Moreover, if human patients who are emotionally and behaviorally disturbed are to be used as a model in studying nonhuman primates, then Chance was entirely right in applying his model to the particular collection of macaques upon which his laboratory observations are based (Schubert, 1983d, 1984a, 1979; cf. Goffman, 1961; and Rubenstein and Lasswell, 1966).

Problems arise, however, when the advertence/abvertence hypothesis is employed to model the modal behavior *either* of feral monkeys (as in the case of Strum's baboons) *or* in regard to what political scientists presume to be the "normal" political behavior of humans (e.g., Barner-Barry, 1978, 1979; Masters, 1979a, 1981b; Wingerson, 1982). Perhaps one should use advertence/abvertence to guide observations of the behavior of putatively psychotic politicians such as those described more than half a century ago by the first political scientist cum lay psychoanalyst (Lasswell, 1930; and cf. Lasswell, 1948: 61–68, especially his contrast between "Judge X" and "Judges Y and Z"). Furthermore, if primates all live now in environments where human cultures result increasingly in such intolerable stresses that there are now more and more crazy monkeys and crazy humans alike (see Foucault, 1965; and Szasz, 1965, 1970, 1973), then Chancian abvertence may well supply the most appropriate model of social behavior. But in a world full of mad primates, who will remain sane to be a scientific observer?

The first two components of Chance's model are familiar to primatologists and social scientists alike. Since I have recently commented on them (Schubert, 1983d, 1983e, 1984a), I shall be succinct here. The first dimension (centric/acentric) describes dichotomously the degree of dominance in hierarchical structure. The role of leaders in centric groups is defense against predators, which surely is reminiscent of the conventional "baboon primatology" of the fifties and before; Chance's only distinctive contribution here was his suggestion that in acentric groups subordinates "escape into the environment" rather than "to the center of the group" where the centric leader can protect them.

The second dimension of Chance's model, which distinguishes agonism from hedonism, is more important. "Agonic" is a more hedonic word than "antagonism," the latter having become common currency among primatologists—to say nothing of psychologists, sociologists, and political scientists. Agonism, like antagonism, denotes leaders who dominate through physical force or threats, whereas hedonic leaders dominate through persuasion and not least because their appearance and manner are appealing to subordinates, who willingly follow (and are led by) such "attractive" leaders.

Hedonism, thus defined, may seem reminiscent of Max Weber's concept of charismatic political leadership; and it is not inconceivable that Chance's development of his ideas might have profited from some familiarity with the considerable literature, in political science and related social science disciplines, that discusses political charisma (e.g., Weber, 1964, part 3: "The Types of Authority and Imperative Co-ordination"; Hummel, 1973). Chance himself never mentions or cites Weber, although the late Ray Larsen, in the chapter that he contributed to the book he coedited with Chance, explicitly advocates that "the Weberian typology" including charisma "should be incorporated into . . . the attention structure model of primate social structure" (Larsen, 1976: 270). Larsen, in explicating Weber's theory, points out that "the external proof of the presence of charisma [in a political leader] is the visible *emotional* state of followers . . . [and that] charismatic leaders. . .are thought to appear most readily in situations of distress and crisis when prevailing institutions are likely to be perceived as inadequate and *followers are inspired with emotions* born of distress" (Larsen, 1976: 254; emphasis added). Larsen was an anthropologist; political scientists Ralph Hummel and Robert Isaak (1980: 178–184, esp. 180; and see Hummel, 1974, 1975) agree that charisma defines the experience of a follower in distress, who projects love through his complete personal devotion to a leader perceived to be extraordinary or supernatural. The follower "resolves personal suffering and agony by subconsciously projecting love onto another individual (potential leader) from whom he or she then perceives his or her love returning in the form of an uncanny attraction." Larsen concludes (1976: 269) that "charismatic leader-follower behavior can. . .be interpreted to be an expression of a primal adaptive process which facilitates group cohesion, integration and stress reduction."

Chance himself treats the hedonic mode of behavior as a positive, affirmative, and desirable alternative to the negativism of agonism, with its impediments of dominance structure, the threats and use of force, and the involuntary insubordination of followers. This undoubtedly explains why a number of political scientists, not excluding myself, were initially attracted to Chance's model of attention structure as a potentially liberating democratic alternative to the intragroup authoritarianism of the older primate model of dominance by force; as applied to humans, it seemed to offer "a breath of fresh air" in a primatological room filled with the stale smoke of aggressive dominance and subservient subordinance.

Three out of every four political scientists identify themselves as either "liberal" or "very liberal" in political ideology (Cattani, 1981), so it is understandable that they should be more interested in a primatological theory of politics that proffers democratic as well as authoritarian alternatives, at least for the purposes of cross-species comparison in which the primate models are going to be applied to human societies, in which political scientists (like other citizens) have to live.

Attention structure "theory" fails to develop the necessary linkages for application to humans, either via the physiological route of the psychology of arousal and brain science (which Chance ignores), or through the political-sociological route of Weber, Parsons, structural functionalism, and ultimately role theories of political leadership (which Larsen recognized as a possible route, but unfortunately did not explore before his premature death). That leaves us with a literature on charisma that describes followership as neurotic behavior, with leaders such as Adolf Schicklgruber who, in the consensual judgment of most appraisers of his political career, was not only highly neurotic but also centric, *both* agonic and hedonic, and abvertent to boot. Attention structure discussion of hedonic behavior says virtually nothing about followers, except that they "look at" and "pay attention to" their displaying leader. Thus the "theory" likewise pays attention only to leaders, which may constitute another example of art imitating life. It may also constitute a major defect of the theory, since it fails to say anything about the motivations of the hedonic followers; the hedonic leaders presumably get psychic income from being looked at. But if hedonic monkey leaders display to the fearful, emotionally disturbed, anomic animals that charisma theory denotes as the human followers of charismatic political leaders, then what we seem to have in Chance's second dimension is a preincarnation of his third dimension: advertent leaders in confrontation with abvertent followers. There seems to be very little that political science can either learn from, or teach to, such a theory of political leadership.

Dominant Chimpanzees

Both Frans de Waal (1982) and his mentor Jan van Hooff (1982) report an unabashedly anthropomorphic analysis of the leadership behavior of chimpanzees of the Arnhem Zoo colony. They employ the flamboyant rubric of "chimpanzee politics," an arresting metaphor suggested to them by Desmond Morris, himself an old hand at that sort of thing (e.g., *The Naked Ape*, *The Human Zoo*, etc.), who served as godfather, of sorts, to the Arnhem chimp project. Since I shall return to the matter of anthropomorphism below, here I deal with the concept of politics utilized by de Waal and van Hooff.

The Arnhem colony consisted of approximately two dozen chimpanzees during most of the time when de Waal's data were collected. Of these, the number of adult males ranged from none to four; adult females, from about six to nine; and the rest were juveniles or infants. The colony was constructed by an aggregation of individuals or pairs or small maternal-family groups of animals, culled from various sources but apparently mostly from European zoos. The original leadership was necessarily female because the first males old enough to be potential leaders (two adults and one adolescent) were introduced together after the colony had become established as a novel synthetic social structure. It was then necessary to remove temporarily the incumbent female leadership in order for the dominance of the oldest male to become accepted by the group.

De Waal and van Hooff describe sexually discrete patterns of leadership motivations, styles, strategies, structures, and aggressive behavior, which they believe are the consequence of genetic/physiological differences between male and female chimpanzees. It is normal among feral chimpanzees for one or more adult males to be associated, for more or less extended periods of time, with several adult females plus the latters' offspring. Under such circumstances, the adult females are born into and tend to perpetuate a highly stable structure of subordinance relationships among themselves. At Arnhem there was one principal female leader but no simple linear pattern of subordinance in the interrelationships of the other adult females. The female substructure was based partly on genetic relatedness, but also partly on affection.

Females are described as being content to accept their respective statuses, as distinct from sexually mature males, who are said to "*strive* for higher status." Moreover, status is determined ultimately by successful aggression—manifest or potential—in dyadic encounters; but in social groups such dyadic conflict usually can, and often does, involve other members of the group. Therefore the social structure of dominance among mature males is highly dynamic and under continuing pressure for change. With three or more males in competition, leadership among them typically requires the dominant male to prevent the subdominants from cooperating in a coalition to oppose him. There is also a developmental dynamic in the process: usually, as in the case of de Waal's data, the pattern of succession is for younger males to displace older ones.

The female dominance structure presumes the coexistence of a male structure, and vice versa. The males do not compete in a social vacuum, and therefore their competition with each other is acted out against the background of the rest of the colony, so that it explicitly involves also the affective relationships between individual males and individual females and their offspring.

For insight into political-science theory of coalitional behavior, de Waal might have consulted such appropriate sources as Riker (1962) or Riker and Ordeshook (1973); instead, he relied on Martin Wight (de Waal, 1982: 177), whom he quotes as advising that "the alternatives to the balance of power are either universal anarchy or universal dominance." Although de Waal himself was evidently surprised that he failed to find Wight very useful, he was looking for a theory of power politics in all the wrong places. In some forty years as a political scientist I had never even heard of Wight, but I was provoked by de Waal's reference to look him up. Martin Wight's *Power Politics* (1946) is a tract of traditional political science by an English writer whose approach is institutional rather than behavioral, and whose implicit theoretical paradigm is Newtonian mechanics—not Darwinian selective evolution (Landau, 1961; ch. 19, below). The "powers" with which he deals are nation states, such as the United Kingdom, the United States, and the Netherlands.

By invoking this outdated and considerably outmoded pamphlet as a source of political theory, de Waal is compounding his anthropomorphic problems, which would be serious enough if he were saying only that the social leadership behavior of chimpanzees can best be understood by discussing chimps as if they are like *individual* humans. Instead, de Waal is implicitly asserting that the behavior of individual chimpanzees can be understood in terms of the competitive/cooperative relationships among nation states that are aggregates of tens of millions of humans (cf. his listing, 1982: 218, of Zinnes, 1970, as a reference). The maximum size of chimpanzee groups—natural or unnatural—rarely exceeds one hundred individuals. Humans in groups of one hundred behave, are organized, and are led very differently from humans in aggregations of four hundred million, or even those in aggregations of fourteen million, the approximate population of the Netherlands at the time of de Waal's research. To compare, explicitly or implicitly, the behavior of an aggregation of even one million persons with that of one hundred chimpanzees is to commit an egregious error of scale (or of "level of analysis").

De Waal's use of Wight as a source of political theory is consistent with the hypothesis that when a person in one discipline attempts to make initial (and unprofessionally guided) excursions into some other discipline, she/he typically cuts into the unfamiliar discipline approximately a generation—three decades, not an *academic* generation—behind its contemporary cutting edge. Wight's pamphlet is typical of the pre-World War II literature of international politics. There are thousands of much better, behaviorally oriented, works on international politics now available (including Zinnes, 1970, which de Waal cites as a reference but otherwise neither mentions nor appears to use); for example, it would have been more useful for de Waal to have cited Quincy (the brother of Sewall) Wright's *Study of International Relations* (1955). But even if de Waal had selected Wright instead of Wight, used Zinnes, or picked a more ethologically oriented political science work on the political theory of international politics (such as Ralph Pettman's *Human Behavior and World Politics*, 1975), his problem of the level of analysis would have remained.

Infanticidal Langurs

Probably the most dramatic and certainly one of the best known empirical examples of sociobiological theory is that provided by Sarah Blaffer Hrdy's hypothesis of regime change in hanuman langur troops. Actually, she has argued on behalf of a very much broader theory of mammalian infanticide (Hrdy, 1977a, 1979) that focuses on the extent to which claims of infanticide among various animals have been made by a variety of observers, informants, and retrospective correspondents; but many of these claims have no manifest nor even discernible latent implication for social structure or political behavior among the animals concerned. The discussion here will focus instead on her own field study of several adjacent troops of hanuman langurs whose ranges overlapped partially with each other, and in some cases considerably with the resort town of Mount Abu in northwest India, about 210 miles south of Jodhpur. In her reports of those studies (Hrdy, 1974, 1977b), male infanticide is a crucial and indispensable element of the theory of change in the "political" leadership of the langur troops that she observed.

The scenario that she sketches is truly elegant in relation to the sociobiological theory that she invokes. Hrdy assumes that langurs act, for genetic reasons, so as to maximize each individual monkey's direct fitness. Both male and female adult langurs improve their reproductive success by producing as many offspring as possible; their success in that endeavor is enhanced by several elements of the social structure of the group. Usually there is only one resident adult male in a hanuman langur troop, although Hrdy does discuss transitory situations in which a coalition of several adult males cooperates to oust an incumbent resident male. But in the latter circumstance, one among the cooperators soon becomes dominant over the other members of the recent coalition and drives them out of the troop. The resident adult male is dominant over all adult females and over younger hanumans of either sex. At puberty, juvenile males are ejected from the troop by the resident adult male. Surplus males, including such juveniles and also adults who are not at the time acting in the role of the dominant resident of a heterosexual troop, combine to form all-male troops. Hanuman langurs occupy well-defined territories, although the range boundaries of both all-male troops and other heterosexual troops not infrequently coincide or intersect slightly with the boundaries of heterosexual troops. As a resident male ages, or becomes injured, or sometimes as a consequence of other circumstances, he can expect to be challenged by another male. Sometimes the challenger comes from an all-male troop; sometimes he is a neighboring resident who undertakes to lead a new band in addition to, or instead of, his present one. The challenger sometimes succeeds (sooner or later) in driving out the incumbent resident.

Most of the time during most of his incumbency—and Hrdy's book mentions many exceptions and explains how and why they happened—the resident has exclusive sexual access to the estrous females in his troop. This means that he sires virtually all of the infants conceived during his incumbency and conversely, troop infants and juveniles of corresponding ages will all (or almost all) be at least half-sibs, in same-mother clusters with full-sibs. Hrdy asserts that the dominance system evolved to guarantee the monopolistic breeding system—from the point of view of the dominant male. The more adult females that he can collect in the troop and the longer he can remain in the position of resident male of the troop, the more he will improve his fitness (in the neo-Darwinian sense). If his incumbency persists for the few years that it requires for a female infant to mature sexually, then producing female offspring is one way to acquire females with which to breed, incestuously.

Langur infants perish for many reasons: the predation of dogs and other carnivores, poisoning and other degradation of the environment, falls, and sometimes aggressive attacks by conspecific males, including, but by no means limited to, the resident male of the infant's troop. According to Hrdy, such attacks by adult male langurs upon langur infants are associated with high states of psychophysiological arousal (cf. McGuire, 1982) that is usually both sexual *and* explicitly aggressive (but see Vogel and Loch, 1984). Hrdy's theory is that when a new resident takes over a troop, he kills off all unweaned infants, which in hanuman langurs has the effect of causing the mothers of these infants to resume ovulation and become estrous within a month or two so that the newly dominant male can impregnate all of the adult females of the troop as rapidly as possible with his own genes (instead of having to wait a year or two for that to happen). Hrdy herself does not assert (1977a) what her sometime mentor and graduate research advisor, Robert Trivers, explicitly claimed (Schubert, 1982b: 228, n.3; ch. 7 in White, 1981c: 225–33) namely, that the new resident continues to kill off all infants born into the troop during the initial seven months of his presence there, but, after that, avoids killing his own offspring. The killer male thereby improves his fitness in two ways: first, he procreates a larger number of offspring than would otherwise have been the case; and second, he avoids wasting his energies providing "protection" for, and engaging in competition (however marginal) for food with, infants carrying genes of other males than himself. But even Hrdy's own theory, if it were supported by the weight of evidence available, would constitute a striking example of the power of sociobiological theory to explain the behavior of primates.

Hrdy's methodology and use of empirical evidence have been frequently criticized (Curtin and Dolhinow, 1978; Boggess, 1979; Vogel, 1979; Eibl-Eibesfeldt, 1980; Schubert, 1982b; Wheatley, 1982); and I do not propose to restate that argument here. What I do want to do now is reexamine Hrdy's purported political theory, which she invokes as an additional prop to help "explain" the data produced by her field investigations of langur behavior.

The short of the matter is that Hrdy's theory of politics is highly anthropomorphic, metaphorical, and ad hoc. She refers to the period of the incumbency of a particular adult male resident, in a specific heterosexual langur troop, as a "regime." When some other adult male drives the resident out of the troop, she describes this as a "political change" involving the "usurpation of power." A succession of male leadership changes in a particular troop is the "political history" of the troop. But there is no politics in what she describes, except possibly in her motivation for inappropriately invoking concepts of politics to theorize about the relatively small-scale and socially uncomplicated relationships among the primates with which she is concerned (at least, as compared to an equivalent number of humans). When behavior results from cultural evolution—rather than high-speed genetic evolution (Lumsden and Wilson, 1981, 1983; ch. 10, below)—and the social group involves interactions among adult males and females alike, who are organized for breeding purposes into a substantial number of only partially related families that must agree upon common social policy in order for the multifamily group as a whole to survive in the available ecological niche, then it may be appropriate to speak of the beginnings of political behavior, at least in prototypical form. For Hrdy, this would require that she had observed a system of social control that was developed and operated *by langurs* to deal with the interrelationships among all of the langur troops in a given locality—not merely the social relations within each langur troop considered individually. A necessary, although by no means sufficient, condition for the construction of such a theory would be systematic observation of the behaviors that occur in the relationships *between* troops—"among" apparently does not occur.

The postulation of a cultural rather than a direct genetic basis for animals to behave politically is in diametric opposition to Hrdy's sociobiological premises (see Manicas, 1983). Such a requirement by no means restricts, by definition, political behavior to humans alone; many other animals, including probably all other primates, act as they do in part—admittedly, in small part, as compared to humans—for cultural as well as for genetic reasons (Bonner, 1980; Mainardi, 1980). Much of species-specific behavior must be learned by wolves as well as by chimpanzees; and Hrdy describes many aspects of langur behavior that must be learned to be acquired. (Maternal care of infants, adequate to permit the survival of the infants, clearly must be learned by most, if not all, monkeys and apes—not just by *human* mothers.) In a later section of this chapter, I shall present a more detailed discussion of what ought to constitute the minimal requirements of a political theory of primate behavior.

Baboon Strategists

For more than a decade Shirley Strum has been (and still is) working with a particular troop of olive baboons in north-central Kenya. There were about sixty individuals in the group when her field work began, including seven adult males and three times as many adult females; the other troop members, a majority of the total, included immature individuals, male and female alike. Female olive baboons remain in their natal troop; males migrate, beginning at adolescence, from troop to troop, although usually with more or less extended periods of residence in the troops with which they become affiliated. Each adult male develops a consort relationship with as many of the adult females as he can manage, although sooner or later changes in consortship are the rule rather than the exception. After seven or eight years of field observations and coding of her data had produced a fairly substantial basis for making statistical analyses, Strum began to ask a number of questions, all drawn from important and persisting issues in primate evolutionary and ecological theory, which could be tested with her accumulated data.

These included: Is there a consistent hierarchy of dominance among adult males in PHG ("The Pumphouse Gang," as she came to refer to her troop)? Is male dominance positively correlated with sociobiological fitness? Do dominant males preempt the meat that becomes available through predation or scavenging? Is the achievement and retention of high dominance status over other males best achieved through success in agonistic encounters?

Strum's answers (1982b) were as follows: Instead of finding a linear dominance hierarchy among males, she found numerous reversals in status and a poor basis (in dominance status) for predicting the outcome of any specific agonistic encounters, or the acquisition of resources. The males who were most dominant were least—not most—successful in establishing consort relationships with females and in gaining access to estrous ones. Neither did dominance rank predict the extent of meat consumption by males. Recently immigrant males outranked in dominance status males with substantial residence; but resident males did attract more consorts than the immigrants, evidently not because of greater dominance status but rather because of the residents' better knowledge and understanding of the social psychology of the group (cf. Reynolds, 1984). Correspondingly, resident males obtained more resources—both females and meat—than immigrants did. However, immigrant males do initiate relationships with troop females; and all adult males develop similar relationships with troop infants. Such social relationships, which are nonaggressive and cannot be based (in this species) on agonism, are much more important to an adult male's access to resources than is dominance rank per se. On the other hand, the agonistic encounters of immigrant males probably accelerate their acceptance by the other males of the troop. Strum concludes (1982b; 199; cf. 1982a) that her findings demonstrate "the limitations in the heuristic value of the concept of dominance hierarchy, at least for male baboons."

Strum therefore proposes that it is essential for males to learn to exercise social skills in circumstances in which agonism and dominance do not assure preferential access to conspecific and environmental resources. She suggests that her data exemplify the behaviors of cooperation, strategy, division of labor, and certain cognitive abilities that would have been essential to a primate preying on other animals larger than the hunter. These skills are found in contemporary nonhuman primates and surely should be hypothesized to have had to evolve among hominids (Strum, 1981; Baldwin and Baldwin, 1979; Durham, 1979). But she does not think that such a premise presumes any model of "*man* the hunter"; on the contrary, her data show that a sexual division of labor appears at the *earlier* (than organized hominid hunting) collector/predator stage. That division could be a consequence of male and female differentials in the social structure and reproductive strategies characteristic of different primate species. Thus a sexual division of labor, with adult males specializing in predatory behavior, may have evolved even earlier than has been thought likely, *before* the shift from collector/predator to gatherer/hunter modes of hominid adaptation. Western and Strum (1983: 25) speculated that natural selection may have acted directly on social skills, accelerating the encephalization of primates (and especially, of hominids). "It is in humans," they remark, "that social manipulation has attained its greatest sophistication, *as evidenced by human political systems*" (emphasis added).

As noted above, Strum's work focused on nonhuman primates in the most natural—that is, relatively least disturbed by humans—setting of the four cases discussed in this chapter. In her analyses and interpretation of her data, she also makes the most sophisticated use of formal primatological theory of the four cases considered here; indeed, there is a perfect, positive, and by no means necessarily spurious correlation between the two scales of naturalness of the primate group studied and of data analysis in terms of primatological theory that is both mainstream and formal. Strum is sufficiently interested in the possible political implications of her work to have presented a paper at the meeting of the Western Political Science Association in 1982, and also to have become further involved with political scientists at a symposium of the International Primatological Society in Nairobi in 1984. Like Chance, de Waal, and Hrdy, Strum makes no attempt to relate her primatological theory to contemporary formal political theory about relevant political behavior; but unlike them, she does not purport to be describing or analyzing political behavior per se in her reports on olive baboons. She is careful to restrict her claims to what is manifestly supportable in her data: that the behaviors she observed among baboons are of social strategies, many of which were nonaggressive and complementary to agonism and dominance (Strum, 1983); and that her data are consistent with the necessarily speculative hypotheses that she suggests about selection for encephalization among primates, about the dynamic relationship between scavenging and predation as primate responses to rapid ecological change, and about sexual selection for a division of labor prior to the emergence of a gatherer/hunter adaptation among some hominids.

REQUISITES FOR A BEHAVIORAL THEORY OF PRIMATE POLITICS

Interpretations of simian social behavior should not, for several reasons, be based on Shakespeare's plays, Machiavelli's advice to Italian princes of the Renaissance, or observations of the anomie and social isolation of human neurotics and psychotics. It is just as biasing of their observations, findings, and interpretations, for primatologists deliberately to anthropomorphize their investigations of simian behavior— "political" or otherwise—as it is for political scientists to "simianize" their investigations of human behavior—political or otherwise—by purporting to understand war, aggression generally (Kortlandt, 1972), the campaigning behavior of candidates for the Presidency of the United States, or sex differences in participational and other political behavior, on the basis of simplistic analogies with the behavior of simians. So we must first examine what "primatizing" involves.

If we want to compare the imputed political behavior of simians with that of humans, analysis should begin with inquiries into parallel evolution (Hall and Sharp, 1978) between hominids and the simian species being compared; not with linguistic analogies with the behavior of whatever species is alternative to the one being studied. We can appropriately begin such a homological comparison by discussing a matter of cultural evolution: the changes in the political structure of human populations and groups that have taken place during the most recent ten to twelve thousand years or so since the transition from gathering/hunting to agriculture/pastoralism for mainstream human populations (cf. ch. 8, below); such a transition was still in process for a few isolated—including some regressed—groups during the present century. Then we can turn to my main thesis: the requisites for the construction of behavioral political theory about simians. In considering the extent to which this can be modeled upon the political behavior of modern humans, surely we ought to understand something about how political behavior developed among humans. Such a phylogenetic model of human politics ought also to be related to the phylogenetic evolution of social behavior in each simian species to which comparison is to be made; but I leave that task to primatologists, whose specialization in one or another of the relevant simian species gives them a head start in pursuing such inquiries.

Primatizing

It can be conceded that simianism in the analysis—and even more so, in the prescription—of political behavior is much more dangerous to humans than anthropomorphism is in the study of primates; but the issue of anthropomorphism is nevertheless an important one.

There are three principal aspects of anthropomorphism in primatology to which I shall direct attention: first, the cognitive consequences, not for the subject animals but rather for their human analyst(s), of the mode of their designation; second, the consequences of näive (and truly uninformed) lay use of professional concepts (or, "Doesn't everybody understand politics, intuitively?"); and third, the methodological consequences of anthropomorphism, with particular regard to levels of analysis, the use (or nonuse) of evolutionary theory, and the dangers of insider analysts becoming blinded by the transitorily dominant paradigm in their field of personal academic socialization. (The latter—knowing too much about too little—is of course the converse of the question of interdisciplinary borrowing, where the problem typically is one of knowing too little about too much.)

First, to name an object is to know it; and to misname it is to fail to recognize it—for what it is in terms of alternative grammars, rhetorics, and systems of thought and discourse (see Peterson and Lawson, 1982b; Graber, 1982b; Peterson, 1982b, 1983a, 1983b, 1986; and ch. 17, below). The failure to recognize such risks and consequences, which is so evident in most of the cases considered above, indicates that many primatologists might profit, in both the conduct and interpretation of their field, from a more intimate acquaintance with contemporary scholarship in cognitive psychology and psychobiology (e.g., Secord, 1982; and especially Manicas, 1982). For starters, they would then certainly be more self-conscious about the consequences for themselves of bestowing the names of human children upon the animals they purport to study. Such names are not scientifically neutral; every one of them comes (for the user) equipped with the experience of the user's lifetime in previous associations with humans, whether they were known personally or vicariously through literature and television. Such associations influence what the user "sees" in the named animal's behavior and also what the user subsequently "finds" to be significant about that observed behavior. It would be impossible for any primatologist to repress such entailments, but there is no apparent evidence that many of them even try to do so. The behavior observed may survive as an electronic impulse keyed into a particular locus in the memory of a portable data-events-recording instrument; but the name, identity, and character of the designated animal live on in a much more complex recording instrument: the memory of the primatologist observer. And this is especially true of the affective, as distinguished from the effective, traces in human memory.

Second, the facile use by some primatologists of political concepts, wrenched out of the context for which they were designed and in terms of which they make some, at least limited, sense, is associated with no supporting evidence to suggest that they have troubled themselves to consult the relevant primary political science research literature through which political scientists themselves become socialized into an understanding of the range of meanings, theoretical contexts, and necessary limitations to the use of political concepts. To facilitate the possibility of such primatological excursions, half a dozen of the best available contemporary introductory undergraduate political-science textbooks on political behavior can be cited: Kavanagh (1983), Manheim (1982), Gianos (1982), Pettman

(1975), Rosenbaum (1975), and Jaros and Grant (1974). These works discuss the relevant questions of theory, methodology, and substance; and they list an extensive research literature in the primary sources upon which they rely.

This process cannot be viewed as a one-way street, of course. Political scientists who study primatology and related fields of animal behavior need a good comprehension of how field ethologists do their research. My own case indicates the kind of interdisciplinary work needed to become less of an amateur in using the research theory of ethology and primatology to help convert political science into a more biologically based (e.g., a less exclusively culture-bound) academic discipline. As an active member of such professional organizations as the [US] Animal Behavior Society, the American Society of Primatologists and the International Primatological Society, I spent a postdoctoral year (1977–1978) at the ethology laboratory at the University of Groningen in the Netherlands, studying communication behavior in birds with Niko Tinbergen's first doctoral student, Gerhard Baerends. A dilettantish social scientist does not bother to devote a year of his life to such research behaviors as getting up at 4 A.M. in order to beat the sun in arriving at a blind on a wet, freezing Dutch polder. What I have learned about the process ought to be worthy of consideration by primatologists approaching political theory from the opposite direction. There are, by now, at least a hundred political scientists with similar convictions and activities with regard to the relevance of the biological sciences to political science (Somit, 1976; Wiegele, 1979b; Watts, 1981; White, 1981c; Schubert and Somit, 1982; Somit, Peterson, Richardson, and Goldfischer, 1980, as updated by Peterson, Somit, and Slagter, 1982, and by Peterson, Somit, and Brown, 1983).

Third, there are the methodological implications of anthropomorphism, of which I shall begin with the question of levels of analysis. Clearly what Ronald Reagan does in both his verbal and nonverbal communication in a virtually worldwide televised "news conference" functions at a different level, for their respective populations, than Luit's grimacing from an electrified tree top in the Arnhem Zoo. Whatever the analogical similarities in the appearance of the two subject primates (cf. Masters, 1976, 1978b), it is näive to assume homology between events of such different scale (see also Peterson and Somit, 1980; and Somit, 1984).

Another question of methodology concerns the use of evolutionary theory, with a mind sufficiently open to be willing to entertain seriously the possibility of alternative explanations of the phenomena observed—even though such alternatives may not support, and might even detract from, the theoretical approach with which one is preoccupied at any given moment. This is a problem for many primatologists just as it is certainly a problem for many political and other scientists (see Schubert, 1982b: 223–35; and Kaplan, 1964: 351). An alternative way of stating this matter is to focus on the emotional implications for any scientific analyst—thereby including, of course, both primatologists and political scientists—of intellectual human bondage to what may be the most pernicious form of mood convection: the dominant paradigm that motivates the observer-analyst's field of inquiry at a particular time (Kuhn, 1970; ch. 19, below).

Politicking

How does political behavior differ from any other aspect of social behavior? It is easy for nonspecialists to define the core subject of any discipline; but it is no more reasonable to expect political scientists to agree upon what politics is than to expect agreement among psychologists about the mind, among sociologists about society, among economists about the economy, among biologists about the nature of life, or among astronomers about the ultimate fate of the universe. A preeminent political scientist defined politics as "who gets what, when, and how" (Lasswell, 1936). A position that many biopolitical behavioralists would accept is that politics deals with the sociopsychological processes of human individuals and groups seeking to influence or control others who are not closely related to them in regard to any questions in which humans are, or might become, interested. The restriction of genetic heterogeneity means that politics does not occur other than metaphorically *within* families. Human intrafamilial relationships are studied by sociologists and cultural anthropologists, not by political scientists; social scientists agree that "families" consist of individuals who live and breed and raise offspring in intimate spatial and psychological proximity to each other (see Spiro, 1975; Tiger and Shepfer, 1975; and Rossi, 1977; but cf. Breines, Cerullo, and Stacey, 1978). Discussions of "family politics" reflect simile and metaphor, like discussions of "langur politics" or of "chimpanzee politics," and substantially for the same reason. When the units of social structure become sufficiently large and diverse to encompass many different consanguinous (extended-family) subunits in cooperation or competition with each other for control over the policies that are important for the group overall, *then* it is possible to recognize the beginnings of political behavior.

The first three cases discussed above involve primatologists who invoke selected political concepts about the behavior of modern humans—modern in the sense of cultures that evolved long, long after the transition to agriculture about eleven to twelve thousand years ago in the Fertile Crescent and Mesopotamia. Whatever may be their cultural differences in politics and *Weltanschauung* (Rosenbaum, 1975), Plato, Machiavelli, and Shakespeare are, like us, modern humans; the relevant time scale is evolutionary rather than historical (Pettersson, 1978), in the sense that "history" is understood and measured by social scientists. Primatologists are surely right in presuming that the genesis of political theory must be sought in the bones of Lucy and the footprints left in the volcanic ash of East Africa by what may have been three of her conspecifics, some 3 3/4 million years ago—not in such latter-day political writings, whatever their merit, as Aristotle's *Politics* (Schubert, 1973: 240, n. 1; ch. 6, above).

Of course, the simians whose behavior is discussed in the four cases here are also moderns, just as modern as we are. Discussing several lines of evidence from different disciplines (primarily the electrophoresis of detectable substitutions in the comparison of polypeptides, and bipedalism versus knuckle walking in the muscle system and skeletal structures of both living and fossil primates), behavioral geneticist Alan Templeton (1984) asserts that chimpanzees and gorillas diverged from a common ancestry with hominids at least five million years ago and subsequently evolved their unique knuckle-walking adaptation. That evolutionary change occurred while hominids were evolving their unique extremities of feet, and brain size and cortical structure, especially after about 1.8 million years ago (Kurth, 1976; Brace, 1979). Whether or not a scientific consensus comes to agree with the details of Templeton's scenario concerning primate evolution (see Cherfes and Gribbin, 1981a, 1981b; Yunis, Sawyer, and Dunham, 1980; Yunis and Prakash, 1982), it is no more justifiable for primatologists to homologize from the political behavior of living humans in order to interpret the behavior of living simians, than it would be for political scientists to do the opposite (but see Willhoite, 1976; Tiger and Fox, 1971; and Loy, 1975). Strum (1981; and cf. Geist, 1978) has demonstrated how volatile such a fundamental change in strategy as a shift from scavenging to much greater predation can be, in response to extremely rapid and extraordinarily extensive ecological change (chs. 12, 13, and 14, below). Whatever may be assumed about progress in the social behavior of modern simians in relation to their respective ancestral species, the authors of the other three cases evidently presume that either social behavior, of the species each has studied, has remained essentially static—at least in relation to human social behavior—or phylogenetic evolution in the social behavior of simian species is irrelevant in making comparisons to the political behavior of modern humans.

To understand the evolution of political society, it is necessary to go back to a time before the earliest classic books of traditional political philosophy, before the earliest surviving political symbols such as the pyramids, and even before the invention of written language in the early city-states of Mesopotamia approximately five thousand years ago. It is necessary to turn to the evolutionary anthropologists and botanists who have specialized in the study of the artifacts of human prehistory (Fried, 1967; Carneiro, 1970; E. Fisher, 1979). The most generally accepted model is that of Kent Flannery (1972; cf. Alexander, 1979a: 250; and Corning, 1983: 346). Flannery, following Service (1962), distinguished four levels of human political organization: the band, the tribe, the chiefdom, and the state. Bands are characterized by local group autonomy, egalitarian status, ephemeral leadership, and ad hoc rituals. Bands typically include only about forty persons, who tend to be relatively highly interrelated genetically, belonging to one—or at most, very few—extended-family lineages. Living examples of bands, during the past two centuries, include mostly gatherer/hunters, such as the Kalahari San. All humans were organized in bands prior to the transition to agriculture (Struever, 1971; Cohen, 1977; and for a good review of recent work on political evolution, see Corning, 1983: 345–75).

A tribe included about one hundred individuals, with more lineages than a band, a lower average level of genetic relatedness, and much more stable and specialized social organization. Tribes evolved in synchrony with the domestication of plants and animals, which so stabilized food supplies that larger populations could be sustained. The economy of tribes initially was mixed, but some of them subsequently became based primarily on the domestication of plants, while others specialized in the domestication of animals for food consumption and developed pastoral economies if the animals they domesticated foraged on wild rather than domesticated plants.

As the density of population increased, so did competition *between* tribes for access to land, whether for use in horticulture or grazing, and this resulted in intertribal warfare (Durham, 1976b; Borgia, 1980: 187). This initiated, possibly as early as six thousand years ago in some desirable Near Eastern locales, the subjugation of some tribes by others; and the labor of the subjugated slaves or serfs shifted resources to their conquerors. The latter became supertribes or chiefdoms with an even more complex hierarchical political structure, corresponding divisions of social and economic status, and populations ranging in the hundreds. Tribes with an agricultural economic base had developed permanent residential sites as early as 9000 BP; and by 5000 BP the earliest states, which were hierarchically organized aggregations of a dominant city with many towns and villages extending over a relatively large area, began to be established. For present purposes I do not need to continue summarizing (but see ch. 8, below) increasingly more familiar political history concerning the establishment of "states" (which happened almost invariably through military conquest and aggrandizement), as recounted in works of scholarship beginning 2 1/2 millennia ago with Herodatus' history of the rise of the Persian Empire and its attempted conquest of the Greek city states and their confederations.

It is evident that the most complex level of social organization found in simians corresponds to the level that humans had achieved one hundred thousand years ago at the beginning of the Wurm-Wisconsin Glacial, not at its ending, which helped to catalyze the transition in the Fertile Crescent and elsewhere (Darlington, 1969: 70). Occasionally in contemporary times chimpanzee bands have been observed to congregate in temporary aggregations of one hundred individuals; and troops of baboons may spend the night in contiguous territories that result in hundreds of animals sharing the same cliffs. But no student of chimpanzee behavior has reported any evidence of a system of social control—other than, perhaps, "mood convection"—by the leadership of such a temporary aggregation of chimps; it does not appear that sleeping baboons require or utilize any system of social control external to the troop groupings, each of which occupies its customary nocturnal territory just as it occupies its customary, though much more widely decentralized, daytime territory (Marais, 1967). Thus the highest level of complexity in social organization and behavior for simians corresponds to the *pre*-political band level among humans.

No doubt it would be better, if possible, to compare early Homo sapiens of the Paleolithic (Geist, 1978: 354–55) with then-contemporary simians. Not having such an opportunity, at least we can restrict comparison with humans to what characterizes the band stage of human social unit organization. And there we find (under "natural" conditions) heterosexual but highly consanguinous groups of perhaps forty persons, living much like the Kalahari San did thirty years ago. For the San we now have continuous multidisciplinary observations, by primatologists as well as by cultural anthropologists, over a period of more than a human generation (e.g., Lee, 1972; Lee and DeVore, 1976; and Konner, 1982: 5–10, 31–32, 301–15, 371–76, 447–50, and *passim*). Such studies of the San and other gatherer/hunter bands suggest a model of small human groups that are egalitarian in their social organization and behavior. There are no dominant, aggressive, male bosses of status hierarchies. There are, therefore, no regimes to be usurped, no tyrants to be displaced, and no strategies necessary to the construction of minimal winning coalitions, so that the spoils of political victory can be divided among an optimally small number of "winning players" in a zero-sum game (Riker, 1962; Riker and Ordeshook, 1973; Axelrod, 1981, 1984). There is, therefore, virtually no political behavior worthy of being imitated by—or, what amounts to the same thing, attributed to—the murderous langurs described by Hrdy, de Waal's power-hungry chimps, or the hedonistic macaque leaders who display to their attentive followers in Chance's scenario. At least prior to fifteen thousand years ago, humans at the band level were not yet filling a niche that would either require or permit them to develop the complex social and political structures and behaviors that stemmed from their successive domestication of plants, other animals, and finally themselves (see ch. 8, below, for details of that process).

Even if primate politics as a subject of scholarly inquiry cannot be pursued by nimbly skipping back and forth between the social behaviors of contemporary apes and humans, this does not signify that the topic is an unworthy one or that primatologists and political scientists have little to share with each other. It clearly remains possible to consider the question of homologies in the prototypical behaviors of humans and of simians, even if it should be the case that these become manifest at different developmental stages in the ontogeny of contemporary simians and humans. Given the neoteny of contemporary humans, it should hardly be surprising that the behavior of our children has the most in common with the behavior of adult chimpanzees or baboons (Barner-Barry, 1977, 1981, 1982; Strayer, 1981; Jones, 1983; Omark, Strayer, and Freedman, 1980). Furthermore, we ought to expect (or to have expected) that the route to our common goals, as students of the roots of political behavior, must be found in the study of the parallel evolution of primate species, and not merely in linguistic exegeses on the apparent social behaviors of living apes and humans. Such observations can just as well denote analogies or metaphors rather than homologies (Hall and Sharp, 1978; Thompson, 1975; Campbell, 1979; Harding and Teleki, 1981; and ch. 6, above).

Nevertheless, there must be many important homological continuities between the substrata of social (including, when and where it occurs, political) behavior in humans and social behavior in simians. These must include many aspects of physiology and behavior genetics. Endorphins must operate similarly for all primates (Tiger, 1979b), and arousal may well be related to attention in the psychophysiology of perception, for simians as well as for humans (Pribram and McGuinness, 1975). Whether anticipatory socialization into leadership roles operates similarly for human males as it does for rhesus males (McGuire, 1982) appears not yet to have been studied (but see Madsen, 1985, 1986). Notwithstanding important sex differences between humans and simians (and among different species of simians) in regard to structural dimorphism, sexuality, and mating behavior (Hrdy, 1981; H. Fisher, 1982; Watts, 1983; Hall, 1982, 1985), there appear to be important questions about sex role differences (in both social structure and social behavior) for which continuities can be observed between humans and simians, as the de Waal and Strum case studies imply.

Of particular importance will be further inquiries into the wellsprings of human speech (e.g., Andrew, 1963); more sophisticated and extensive studies of human nonverbal communication (e.g., van Hooff, 1972); and the cognitive primatology of both emotions and conscious thought. In time these will enhance considerably our understanding of the psychobiology of simians as well as of humans, notwithstanding the extent to which the respective brain sizes and cortical structures constitute what is probably the most important difference between us and other primates. After all, strong selection on the human brain is considered to have stopped twenty thousand years before (Kurth, 1976) politics began, as political science understands and studies it.

Primatologists and political scientists are most likely to discover common ground for their mutual interest in the genesis—and epigenesis (see ch. 14, below)—of political behavior through the more deliberate use of evolutionary theory to keep their ethological observations in better perspective.

8
Religious

THE CONSTITUTIONAL STATUS OF SCIENTIFIC EVOLUTIONISM AND CREATIONISM

Almost six decades have elapsed since the Scopes trial, and we tend to forget now that Scopes was *convicted*: the Tennessee antievolution statute was upheld in the state courts (*Scopes v. State*, 154 Tenn. 105, 1927), and the case did not reach the United States Supreme Court. This is a not atypical example of the law going one way, looking backward in a way very different from that of Edward Bellamy, while society and the American national culture in time attach opposite meaning to the significance of such a cause célèbre. Indeed, a constitutional challenge to the descent of man did not get to the Supreme Court until *Epperson v. Arkansas*, 393 U.S. 97 (1968); and by then, public attitudes had temporarily shifted away from bedrock fundamentalism to such an extent that there was no question—as there certainly would have been if it had been the Taft or Rehnquist instead of the Warren Court doing the deciding—of *not* striking down the Arkansas statute that prohibited the teaching of the Darwinian theory of evolution and, hence, of the modern science of biology in the public schools. The average member of the Warren Court was twenty-two years of age, when Clarence Darrow confronted William Jennings Bryan in the Tennessee "monkey trial"; but at that time the

---

This chapter is a revision of "Scientific Creation and the Evolution of Religious Behavior," *Journal of Social and Biological Structures*, 9: 241–60 (1986), which in turn was a revision of what initially was presented as part of the Third Annual Rocco J. Tresolini Memorial Lecture on Law and Legal Institutions, Lehigh University, April 27, 1980; most of the remainder of that lecture was published in revised form in Schubert (1985a). Reprinted with permission of the publisher.

average age of members of the Taft Court was that of one who had been born before the outbreak of the Civil War. *Epperson* was an unanimous decision; and the author of the opinion of the Court was the late Abe Fortas—one of the three most radical justices (with John Clarke and William O. Douglas) ever to serve on the Supreme Court of the United States of America. The decision came from one of the two most stable and relatively extremely liberal majorities of justices the Court has ever had,[1] including Douglas, Fortas, Marshall, Warren, and Brennan (and in that order: Schubert, 1974: 60, 159). We can take Hugo Black (Schubert, 1984b) as an example of why the other four justices accepted the decision: even though he had been socialized as a youth into the very kind of political and religious culture that was later to spawn the pro-"Scientific Creationism" statutes (O'Neil, 1982; Wood, 1982: 234), he was ready in 1968 to acquiesce in the tutelage of the American Civil Liberties Union, the National Educational Association, and several other proponents of the postulated constitutional right of "academic freedom." By 1982, the judicial climate of opinion had become sufficiently liberalized generally (Schubert, 1985a) to make both possible and probable the decision of federal district judge William Overton, declaring unconstitutional the Arkansas Scientific Creationism statute, in *Rev. Bill McLean vs. Arkansas Board of Education.*[2] The prevailing American constitutional policy (Schubert, 1960, 1970) postulates both a constitutional right to teach scientific evolutionism and also the denial of Scientific Creationism's claim to a constitutional right to separate-but-equal status to be taught as a theory of scientific evolutionism in the public schools.

MODELS OF HUMAN EVOLUTION

Evolutionary Biology

What model of human evolution does modern biology accept (Williams, 1966; Futuyama, 1979; McEachron and Baer, 1982; McEachron, 1984)? Speaking as one who, as an adolescent, first read Bertrand Russell's "A Free Man's Worship" (1918 [1903]), I deem it fair to report that the intervening unfolding of the twentieth century to date has done little to change the account of creation, at least so far as concerns the earth, reported by Russell in 1903, which in turn was based on Goethe's *Faust* of the early part of the previous century. Radio-astronomers have learned more about quasars and black holes, and "Big-Bang" cosmology has long been controversial but remains the most probable and widely accepted explanation of the physical dynamics of the universe (Jastrow and Thompson, 1974). According to that theory, the universe is about thirteen billion years in age; and independent evidence dates earth at about a third of that—four and a half billion

[1]The other period was during the latter half of Stone's Chief Justiceship, 1943–1946, when he, Murphy and Rutledge, together with Douglas and Black, made possible a solid liberal majority on most issues of civil liberties; see Pritchett (1948) and Schubert (1970).

[2]Reproduced in full in Overton (1982) and also in Murphy (1982, pp. 117–148).

years. The displacement of dinosaurs by mammals is estimated to have occurred about seventy million years ago, with the earliest primates diverging soon thereafter. The early higher primates are traced to about thirty million years ago, and our earliest hominid ancestors became differentiated from pongids perhaps five to eight million years BP (Hall, 1985: 135; ch. 7, above).

During the past two decades particularly, explicit fossil finds in both Africa and Asia are identified as protohominid and dated at fourteen and one half million years BP. The expansion in both the size and convolutions of the neocortex and cerebellum of the human brain, transactionally with the probable development of speech, social cooperation and organization (Boyd and Richerson, 1985), laterality, and other distinctive attributes of the species as we know it now, have occurred primarily within the last million years; and Cro-Magnon *Homo sapiens* radiated into Europe no more than fifty thousand years ago, apparently to precipitate the gradual but complete extinction or absorption of *Homo sapiens neanderthalis* (but see Stanley, 1986) within thirty thousand years BP (for imaginative reconstructions, see Golding, 1955; Kurten, 1980, 1986; Auel, 1980, 1982, 1985). Our present subspecies *Homo sapiens sapiens* evolved about twenty thousand years ago (Washburn, 1978, pp. 146–54; Campbell, 1966; Darlington, 1969; Birdsell, 1972; Braidwood, 1975; Lancaster, 1975; Geist, 1978); and by ten thousand years later had begun the fateful transition from hunting and gathering to pastoral and then to agricultural economies (Cohen et al., 1980). The latter led on to the development of villages by 9000 BP, towns by 6000 BP, and soon thereafter the great cities of the Mesopotamian basin (Trump, 1980: 7–28, especially 25, 28; Braidwood, 1975; Birdsell, 1972; Fisher, 1979: 215–33).

The earlier Middle Pleistocene increase in the size of the brain of *Homo erectus*, in counterpoint with, and certainly in substantial degree due to, the development of nonverbal and spoken language in that precursor hominid species (Crook, 1980: 120–49; Sagan, 1977; Geist, 1978: 241–68), undoubtedly was due primarily to genetic evolution, even though its major consequence was to make accelerated cultural evolution possible (Waddington, 1957, 1960; ch. 14, below). The more conservative and classical population-genetics argument is that the 300 (minimum) to 350 (maximum) human generations of the eleven thousand years of the recent geological era that have elapsed since the transition to agriculture began are utterly inadequate for credence to be placed in a genetic explanation for the changes in human adaptation since then (Kurth, 1976: 508–9); but in several radical recent works, Charles Lumsden and Edward Wilson have urged strongly a "thousand-year rule" to the contrary.[3] Except for speech, the transition was without any doubt the most important change, not only in our hominid adaptation but also in our social and psychological structure, that occurred in half a million to a million years (Keith, 1949: 267–77; Pettersson, 1978; Fisher, 1979: 177–89). It

[3]See Lumsden and Wilson (1981) and the peer commentary on their precis of the book (Caplan et al., 1982). Similarly, see Lumsden and Wilson (1983), and the precis plus peer commentary on that book (Barash et al., 1984). See also Kort (1983) and Cavalli-Sforza and Feldman (1981).

was accomplished by means of phylogenetic epigenetic manifest cultural evolution that undoubtedly entailed latent consequences for genetic evolution (Webster and Goodwin, 1982; cf. chs. 10, 13, and 14, below).

Domestication

The transition was based upon the domestication, by humans, of selected plants and animals (Ucko and Dimbleby, 1969): "domestication" means that humans exercised control over the reproduction of the subject flora and fauna. Humans did this in order both to manipulate the subject species' phylogenies so as to enhance those phenotypic characteristics which were deemed desirable by the human manipulators and also to accelerate the rate of genetic change for such experiments. But no such changes in the morphology of domesticated plants or animals or in the behavior of domesticated animals could take place without reciprocal changes in the behavior of the human domesticators. The herding, feeding, and protection of domesticated animals placed pervasive constraints upon the herders, just as the preparation, sowing, protection, and reaping of crops constrain the possible range of behaviors of farmers, who labor under the additional limitation of residence adjacent to the plantings—at least until after harvesting, and perennially if bulky crop surpluses can and must be stored. Domestication involves, therefore, a combination of cultural and genetic selection—human culture determines the rules and processes, and genetic change defines the results—in plants and animals, both those selected for experimentation and those that are not target species for domestication but nevertheless are affected ecologically by the processes. At the same time humans domesticated themselves, by means of cultural rather than genetic evolution (as indicated in the preceding paragraph), as another unintended but inescapable consequence of the transactional nature of the domestication relationship and process. Of course, an equivalent argument can and ought to be made concerning the entailed effects of industrialization upon nineteenth-century humans and of nuclear fission upon the people of the twentieth century; those developments are beyond the scope of the present article (but see Green, 1986).

Elizabeth Fisher (1979: 132) suggests that "the Near East was the crossroads of the three inhabited continents: Asia, Africa, and Europe. It would therefore be the natural meeting ground for peoples, genes, and ideas—logical sources for the greatest diversification as well as the most efficient use of possibilities—so that new human species and new inventions would indeed be most likely to arise in that area." Lewis Binford, who hypothesized that "cultivation arose as a response to similar pressures many times and in many places," expected that evidence "for the initial domestication of plants and animals in the Near East will come from areas adjacent to the Natufian settlements in the Jordan Valley" where the Jordan River flows south from the Sea of Galilee to the Dead Sea (Binford, 1971: 48–49). Binford anticipated that similar evidence should be found in European Russia and south-central Europe, adjacent to the rivers flowing into the Black Sea. Fisher (1979: 131) has commented that "in the Judeo-Christian tradition, which has shaped much of Western civilization, the Near East plays an important role. The Garden of Eden, first home of our mythical ancestors, and Mount Ararat, on which, at a later date, Noah's ark came to rest after the other inhabitants of the world had been destroyed by flood, are both situated in the general area. It is curious, then, that twentieth-century studies of our beginnings, and possibly those of all modern humans, keep bringing us back to the same area, for the origin of *Homo sapiens sapiens* and again at later stages in our story" (and cf. Steiglitz, 1982). The Natufian ancestors of the Israelite tribe were in one of the right places, at the right time, not only to have been participants in the transition, but also for proof of their presence there to be unearthed.

An abridgment (Davis and Dent, 1968: 9) of Frederick Zeuner's classic study asserts that "so far as the present state of knowledge goes, the earliest signs of domesticated animals are to be found in remains of the ninth millennium B.C. and in the neighborhood of Jericho in Palestine. Besides the dog, these people of Jericho seem to have kept the goat as a domestic animal, and perhaps to have embarked on the taming of another grazing animal [the gazelle] that proved after all a blind alley." It seems most likely, however, that the domestication of dogs came first, possibly unintentionally, and independently among widely separated hunting and gathering peoples in several different continents—although everywhere from wolves (Hall and Sharp, 1978; ch. 6, above). Evidence for the domestication of dogs indicates Iran, 10.5 thousand years ago (Davis and Dent, 1968, figure opposite p. 1; Zeuner, 1963) and 10.4 MBP (millennia before present) in what is now the USA, 9.5 in England, and 9.0 in Turkey (Reed, 1971). Sheep are associated with Iraq by 10.8 MBP, and goats with western Iran at 9.0, but, unlike the situation for dogs, the initial domestication of goats and sheep is limited to Mesopotamia and the Levant—the "Fertile Crescent," where the domestication of plants is thought to have begun, about 11.0 MBP, with the emmer wheat and two-rowed barley that grow wild there and in Anatolia (Butzer, 1971). Domesticated pigs are found in Anatolia at 9.0 MBP, cattle in Greece at 8.5 MBP, onagers in Mesopotamia by at least 4.5 MBP, the horse from 4.5 to 5.0 MBP, originally (like the reindeer) in North America. Domesticated cats are first found in Egypt at about 3.5 MBP. It is important to note (Davis and Dent, 1968, p. 23) that "as one might expect when an animal no longer has to fend for itself, its brain changes in shape and function. . . . [T]he parts of the brain which tend to shrink as domestication progresses are those controlling sight, hearing and scent." In other words, use it or lose it. Also, "except for those animals which man has put to work, such as the horse, and formerly the plow ox, the tame variety is always less muscular though possibly fatter than the wild. In almost all animals, the muscles which work the jaw tend to become smaller and weaker with domestication, and this in turn affects the shape and size of the bony ridges on the skull to which they are attached." Evidently, the described effects are due to epigenetic causes that have developmental consequences and, given enough time overall and a sufficiently short species-generational time, they will also select for genetic change. Domesticated humans certainly are subject to the epigenetic effects, and possibly to the indicated sensory deficits as well (although we should expect that sensory deficits due to the transformation of the human niche and habitat will be even greater, as a consequence of the industrial and nuclear transitions, than can be attributed to the agricultural transition; ch. 11, below).

Thinking Animals

Paul Shepard (1978) has explored both extensively and intensively the relationships of humans with other animals, in both the phylogenic and ontogenic development (Gould, 1977a) of human cognition (Peters, 1978b; Porter and Russell, 1978; White, 1974). Shepard (1978: 164) points out that "wild animals are genetically more diverse than appearance would indicate. Intelligence, the capacity for learned behavior, synchronizes behavior and thereby compensates for the hidden genetic polymorphism that keeps a species young and adaptable." The effect of domestication upon animals is, of course, to make them into stupid freaks, when compared with their wild progenitors. Shepard says (1978: 200; cf. ch. 11, below) that "the captive is a sick creature, the domesticated animal a monster; that we find them remedial [when we make them into our "pets"] is a comment on the desperate status of our human ecology"; he continues (1978: 250) that "minding animals has its pathology. The most widespread is the yoking of non-human creatures into human society. . . . That perversion was first unleashed by the keeping of animals in agriculture, but today is increasingly the result of keeping animals as objects of leisure or toys. . . . As the farm animal disappears, the pet population redoubles." Shepard's thesis (and cf. Levinson, 1980) is that the presence of other animals, as parameters that define the human habitat, is for genetic reasons (that extend back throughout hominid and earlier primate evolution) essential to normal human cognitive development—and hence necessary to the continuation of our present species, as such.

Ecologies with "animal-defining" environments suitable for humans were of course characteristic of the hunting and gathering niche; and, albeit in increasingly warped form, they persisted throughout the last eleven thousand years until industrialization combined with escalating human population growth (during the present century) to create present conditions, which can be expected to result in the extinction of all nonhuman forms of mammalian life before the middle of the twenty-first century. Humans always have exterminated competing animals to the greatest extent possible, usually with considerable success; but now human pressures upon limited space, water, and food place in competition with us all other mammals (at least). But as thinking animals ourselves, we cannot lose them without losing also an important component of our own humanity. To understand "minding animals," it is necessary (Shepard, 1978: 249, and 2–3, 253)

> to identify and explore the ways in which the human mind needs animals in order to develop and work. Human intelligence is bound to the presence of animals. They are the means by which cognition takes its first shape and they are the instruments for imagining abstract ideas and qualities, therefore giving us consciousness. They are the code images by which language retrieves ideas from memory at will. They are the means to self-identity and self-consciousness as our most human possession, for they enable us to objectify qualities and traits. By presenting us with related-otherness—that diversity of non-self with which we have various things in common—they further, throughout our lives, a refining and maturing knowledge of personal and human being.

It is well known that the humans called Neanderthals observed complex rituals for the burial of their dead; and scholars have interpreted this to signify the awareness of mortality, concern about a hereafter, and the practice of religion, at a time ranging from thirty-two thousand to seventy thousand years ago. Cro-Magnon humans frequently occupied huge limestone caverns, scattered throughout Europe, during the Main Wurm glacial stage of twenty-five thousand to fifty thousand years ago; and in many of these caves the walls of the innermost sanctums are still decorated with the brilliant sketches of the animals upon which they preyed, or with which they competed for food and shelter. Some scholars (Tiger, 1979b: 71–79; Marshack, 1971) believe that many of these exceedingly lifelike drawings represent the totems of the animal gods that these humans worshipped. Shepard's comment (1978: 30; and cf. Marshack, 1971, and Geist, 1978: 301–53) is that "this biology of art is most clearly seen in some of its oldest surviving expressions, the fabulous, worldwide cave painting and etching which span more than twenty-five thousand years. These cave and rock-shelter sanctuaries were the antecedents of temples, where the significant passages of human life were ceremonially represented. The figures in them, however, are mostly animals, a fact that misleads the modern viewer into thinking they are about animals. They are fossil forms of thought in which the act of imaging and symbolizing is embedded in its maternal substance, the hunt."

Cave art constitutes, indeed, what in cultural evolution is the ancestral prototype of the totemic art and cultures that have been so widely reported and discussed by social anthropologists of the nineteenth and twentieth century, based on their field observations of remnant neolithic hunting and gathering peoples that survived into—or almost into—the contemporary era (Haddon, 1895: 306–20; Boas, 1927; Marshack, 1971; Geertz, 1973: 421–25).

Creation Myths

According to Shepard (1978: 125), "the core of totemic culture is a set of myths or stories about creation. These tales of how things come to be as they are narrate events in the first society of beings, the ancestral ones, which includes some humans and a variety of other creatures, all speaking and behaving as though they belonged to a single sociopolitical unit."

The fundamentalist Christian view of religion presumes a time frame of barely two thousand years since the initial Coming of the Father's only begotten Son, to make possible everlasting life for humans (or at least for some suitably eligible fraction of the species), and not much more than twice that time (Jones, 1968: 343, 6) since the Creator "fashioned man of dust from the soil," supplied him with a helpmate, and then proceeded to tempt her with a phallic symbol in the form of a serpent—with consequences for human mental health, adornment, suffering, female subordination to males, and the reproduction of the species, with which we all are only too familiar (but see Paige and Paige, 1981). Evidently, too, this was a time when human problems with the environment accelerated because denial of access to the oasis of Eden left only the alternative of the desert and wasteland surrounding it—the Palestine that many contemporary peoples still remain willing to fight to the death over today. We need not be concerned with any preceding stages of earth development, according to this literalist model; Yahweh accomplished that prior task in six allegorical days; and even if the diurnal cycle then was a much longer one than now (Newman and Eckelmann, 1977), we are left with a portrait of human existence that seems reasonably congruent with the probable latter-day trials and tribulations of the Israelites as a distinctive ethnic tribal aggregation among dozens of desert nomad primitives descended from the Natufian aboriginals in the Neolithic Levant of 9000 B.C. (Butzer, 1971: 223–37; Keith, 1949: 379).

The preceding synopsis of Genesis incorporates both of the two quite distinctive, in terms of both content and time of composition, creation myths that characterize the Hebraic and the Christian religions. The older of the two (Adam and Eve) is dated at approximately 900 B.C., while the younger (Sabbatarian) version is dated about five hundred years later (Sproul, 1979: 4, 123; for a comparative analysis of both versions, see Van Over, 1980: 232–33). There are many indications that both Judaic myths were derived from older Mesopotamian cultures, the early Sumerian (prior to 2000 B.C.) and the Old Babylonian (1750–1550 B.C.), and that these were transmitted in a parallel line of descent to a contemporary (of Genesis) Mesopotamian culture, the Neo-Babylonian culture of 600 B.C. (Sproul, 1979: 91, 114, 120). Raymond Van Over (1980: 171; and cf. McClain, 1982) remarks that "it is clear that there are traces of . . . early Ugarit or Canaanite myths in Hebrew mythology. . . . [For example] the Hebrews took over much of the Baal myth, and transferred it to Yahweh when the Hebrew peoples settled in Canaan. . . . There is obviously great intimacy and cross-influencing among the several Mesopotamian cultures. While the Hebrew religion remains unique in the Semitic tradition, it was influenced by the other groups in numerous and subtle ways." Barbara Sproul (1979: 123) agrees:

> The opening book of the Old Testament was called Genesis by the third-century B.C. Greek translators who made the Septuagint version of the Bible; its proper Hebrew name was Bereshit ('In the beginning . . .'), referring to its opening words in the same fashion as the Mesopotamian creation epic, the *Enuma Elish*, is named after its first words. The par-

allels between the first creation account in Genesis and the Mesopotamian epic are not confined to their naming process. Not only are there marked similarities in specific details but also the order of creation events is the same, leading many to presume a dependence of the Old Testament account on that of the *Enuma Elish* or similar Babylonian documents. (The relative ages of the documents and their civilizations preclude dependence of the *Enuma Elish* on Genesis.)

Among the other creation myths described and dated by Sproul (1979) are those from Egypt (Ptah [Memphis], *c.* 1400 B.C., p. 10; cf. Van Over, 1980: 249–50), India (Rig-Veda, *c.* 1200 B.C., pp. 6, 179); Kena Upanishad, 800–400 B.C., p. 8; Chandogya, *c.* 700 B.C., p. 9; and Jinasena [Jain], *c.* A.D. 900, p. 10), Assyria *c.* 800 B.C., p. 15), China (Lao Tzu, *c.* 600 B.C. [?], p. 12), and Central America (Mayan [Popol Vuh], *c.* A.D. 1600, p. 18). Sproul describes also the creation myths of several contemporary (nineteen- to twentieth-century) aboriginal cultures, including Hawaiian (pp. 358ff.), Mali (pp. 49ff.: African), and Eskimo (pp. 220ff.).

Even Noah's flood has an explicit and appropriately timed geologic precedent, in the precipitous transition from the Wurm glacial to the present postglacial, approximately thirteen thousand years ago—a date only two thousand years prior to the earliest evidence of domestication in the Jericho mound; and Noah's passengers patently were a sample representation of animals associated with the human hunting, rather than pastoral or agricultural, cultural environment. Clearly, Jericho itself places the Natufian aboriginals of the Genesis patriarchs at the very threshold of the transition to domestication. By then we are back in touch with all except the Paradise Lost stage of the Genesis creation myth.

Integrating Biology and Culture

The factor by which the geological and the biblical accounts of the formation of the earth differ is 750,000:1; and even the respective estimates of the age of humans as a distinctive type of creature differ by a factor of 2400:1. If we take account of our primate ancestors (as we ought to do because it is relevant to understanding the importance of religion to us), then the factor becomes 5000:1. But in a qualitative sense, the cultural model of the creation myths fits like hand-in-glove into the biological model of human evolution.

Generally in the Near East as well as elsewhere,[4] just as in the more particular image of it reflected in the Pentateuch, the historical record of human cultural evolution begins with queens (Fisher, 1979: 267–80; and cf. McClain, 1976: 150) or kings who themselves are the living Gods worshipped by their subjects,[5] or else they are their people's living spokespersons to communicate with supernatural Gods (Jaynes, 1976).

[4]Wheeler (1975: 12) has noted that "Wheatley argues that ancient cities, like those of the ancient Mediterranean world, were constructed as a sacred precinct, a residence of the gods, and an *anis mundi*. . . ."

[5]As was, of course, the then incumbent (and still-living) emperor of even such a modern nation as Japan, until less than forty years ago.

Julian Jaynes (1976) claims that the mind (and hence consciousness and capacity for reasoning and memory) of even such relatively modern people as Periclean Greeks and Pentateuchal Hebrews was fundamentally different, *in its mode of physiological operation*, from that of contemporary educated Western persons. Jaynes asserts that gods spoke directly *to and through* not only the ancient Greeks and Hebrews but also that gods *continued to speak thus* to selected humans, at least until the end of the Middle Ages in Europe. Indeed, they still do so although perhaps more selectively; now we consider those who hear such voices to be deranged: paranoid or schizoid (in the controlling language of our contemporary psychiatric myths: Szasz, 1965, 1970, 1973; Foucault, 1965; Goffman, 1961).

The need to construct an account of the beginnings of human life, an account that is credible in terms of the dominating characteristics of the particular environment (both physical and cultural) in which people find themselves, appears to be universal because of biologically based, genetically transmitted, neurological responses to the psychophysiological demands of human needs essential to both development and survival.[6] Barbara Lex, for example, asserts (1978: 300) that "upon analysis distinctions between 'political' and 'religious' power are shown to be more apparent than real. If supernatural sources are thought to cause unusual psychophysiological states, then the aggregate prophets' pre- and proscriptions, in addition to their instructions regulating and curtailing conditions which threaten the life processes and continued effective functioning of individuals, can be understood as mechanisms which ordain the establishment of precisely defined, essential, and sanctioned emotional feeling states in individuals and among groups."

When we get to the first cities in Mesopotamia eight thousand years ago, the central structures are temples, with priests who, until a much later time, remained also scribes with a monopoly over writing. This gave them control of the past, as well as of the future (Orwell, 1949); and they were in addition political leaders who thus controlled the present (d'Aquili et al., 1979; Pfeiffer, 1977; Fisher, 1979). Shades of the Ayatollah Khomeini! Shades of Moses, too—at least until his appointment of the judges! But in the beginning there was no word, no word at all, only gestures and nonlinguistic vocalizations (Corson and Corson, 1980; Key and Preziosi, 1982; Calvin, 1982); there were priests long, long before there were judges. The political function of control of the present, and the religious functions of the control of both past and future, were united in the same person, or elite, for eons before cultural evolution had, by a few thousand years ago, separated church and state (in this sense) by ascribing them as the statuses of separate individuals within the group—or more likely within some groups of certain populations.

At a time when writing remained a novel invention that was restricted in its knowledge and use to a minuscule elite of religious leaders, the remembrance of things past depended, for all humans, upon the constraints imposed psychophysiologically as well as culturally upon human memory (see de Nicolás, 1982).

[6]As asserted by both Shepard (1978), and Laughlin and d'Aquili (1974); see also Davies (1963) and Schubert (1976).

Neither then (Restivo, 1983: 170–71), nor even earlier at the threshold of the transition, still several thousands of years before the beginnings of written language, could any humans be guided by an understanding of the paradigms of modern science. (I refer here to those attributed to Copernicus, Newton, Darwin, Mendel, Einstein, and Bohr—to toll merely some of the most obvious progenitors of the relevant theories upon which modern evolutionary biologists build in reconstructing the biological model of human creation.) Even the classical Hellenic science of Pythagoras, Plato, and Aristotle surely was unknown to the scribes who wrote down the creation myths of Genesis.

Constructional biologists (Danielli et al., 1982; Webster and Goodwin, 1982; Gray, 1983) seek to develop biocultural models of human evolution explicating modern physics and brain science, and they include some (Wheeler, 1982; and cf. Comfort, 1982, and Steiglitz, 1983) who mediate ancient with modern wisdom to reconstruct a "mathopoeic" model of human creation myths, as exemplified by the work of McClain (1976, 1982)—which Sacksteder (1982) describes as "numerology."

Creation myths that are both the expression and the byproduct of oral traditions of cultural transmission necessarily deal with past time in a very different way from that possible for modern astronomers, geologists, and archaeologists. And even they, within our own lifetime, treat and certainly understand it (Reichenbach, 1956; Grunbaum, 1973; Salmon, 1980) quite differently from the way in which they could have before the invention of such methods as carbon dating. The scribes of Genesis were in a position more akin to that of modern quantum-particle physicists (for whom time cannot be related to either macro- or microcosmic physical events except through the mediation of human perception and cognition: Bohm, 1980; P. Davies, 1980, 1982; Wolf, 1981; Pagels, 1982; ch. 19, below) or psychobiologists (for whom time is a phenomenon dependent upon the neurophysiological structure of the *left* hemisphere of the human neocortex: Ornstein, 1972; Jahn, 1981; ch. 17, below). The scribes did remarkably well in foreshortening into an allegorical week a billion or more years of earth development and human prehistory relating to the evolution of life; and they did even better in compressing the most relevant (to them) nine thousand years (since the transition) into slightly less than half of that (if we are to accept as authoritative Archbishop Ussher's [1650–1654] calculation of 4004 B.C. as the creation date for Genesis, 1–2: 3). From any reasonably nondogmatic perspective of either human biology or human culture, the biological and cultural models of creation do not contradict—they complement each other.

EVOLUTIONARY MODELS OF RELIGIOUS BEHAVIOR

Prologue

As a prelude to his chapter on the role of families in the development of species consciousness and on the familial role in shaping concepts of the future, Tiger

(1979b: 82) presents the following version of the Christmas legend, as the parable central to the religion of Christians.

> St. Luke describes the infant as lying in a manger of straw in a stable, like any other newborn infant animal. . . . ([But] the central fact of the account . . . [is] that Jesus is, although an infant and an animal, the incarnation of God. It is also probably important that he becomes a shepherd.) The loving and constructive members of the community welcome the birth. Stars of wonder, stars of light and other elaborately imagined signs of general interest attend the event. Their very exoticism betrays the deep importance that is attached to it, to birth. With characteristic *elan* the central mammalian process becomes the formative religious idea of a massive cultural tradition. The symbols of Christ, his birth, the manger, the animals, the stars, the parents, the uphelping innkeeper—all reflect a generalized adult concern with the birth and protection of infants, this birth, of Jesus, reenacted annually, represents the core of the contract annually renewed which demands protection and love for the needy innocence of children.

Paul Shepard surely would agree.

Primate Dominance

In a chapter entitled "Fighting" in a book that discusses human behavior from the perspective of comparative primatology, Desmond Morris (1967) has proposed that

> in a behavioral sense, religious activities consist of the coming together of large groups of people to perform repeated and prolonged submissive displays to appease a dominant individual. The dominant individual concerned takes many forms in different cultures, but always has the common factor of immense power. Sometimes it takes the shape of an animal from another species, or an idealized version of it. Sometimes it is pictured more as a wise and elderly member of our own species [see Guthrie, 1980]. Sometimes it becomes more abstract and is referred to as simply "the state," or in other such terms. The submissive responses to it may consist of closing the eyes, lowering the head, clasping the hands together in a begging gesture, kneeling, kissing the ground, or even extreme prostration, with the frequent accompaniment of wailing or chanting vocalizations. If these submissive actions are successful, the dominant individual is appeased. Because its powers are so great, the appeasement ceremonies have to be performed at regular and frequent intervals, to prevent its anger from rising again. The dominant individual is usually, but not always, referred to as a god.

Morris then goes on to explain the then-accepted (in the mid 1960s) model of primate social organization based primarily upon field studies in a "baboon primatology" (see also Chance and Jolly, 1970; Omark et al., 1980; Kortlandt, 1972; Schubert, 1982b; and ch. 7, above). Why, he asks, were such gods invented? The answer, to him as a zoologist, must be sought in the ancestral origins of the human species. There was a time, he postulates, before humans were capable of joining together in cooperative hunting bands—more than a million years ago (and perhaps four to five times that, as we can now add), according to the model of hominid evolution sketched above. Then hominids were bipedal, savannah dwelling omnivores; but necessarily their social organization must have been like that of presently contemporary pongids dwelling in similar habitats—even if the latter occupy different niches because of their substantial confinement to a vegetarian diet (Teleki, 1975; Freeman, 1981; ch. 6, above). The social organization of other species of apes and monkeys, in typical cases (according to Morris), is based on the absolute rule of a single dominant mature male, who both maintains order within the group and protects it from external dangers, so that "his all-powerful role gives him a god-like status." Such an authoritarian pattern of leadership was inappropriate for the development of the cooperation so essential to successful group hunting, and hence it was displaced as hominids moved into the latter niche. But there remained a "need for an all-powerful figure who could keep the group under control," and that role came to be filled vicariously by supernatural (as well as superhuman) gods. Therefore, Morris concludes (1967: 179–80), religion's "extreme potency is simply a measure of the strength of our fundamental biological tendency, inherited directly from our monkey and ape ancestors, to submit ourselves to an all-powerful, dominant member of the group" (cf. Day, 1956 [1920]; and Marais, 1967 [1934]). It should be perfectly clear, however, that the god described is a monkey-god; and to the extent that similar relationships are institutionalized by Christian or Hebraic or Islamic or other contemporary orthodoxies, it is a mistake to conceptualize or to refer to them as "anthropomorphic"—if by that it is meant to imply "in the image of a human being."

As applied to humans, the Morris model explicitly separates political and religious functions while deriving them from the same source. It reflects the state of the art in field primatology two decades ago, just when that field was about to experience tremendous growth, diversification, and complication in both its theory and methods, as well as its empirical data base (De Vore, 1965; Lancaster, 1975; Tuttle, 1975; Berstein and Smith, 1979; Schubert and Somit, 1982; ch. 6 above). We now know, for example, that the monkey tyrant model does *not* apply to the social group relationships among contemporary gorillas or orangutans and that it applies only under conditions of resource scarcity to wild chimpanzees (Hamburg and McGowan, 1979); on the other hand, it continues to have broad application among humans, today as in our written, historical past (van den Berghe, 1978; Chagnon and Irons, 1979; Lockard, 1980; Omark et al., 1980; Baer and McEacheron, 1982; McEacheron and Baer, 1982), particularly in face-to-face groups or assemblages, and no less in industrial than in agricultural societies. This suggests (as some human ethologists have argued: Tiger and Fox, 1971; Hamilton, 1975; Tiger, 1979a) that our "progress" toward civilization during the past ten millennia has been regress from the point of view of the sort of social relationships for which we had become genetically adapted during the preceding thousand millennia.

Hemispheric Dominance

This model deals with two interrelated themes: the human, genetically based neurobiological need to create myths about supernatural beings (d'Aquili, 1978) and the neurobiological *effects* of ritual in the form of socially prescribed psychophysiological behaviors (Lex, 1978). Both hypotheses derive from the theory of biogenetic structuralism (Laughlin and d'Aquili, 1974), and the hypothesis about ritual (d'Aquili et al., 1979) builds upon Eliot Chapple's work (1970) on biorhythms. Eugene d'Aquili (1978: 269) asserts that

> the concept of some form of supernatural (inferred rather than naturally observed) power, be it God, demons, spirit, or whatever, is required by the internal dynamics of myth. . . . This supernatural force is essential to give man the power to resolve the other polarity, which is the problem of the specific myth. The problem can be the life-death polarity, health-sickness, good-evil, or any other existential polarity. . . . The internal dynamics of myth therefore require the counterbalance to man by a polar opposite or "being of power." The presumptive relation of man and the power being as man's relation to some sort of god-king or divine man then permits the resolution via this divine and human personage of the existential problem presented by the other existential polarity. . . . [T]his is simply the way the [causal] operators work. We have no choice over their internal dynamic [which] represents necessary neurological functioning, the subjective correlates of which necessarily are produced, namely, power beings or gods.

Elsewhere, d'Aquili and Laughlin (1979: 154) present the ritual hypothesis, not as scientific dogma but rather as the most probable neural model; and they compare (p. 158) the ecstasy states produced socially in ritualists, as a response to rhythmic auditory or visual tactile stimuli, to the feeling of union experienced by a sexual couple during orgasm. The goal of religious ritual is the "ultimate" union of opposites—that of vulnerable humans with the power, perhaps omnipotent force of deity (d'Aquili and Laughlin, 1979: 162; Geertz, 1973: 104, 105). Three higher cortical functions—conceptualization, abstract causal thinking, and "antinomous" thinking—are prerequisites for the ability to create myths. These brain functions are hypothesized to involve a specific area of the human brain, which "can best be visualized as the area of overlap between the somesthetic, visual, and auditory association areas . . . [which] allows for direct transfer across sensory modalities without involvement of the limbic or affective system" (d'Aquili and Laughlin, 1979: 163). The rhythmicity of religious procession, chanting, and prayers acts as a physiological energizer, driving the cognitive ergotropic functions of the dominant hemisphere independently of the meaning of the words employed, producing the same end state as meditation, even though proceeding from the opposite neural starting point. Such ecstatic states, even though extremely brief, are shared by many or most of the participants in the ritual; indeed, according to d'Aquili and Laughlin (1979: 178)

> numerous reports from religious traditions point to the fact that such states yield not only a feeling of union with a greater force or power but also an intense awareness that death is not to be feared, accompanied by a sense of harmony of the individual with the universe. This sense of universal harmony may be the human cognitive extrapolation from the more primitive sense of union with other conspecifics that ritual behavior excites in animals. In point of fact the feeling of union with conspecifics carries through the human ritual as well. Even if the feeling is elaborated on a higher cognitive level to become a feeling of harmony with

> the universe (and a lack of fear of death), most human religious rituals also produce an intense feeling of union with the other participants. This oneness has contributed to the feeling of "a holy people," "a people of God," "a people set apart."

These authors conclude (p. 180) that the function of ceremonial ritual "cannot be adequately or fully understood apart from the evolution and function of those neural structures underlying ritual behavior, many of which arose far back in man's phylogenetic past, and some of which are responsible for his becoming *Homo sapiens.*"

An alternative explanation and description of the same phenomenon is provided by Lex (1978: 144–45; citing with approval both Chapple, 1970, and Ornstein, 1972), who suggests that

> individuals, eager or reluctant, are integrated into a group, not only by the sharing of pleasurable emotions through participation in formalized, repetitive, precisely performed interaction forms, but also by a mode of thought that reinforces feelings of solidarity. The *raison d'être* for rituals is the readjustment of dysphasic biological and social rhythms by manipulation of neurophysiological structures under controlled conditions. Rituals properly executed promote a feeling of well-being and relief, not only because prolonged or intense stresses are alleviated, but also because the driving techniques employed in rituals are designed to sensitize or "tune" the nervous system and thereby lessen inhibition of the right hemisphere and permit temporary right-hemisphere dominance, as well as mixed trophotropic-ergotropic excitation, to achieve synchronization of cortical rhythms in both hemispheres and evoke trophotropic rebound. [But] it is difficult to separate the impact of repetitive behaviors on the brain from their influence on the rest of the nervous system because the various driving techniques simultaneously excite numerous neural centers. In a given ritual one specific practice alone may be sufficient to establish a state of trance; that several techniques are engaged concomitantly or sequentially indicates redundancy, to guarantee reliability, potentially affecting the entire group of participants . . . [and] manifold driving techniques accommodate individual differences in experience and genetic makeup.

Lex concludes her discussion (p. 146) with the postulation of ten hypotheses derived from general principles of neurophysiological function and the limited observations that neuroscientists have reported of sedentary ritual trance, of which the most relevant for present purposes is the eighth: "Driving techniques facilitate right-hemisphere dominance, resulting in *Gestalt*, timeless, non-verbal experiences, differentiated and unique when compared with left-hemisphere functioning or hemisphere alternation."

The Dominance of Hope

> Hope springs eternal in the human brain:
> Man, never is, but always to be, free of pain.
> —Apocryphal

Lionel Tiger, an anthropologist with a patently feline name, informs his readers (1979b: 37) that he grew up in a "cynical ghetto community" in Montreal, as a member of one of that city's political and ethnic, as well as religious, minorities.

His model is anchored in the phylogeny of hominids and their gradual development into the human species with their gathering-scavenging-hunting-band mode of life, during 99 percent of the past million years (e.g., Lee and De Vore, 1968; Ardrey, 1976); so Tiger has no doubt (1979b: 43) that "religion taps political responses of acceptance of authority central to evolution." But he takes full cognizance also of theory and empirical knowledge in contemporary primatology and evolutionary biology. Tiger's concern is with a much broader range of human beliefs and behaviors than either the religious or the political; the latter he develops in a paper (1979a) on the political ideology of progress (with particular reference to such capitalistic theories as Marxism).

Tiger (1979b) starts with the premise that for hunting bands generally and any specific hunting band in particular to survive, at least some of its hunters must have been successful during some minimal frequency of quests and at some critical interval. Even with the assumption of cooperation and food sharing within the larger group (which must have been no more than an extended family throughout virtually all of the relevant timeframe), hunters who expected to be successful were more likely to survive, individually, to reproduce, and to support wives that did reproduce and children who survived to become (if male) successful hunters themselves. The optimistic hunters were more likely to behave in such a manner that they *could* become actors in self-fulfilling prophecies; and they were more likely than pessimists or those who failed in the hunt for other reasons (such as recent wounds, accidentally ruptured eardrums, flat feet, and a large array of other possible physiological as well as psychological deficiencies) to procreate offspring who themselves carried genes necessary to the complex of physiological, psychological, behavioral, and cultural factors that made possible an optimistic stance toward hunting. According to Tiger, the inevitable consequence in terms of the neo-Darwinian modern synthetic theory of evolution accepted by most biologists today (Futuyama, 1979; but see Webster and Goodwin, 1982, and McEacheron, 1984) must have been that optimistic (i.e., good, successful) hunters were selected for, to such an extent that optimism became fixed through time—that of a million years—as a species trait. If a more technical statement of the point be preferred, the ratios of relevant genes tended to increase within the gene pool of the species to such an extent that it became likely that, in any given hunting band of protohumans or humans, there would be enough optimists produced to assure the survival of the group—disregarding, of course, the ecocatastrophes (of which Noah's flood can serve as a latter-day example) and other forms of bad luck and the warfare that must have been continuously eliminating certain individual bands and clusters of bands, though not all of the ones that by this time had radiated over a substantial portion of Asia as well as Africa (Wilson, 1980; Eshel, 1972; Keith, 1949; McEacheron and Boyd, 1982).

Beyond the decision whether to hunt today lies the question of future hunts and, indeed, of future habitats; but these are questions that, so far as we know, trouble (in the cognitive sense of self-conscious thought) no other animals than ourselves (Silverman, 1983; chs. 16 and 17, below). One can be empathetic (Shepard, 1978; Crook, 1980) to the possibilities of what cognitive ethologist Donald Griffin (1978, 1981, 1982, 1984) has described as "the evolutionary continuity of mental experience" across species and yet agree that the available evidence supports Tiger's remark (1979b, p. 54) that "humans have religions and so far as we know no other animals do, and this is the distinction which has got to be explained."

Tiger assumes that the phenotypic composition of hunting bands must have been heterogeneous, notwithstanding the relatively close genetic relatedness of individuals in any particular band. The fixation of the gene complex—for surely many genes interacting with pleiotropic effects and subject to ecological-ontogenic influences, and not any single "marker for optimism" (Wright, 1940; ch. 9, below)—in the species pool and in any particular subpopulation of it in no way should be understood to imply that all, or even most, adult males in a group were optimistic hunters. Tiger discusses for example, the important social role probably played by manic diabetics (1979b: 162), even though their particular form of the disease was relatively rare earlier just as it is nowadays. "The strident exuberance of the manic phase is an energizing and unsettling stimulus to communities otherwise too complacent and unchanging . . . as . . . in the religious field, where certain individuals with improbably grandiose ideas coupled with very imperious attitudes toward other people may produce charismatic religious or political groups which often have extraordinary effects on formerly settled social systems." A manic diabetic "typically suffers from great fluctuations of mood from the manic to the depressed. When requiring meat the [person] is "high"—supremely confident, untroubled by discordant circumstances and in general prone to decisive and grand action . . . [and] such people carrying such genes would be inclined to stimulate hunting or gathering for the particular foods they can tolerate." So as sometime super-optimists, they would goad the group into action, although, and this point appears not to have been noticed by Tiger, with consequences that may well be unfortunate for the group. There may have been sound reasons apparent to nonmanics why hunting at a particular time or for particular quarry might have been socially ill-advised, which may in turn help to explain why the genes for manic diabetes are fixed but at a low frequency ratio within the species pool.

Perhaps a better example, which is taken from a recent work on human sociobiology by clinical psychologist Daniel Freedman (1979: 6–7), concerns schizophrenia, the occurrence of which in about 1 percent of all human populations throughout the world has long baffled behavioral scientists. The recently discovered explanation is that the offspring and siblings of schizoids are often exceptionally creative persons (cf. ch. 18, below), so that schizophrenia constitutes "a price paid by the family and the breeding population for the creative phenotypes associated with it"—a price paid by the schizoids also, of course. So the hunting bands very occasionally included cyclically depressed diabetics (during nonmanic phases of their disease); but there was also a relatively more common representation of schizoids, of whom some would, at any given time, be chronically depressed while others were hyperactive.

While Tiger's book was still being written, cognitive psychologists (Iverson, 1979) reported a new discovery that impinged directly upon his concerns: namely, that the brain itself is the site and target for endorphins, naturally produced opiates which perform the critical function of reducing both the perception of pain and also that of threat and danger for an appropriate stimulated (that is, stressed, threatened, and aroused) animal, whether mouse or man. Not only does pain become more tolerable but also, by distorting the "organism's cognitive assessment of the survival situation," such endorphins "permit the animal to obscure the understanding that its situation is dire when it is in a bad fight or a fire or already wounded and allow it to continue to function as if it were not operating under such adverse circumstances" (Tiger, 1979b: 168). In short, by anesthetizing the animal during crises, endorphins make optimism possible when pessimism would be fatal.

Operating at the social and cultural, as well as at the individual physiological level, religion represents the institutionalization of hope;[7] and, like endorphins, it has the effect of biasing positively (i.e., in the direction of aspirations) cognitions about future events, the otherwise observable facts to the contrary notwithstanding. Tiger does not say so, but given his longstanding interest in the historical relationships between Darwin and Marx, it might not have been inappropriate for him to have noted this lack of support from modern brain science for one of Marx's better known dicta, that "religion . . . is the opium of the people" (see O'Malley, 1970). Ralph Burhoe (1979: 151) has clarified this point, observing that religions are cybernetic mechanisms "and not simply the opiates, as Karl Marx correctly suggested," and that "religions present the other side inherent in any control mechanisms: the stimulus to action as well as the prevention of overloads that terrorize and immobilize."

Tiger thus explains religion as role specialization in what he calls "the control of the future," by which he means the future as prospect and perceived possibility (cf. Bohm, 1980; Wilber, 1982)—not at all in the sense in which the word *control* is used to describe causal determinism in statistics or in the pre-twentieth century physical and biological sciences generally. So politicians and priests now share with astronomers and biogenetic engineers the legitimacy of soothsaying (Gray, 1983).

CONCLUSION

Even though biology should be taught as biology, and religion as religion—and for sound pedagogical as well as for constitutional reasons (at least, in the USA)—

[7]E. P. Odum (1971: 240, 242, in his ch. 8: "Energetic Basis for Religion") states: "What has been energetically system-rewarded is taught as religious truth to strengthen faith, for it leads to further reinforcement and helps the [energy] network survive. Faith in a religion that supports a way of life is necessary to the group's coherent survival and continued efficiency."

they individually proffer cultural perspectives that reflect substantially different stages in very recent human cultural evolution (i.e., that of the most recent two thousand to five thousand years). Nevertheless, it has been barely a century since the evolutionary standpoint displaced the theological one *as science* in this or any other country. So it is hardly remarkable that the twain are considered to be in antagonism, by persons with either short memories or deficient education in biology and science, or in religion.

Three models of the evolution of religious behavior are examined. The first, based on field studies of contemporary social behavior in other-than-human primates, explains the gods of human religions as the role surrogates for the male autocrat of a monkey troop. The next is an anthropological model based on field observations of indigenous human groups found in hunting-band, pastoral, or primitive agricultural adaptations; but it is asserted that this model applies also to contemporary humans in industrial-society populations. This is an explicitly psychobiological model of hemispheric dominance, explaining religious experience as the fulfillment of right-brain, limbic system needs, with emotion driving both the psychological and the social aspects of that experience. The third model likewise is bioanthropological and focuses on heterogeneosity among hunting-band populations, relating group survival to extremes in psychophysiological diversity among the group members. According to this view, religion evolves as the specialized province of physiologically driven visionaries, whose goadings expand the scope of possible futures available to their groups. All three models constitute examples of the diverse functions performed by religion in the later stages of human evolution (i.e., at about the time of the transition of *c.* 12,000 B.C.).

The creation myths of the Near East (where the transition came first) came to be written down in terms of the available science knowledge at the disposal of their compilers, two thousand to three thousand years ago, as the products of a much longer period of repetition first through oral and then through subsequently invented audial cultures. Those myths project a view of human social and psychological development not at all incompatible with the scenario proffered by modern science on the basis of the empirical evidence for the past twenty thousand years. The scientific scale continues to be substantially longer than the religious mythical one, but for this period, by a ratio of only two or three to one (instead of the hundreds of thousands to one that would apply to the beginnings of life in any form on earth).

The biological model of human evolution indicates that the origins of life out of which humans evolved must be scaled in terms of billions rather than thousands of years; but the most relevant portion of that immensely long history—the time during which modern *Homo sapiens sapiens* developed as a distinctive hominid species—can indeed be measured in terms of the few tens of thousands of years since the denouement of the last ice age and the advent of the present interglacial. Surely, the most recent ten thousand to fifteen thousand years are of greatest interest in terms of the species-wide shift in niche, habitat, and adaptation; but these dramatic changes were made possible only by the enslavement (domestication) of highly selected species of both plants and other animals (Ucko and Dimbleby, 1969), as well as of targets of opportunity provided by disadvantaged human populations.

The symbiotic relationship between humans and other animals is a central feature of the species genotype for brain development; therefore it necessarily is a central feature of the creation myths and also of the brain structure of contemporary living humans. At the level of the psychobiological philosophy of science stance taken in this paper, the creation myths ("creation science") and modern evolutionary science ("scientific evolutionism") purport to provide alternative explanations for the origin of the species—with each explanation grounded in the science theory and knowledge of its respective time of construction. Certainly the older ("religious") view rests on the science of its day; but it is likewise true that the modern ("scientific") view functions as a religion too for its own true believers (see ch. 19, "Natural Science Macroparadigms," below). And a cosmology broad enough to encompass both the theological and the scientific myths of human creation indicates that there is much more of intellectual interest to be found in their harmonies than in their manifest dissonances.

PART III

Evolutionary Theory

What political scientists generally need most, but understand least, is evolutionary as distinguished from historical theory (but see Corning, 1983; and Kaufman, 1985). Political scientists do not need to understand evolutionary theory because it explains everything; it certainly does not do that, and besides, it can predict very little—in the sense in which classical physicists think of prediction. Nor is it for the reason that many, and doubtless most, academic political scientists (conservative and liberal alike) would expect to get out of it: that evolutionary theory would justify the slow, gradual, peaceful (at least, for reigning political elites and their academic acolytes) non-boat rocking stasis that is their ideal for "liberal, constitutional democracy." Evolutionary theory should be important to political scientists because it is directly and highly relevant to an adequate understanding of the behavior of all animals, not excluding the political behavior of humans. Political scientists can and do understand static slices of culture, stained by survey research for their inspection on the functional equivalent of a glass slide with computer output substituting for a microscope; but they cannot deal effectively with cultural dynamics except as an aspect of cultural evolution. And to understand cultural evolution other than dilettantishly, it is first necessary to comprehend the biological evolution that is the template for cultural evolution.

There are four chapters in part 3, and the first of these examines in considerable detail the principal components of the scientific ideology of neo-Darwinism-Mendelism, the conventional "modern synthesis" of nonteleological, gradualist, incremental change due to the operation of natural selection on surviving individuals of all species, whose reproductive efficacy determines both their own "success," as life forms, and the phylogeny (future genetic potentiality) of their respective species. The adaptation of a species to a niche assures only that competition among its individual members will soon lead to a marginally sustaining existence for all of them.

The next chapter evaluates the most popular (at least, among biologists who are not themselves evolutionary theorists) version of cultural "evolution" that has been thrust upon the attention of biosocial scientists during the 1980s; but the process of criticism of that particular variant of cultural evolutionary theory raises questions of much more general import, in considering the differences (which are obvious) as well as possible similarities (which are more subtle to detect) between the processes by means of which genetic, as distinguished from cultural, information is transmitted from one human (or generation of humans) to another.

Chapter 11 is a precis on the proposed experimental manipulation of the human species genotype (i.e., teleologically directed and controlled unnatural selection, reflecting public or private policy choices instead of "blind" mutation, genetic drift, or ecological change), as appraised from a perspective informed by the analogous miracles humans have wrought (in the space-time of barely twelve thousand years or so) in their domestication of plants and other animals.

Chapter 12 examines a new evolutionary paradigm that is already in the process of displacing neo-Darwinian orthodoxy as the major theory of selection and change, certainly at the level of species and higher taxa (cf. part 4: "Evolutionary Development"): punctuated equilibria catastrophe theory. Punctuationalism asserts that almost all natural selection takes place suddenly, rapidly, and with drastic transformation of a former species into a novel one, because of the influence on a small, isolated remnant population of a well-established species, in a "quantum leap" into what, through rapid reorganization of gene *expression* (in response to novel environmental stress and opportunity), becomes a new species, either as a continuing clade of the founder species (but in a different niche) or as its successor. Most species, indeed, did become extinct either under such circumstances, or else through mass extinctions of as many as 90 percent of all species extant at the time of some catastrophic change in the earth's climate and surface environment. Only the survivors of either such "background," or else of major mass, extinctions could contribute to the genomes of presently extant species such as humans. After explicating the rudiments of catastrophe theory, as one of several contemporary alternatives to neo-Darwinian gradualism in natural selection, the chapter discusses, and then exemplifies some of the kinds of applications—and implications—that this kind of evolutionary theory might have for modeling political processes, change, and behavior.

9
Genetic

"Sociobiological theory" reintroduces—albeit with a different false face—Economic Man, that familiar bogeyman of the social sciences. Economic Man is a specter whose arid hyperrationalism continues to haunt the social and political ideologies—and their subsumed institutional embodiments—that predominate in our own time. These entail a pedigree evolving culturally (but overlapping phenotypically) from Smith through Ricardo, Marx and Lenin to Friedman.[1] Hardcore[2] sociobiology proposes the substitution of Bionic Man, as Economic Man's surrogate: equally devoid of flesh (and bones, too, for that matter)[3], single-

---

This chapter is a revision of "The Sociobiology of Political Behavior," in White (1981c), pp. 193–238.

[1]Rooted in game theory and other econometrics, economic rationalism has reentered political science during the past two decades, in the guise of "positive theory." In that form it is no less, but certainly no more, in touch with empirical reality than the legal positivism from which, no doubt, its name is borrowed. For a seminal work see Riker (1962); for a more recent example, see McKelvey and Ordeshook (1976).

[2]Credit for having suggested this term must go to Wilson (1978: 155), who writes: "I have called . . . 'hard-core' altruism, a set of responses relatively unaffected by social reward or punishment beyond childhood. Where such behavior exists, it is likely to have evolved through kin selection or natural selection operating on entire, competing family or tribal units. We would expect hard-core altruism to serve the altruist's closest relatives and to decline steeply in frequency and intensity as relationship becomes more distant." Wilson then (1978: 159) further restricts the concept, as applied to humans, by remarking that "human altruism appears to be substantially hard-core when directed at closest relatives, *although still to a much lesser degree than in the case of the social insects and the colonial invertebrates.* The remainder of our altruism is essentially soft" (emphasis added). According to this distinction, genetically determined altruism is "hardcore"; culturally determined, it is "softcore."

[3]George Williams (1966) says that "in its ultimate essence the theory of natural selection deals with a cybernetic abstraction, the gene, and a statistical abstraction, mean phenotypic fitness. *Such a theory can be immensely interesting to those who have a liking and a facility for cybernetics and statistics*" (emphasis added).

minded in its monotheism,[4] and unconcerned with the extent to which human behavior everywhere is embedded in a matrix of human culture.[5]

Because many critics confuse sociobiology with social biology, it is important to distinguish the theories of genetic determinism of behavior,[6] sponsored by a handful of evolutionary biologists and population geneticists,[7] from the mainstream of research in mammalian biology and especially in primatology. Inclusive fitness, kin selection, and altruistic behavior have emerged during the midseventies as conspicuous "sociobiological" topics for popular discussion and public debate;[8] but the vastly larger and more important body of research in social biology incorporates many other disciplines and new syntheses that *do* portend direct and important implications of the study of political behavior (Hinde, 1974; Reynolds, 1976; Schubert, 1979; chs. 4 and 5, above, and 15, below). This main body of relevant research includes the cross-disciplinary field of animal behavior research, described by Hinde (1966) as a new synthesis of ethology and comparative psychology; primatology, representing an even broader cross-disciplinary synthesis; behavioral ecology; physiological psychology, including neurophysiology and what generally have come to be called the brain sciences; and behavioral genetics. All these fields are concerned with cross- species analysis of the social behavior of animals; some of them have also become involved in attempts to collect evidence bearing upon certain aspects of hardcore sociobiological theory. On the other hand, Wilson (1975: 4) claims that his book "makes an attempt to codify sociobiology into a branch of evolutionary biology and particularly of modern population biology"—just as he evidently has spoken frequently of sociobiology's mission to "cannibalize" the social sciences in general, and sociology in particular. But even he apportions to the hardcore theory only one chapter (ch. 5 on "Group Selection and Altruism") and part of another (ch. 1, on "The Morality of the Gene") of what to friend and foe alike has become the Koran of sociobiology. The remaining twenty-five and one-half chapters, including about 550 out of 575 pages

[4]Williams (1966: 8–9) has also said that Darwinian "natural" selection is the "sole creative force in evolution . . . [and] there is no escape from the conclusion that natural selection, as portrayed in elementary texts and in most of the technical contributions of population geneticists, can only produce adaptations for the genetic survival of individuals." Unfortunately, "many biologists have recognized adaptations of a higher than individual level of organization. A few workers . . . have urged that the usual picture of natural selection, based on alternative alleles in populations, is not enough. They postulate that selection at the level of alternative populations must also be an important source of adaptation, and that such selection must be recognized to account for adaptations that work for the benefit of groups instead of individuals. . . . [But these] higher levels of selection are impotent and not an appreciable factor in the production and maintenance of adaptation."

[5]"From the standpoint of the main issues treated in this book," Williams (1966: 197) eventually concedes, "there is no more important phenomenon than the organization of insect colonies."

[6]Wilson, for example, has stated (1978: 140) that human marriage is based on "genetic hardening" of the synapses.

[7]A fair roundup of the principals would include Richard D. Alexander, Richard Dawkins, William D. Hamilton, Robert L. Trivers, George C. Williams, and Edward O. Wilson, all zoologists.

[8]In addition to *Biology as a Social Weapon*, see vols. 7–9 (1975–77) of *Science for the People*, a bimonthly journal of social policy critique, which in more recent years has broadened its concerns considerably beyond the paper tiger of hardcore sociobiology; Barlow and Silverberg (1980); Caplan (1978a); and Sahlins (1976).

and clearly at least 95 percent of the text, are devoted to a discussion of the field of animal social behavior more broadly conceived.

HARDCORE THEORY

Although the roots go much further back, as any perusal of Sewall Wright's four-volume treatise (especially 1968, 1977, and 1978) makes abundantly clear, one issue of basic importance to our discussion here is whether genetic selection has, at least among humans, important effects in aggregations of breeding populations (demes) that transcend the individual and his closest blood relatives. Another issue concerns the relative importance, to humans today, of cultural, as distinguished from genetic, selection. Wynne-Edwards (1962) protagonized the current revival of group selection theory in a forceful book that argues several other issues (such as genetically controlled population-reduction mechanisms in many types of social animals) with which we need not be further concerned here. His antagonist became William D. Hamilton (1964), who sponsored an alternative theory of "inclusive fitness," which extended the concept of fitness to apply beyond the individual organism, but at the same time limited it to the individual and its closest lineal or collateral genetic relatives—hence the term *kin selection*, as proposed by John Maynard-Smith (1964). Accordingly, an animal enhances its total genetic impact by behaving supportively to increase the survival prospects of conspecifics with which it is related, especially those closely related. Certainly there is a great deal of empirical evidence, concerning humans, that is consistent with the theory of kin selection; but there is also a fair amount of inconsistent evidence, given the high frequency with which assault, murder, and rape occur between nuclear family members, to say nothing of the thefts, robberies, and other misappropriations that such persons make of each other's property.[9] It is also notable that there is extraordinary variance in the extent to which the rules of different cultures encourage or discourage behaviors that tend to reduce the fitness of family members, either phenotypic or genetic, with respect to each other. No single generalization can be made about the extent to which cultural norms are supportive or in contradiction of the model of kin selection; but the very fact that many cultures, both past and present establish norms that do contradict the model suggests that either kin selection is an inappropriate model for human behavior generally, or else that the model is appropriate, but the genetic norms are much weaker than, and are frequently overridden by, cultural norms that exert a much more direct and stronger influence upon human behavior (Richerson and Boyd, 1978). Kin selection seems to be best supported by the ethnographic data for

[9]West Eberhard (1975: 30) points out that "the majority of social interactions, even among close kin, are probably competitive rather than beneficent in nature. Indeed, as Alexander (1974) has pointed out, an individual's closest relatives are his closest competitors because of their proximity and dependence on the same, often limited, resources." But this seems to contradict the very premises of kin selection theory!

human primitive societies and least supported by the sociological data for contemporary urban industrialized societies.

Probably the most austere defense of hardcore theory is that of George Williams, who begins his book on group selection (1971: 2) with an appeal to scripture: "Darwin's great achievement was to show that both the mechanisms of self-preservation and those of reproduction are explained by a more basic principle of natural selection, the reproductive survival of the fittest." In an earlier statement, soon after Wynne-Edwards and Hamilton, Williams (1966: 22, 24, 26) had said:

> The essence of the genetical theory of natural selection is a statistical bias in the relative rates of survival of alternatives (genes, individuals, etc.). . . . If there is an ultimate indivisible fragment it is, by definition, "the gene" that is treated in the abstract discussions of population genetics. . . . [The] maximization of mean individual fitness is the most reliable phenotypic effect of selection at the genic level, but . . . a gene might be favorably selected, not because its phenotypic expression favors an individual's reproduction, but because it favors the reproduction of close relatives of that individual. . . . [T]he theoretically important kind of fitness is that which promotes ultimate reproductive survival.

Williams (1966: 19) invokes "the principle of parsimony" as the authority that compels him to sponsor this extreme form of biological reductionism: "It is my position," he says, "that adaptation need almost never be recognized at any level above that of a pair of parents and associated offspring."[10]

Curiously enough, it was Williams himself who, in his next two books, provided an argument that undermines much of the basis that he had previously laid for the hypothesis of genetic control over individual animal altruism—at least among diploid sexually reproducing species such as humans. First he points out (1975: 150) that "sexuality in evolution . . . greatly retards the final stages of multilocus adaptive change and severely limits the attainable precision of adaptation." Moreover (1975: 150–51) "sex generates recombinational load, and this largely annuls the effect of selection in each generation." Indeed (1975: 12), "there is near unanimity on the point that sexuality functions to facilitate long-range evolutionary adaptation, and that *it is* irrelevant and even *detrimental to the reproductive interests of an individual*" (emphasis added). With that many faults against it, one can only echo Maynard Smith's (1971) plaintive question: "What use is sex?"

An animal's inclusive fitness, which is to say its relative success in genetic competition, is in principle reckoned—it would be improper to say "measured" or

[10]But this is simplistic philosophy of science: Occam's razor is at best a conventional, not an obligatory, criterion of relevance; and it is certainly not the only index to the cutting edge for maximodelling empirical knowledge. Here the parsimony argument makes sense only if we restrict our focus of interest and attention to the level of individual organic adaptation. But why *not* biotic adaptation? The world today is full of humans whose organic adaptability is being sacrificed on the altars of other people's notions of social policy supposed to be adaptive at levels ranging from the species to that of local communities. And even at the individual level, it is the biosphere that determines (inter alia) the environment to which the individual adapts. It is the interaction between groups of individuals and their respective environments with which any viable theory of genetic influence upon human behavior must be concerned, to have any relevance to the concerns of political science.

"scored" for reasons that will be discussed below—by the total relative contribution that its genes make, in competition with conspecifics, to the sum of all genes for all animals of the species (the "gene pool"). And, so far as one can infer from the literature, this applies for the entire time that the species is recognized as such. I have seen no reference to logical positivism or other problems of the philosophy of science in the expositions of hardcore theory; and this may explain the apparent unconcern about the easy shifting back and forth, which characterizes many of these discussions, between inclusive fitness as an empirical statement about the consequences of genetic natural selection (under the assumptions that only the Darwinian paradigm is relevant; of the classical gene; and that measurability in fact is possible), and *normative assertions about animal behavior*. There is of course no objection to the latter as deductions from neo-Darwinian theory; but much of the discussion could profit immensely from a sharper discrimination between what are intended to be verbal translations of mathematical statements about observations of the genes and the behavior of phenotypes. Consider, for example, Wilson's admission (1978: 176) that "there [is] no rational calculus by which groups of [*sic*: or ?] individuals can compute their inclusive fitness on a day-to-day basis and thus *know* the amount of conformity and zeal that is optimum for each act." Putting aside for the moment the question whether individual humans are in any better position to make such a calculation than are groups of humans, it should be clear that Wilson is speaking here in a normative sense: to act rationally, people ought to behave according to the rules of inclusive fitness—if only they could figure the odds! But even more paradoxical is why they should need to worry about it: if inclusive fitness is under genetic control, all they should need to do is what their genes tell them to do, and there should be no need—or indeed, use—for conscious thought, in order to maximize fitness.

In any case, the prospects for quantitatively testing inclusive fitness theory seem remote indeed. Wright has observed (1968: 55; and cf. Layzer, 1978) that

> natural selection within a population necessarily operates directly on individuals and thus on genotypes as wholes rather than on allelic genes. The most significant "character" in this connection is by definition the "selective value" of the genotype. This, however, is an abstraction that is practically impossible to measure. It becomes necessary to focus on less comprehensive characters such as viability and fecundity or, still more narrowly, on specific internal adaptations such as the various aspects of metabolism, morphology of internal organs, homeostatic mechanisms or on specific adaptations to the external world such as instincts, intelligence, size and form appropriate to a particular niche, strength, speed, weapons, armor, concealing coloration. As analysis carried down toward the immediate effects of gene replacements, the relation to selection value tends to become more remote and contingent.

West Eberhard (1975: 13) notes:

> The cost of altruism . . . is . . . clearly a function of the intensity of competition between donor and beneficiary. In general, we should find that *the greater the intensity of reproductive or ecological competition between two individuals, the less the probability of altruism*

> *between them.* The probability of altruism thus depends on such ecological parameters as the so-called carrying capacity of the environment, population size, and the population (or social) structure (since not all age or behavioral classes are equally competitive). This consideration immensely complicates the determination of *K* in nature. There is certainly no species for which the total ecological cost to conspecifics of adding another individual to the population (or subtracting one from it) is known. (emphasis added)

Even Williams (1971: 5) admits that,

> unfortunately, it is unlikely that really decisive evidence for choosing between group selection and kin selection, either in general or even for specific examples, will be forthcoming in the near future. It is difficult to determine the degree to which nonbreeding social groups are made up of close relatives, and almost impossible to measure the cost of altruism to the donor or the benefit to the recipient.

On the preceding page he had claimed that "the most serious recent thinking" is on his side; but our conclusion for present purposes has to be that if even the *protagonists* of hardcore theory don't know how to operationalize it, then no matter how hard or seriously they may think about it, there remains no persuasive evidence in support of the theory.

A related question concerns the concept of gene on which theory is based. Wright has remarked (1968: 22, and ch. 3) that "the question whether the genes. . .are discrete entities or merely regions in a continuum is an old one." And he makes his own position clear by pointing out that "genes are merely regions of the chromosomes within which crossing or other breakage has so far not been observed to occur." Williams (1966: 38) is quite aware how simplistic the classic gene of Mendelian selection is from the perspective of modern molecular genetics; and he does concede that "the genetic environment can be considered to be all the other genes in the population, at the same and other loci." Nevertheless, he concludes (1966: 60) that "the fact that this genetic environment is really made up of an astronomical number of genetic subenvironments, in each of which the gene may have a different selection coefficient, *can be ignored at the level of general theory.*" Very much to the contrary, Lewontin (1974: 218) concluded his book with the following emphasis:

> The fitness at a single locus ripped from its interacting context is about as relevant to real problems of evolutionary genetics as the study of the psychology of individuals isolated from their social context is to an understanding of man's sociopolitical evolution. In both cases context and interaction are not simply second-order effects to be super-imposed on a primary monadic analysis. Context and interaction are of the essence.

Indeed, Robert Trivers, speaking before an audience of professional zoologists and comparative psychologists at the 1978 annual meeting of the Animal Behavior Society, stated flatly that the notion of "altruistic genes at alleles" is theosophy.

Trivers' best and most original work (1974) was done on the genetic basis for interest-behavioral conflict between reproducing animals and their direct progeny, in sexually reproducing species. His contributions there may be of considerable value to attachment theorists and others interested in child development. But very little of the transactions that take place between human parents and their infants has much effect on the practice of either politics or political science, as those presently are understood. Trivers' theory (1974: 249) constitutes an extension of inclusive fitness and kin selection to the special circumstances of generational relationships between appropriate sets of consort males, breeding females, and their offspring. "In particular, parent and offspring are expected to disagree over how long the period of parental investment should last, over the amount of parental investment that should be given, and over the altruistic and egoistic tendencies of the offspring as these tendencies affect other relatives." Moreover, the theory deals with frequent conflict between parental preferences for the sex of offspring and the sex of the offspring in fact born to them (among mammals); it also discusses variation in parental experience, as well as the offspring as psychological manipulators of their parents.

SOFTCORE THEORY

Reciprocal altruism, the theory for which Trivers is perhaps better known, is quite another matter. Here Trivers presents a model of cooperative social behavior within and between animal populations; and with whatever virtue can be claimed for consistency (in the light of Samuel Johnson's aspersion of it), Trivers claims (1971: 18) that "no concept of group advantage is necessary to explain the function of human altruistic behavior."

The vocabulary that Trivers (1971: 15)—not Nature—selected for the construction of the model is fraught with morality and emotionality to such an extent that even his abstract is apocryphal: it is in many respects indistinguishable in content from what one might expect to find in an equivalent precis of any of several books of the Old Testament: "Specifically, friendship, dislike, moralistic aggression, gratitude, sympathy, trust, suspicion, trustworthiness, aspects of guilt, and some forms of dishonesty and hypocrisy can be explained as important adaptations to regulate the altruistic system." But the tone and level of the argument of the text are more reminiscent of the argument between Socrates and Glaucon, in a somewhat similar but older model of human nature and behavior (cf. Masters, 1978c). For present purposes, we are less concerned with Trivers' use of his model to "explain" why certain big fish tolerate being groomed by certain little fish and why some birds simultaneously advertise their own presence as well as that of predators. Whatever the merits of an inclusive fitness explanation of interspecific symbioses, and of their own reciprocal (i.e., interspecific predatory-prey relationships), our concern must be with Trivers' application of his model to human behavior.

Trivers begins, appropriately enough, with an appeal to cultural rather than to genetic authority, by citing sociologist Alvin Gouldner—out of the hundreds of social scientists whom he might have selected—in support of the assertion that "reciprocal altruism in the human species takes place in a number of contexts and in all known cultures." But of course. And then he offers as examples of such human reciprocity: helping in time of danger; food sharing; helping those who are temporarily or permanently physiologically impaired; tool sharing; and exchanging knowledge. His discussion does not emphasize the equally universal antonyms of those behaviors: war and pillage; looting and robbery; the systematic starvation of most populaces of the world (today, during the past few millennia, and probably during the next few, if such there be) by their ruling classes; capital growth and concentration; and the obligatory illiteracy of more contemporary humans than ever lived up to a century or two ago. These latter phenomena are seen in moralistic terms as bad things that cheaters do, and are—for genetic reasons!—punished by their more righteous brothers (or at least, fellow reciprocals). Trivers' presentation projects, therefore, an aura of Puritanism that entitles it—whatever its merits as behavior genetics theory—to a place in the ideological mainstream of American literature: if Vernon Parrington had survived to undertake a revision for the seventies of his *Main Currents in American Thought*, Robert Trivers' essay on reciprocal altruism would certainly be worthy of a footnote in it. It is, indeed, altogether meet and fitting that Trivers should have spawned his theory in the same habitat that Hawthorne drew upon for his *Scarlet Letter*, a work in the same tradition of preoccupation with moralistic aggression. But for these very reasons, there are no new revelations about human behavior in Trivers' paper, for readers already familiar with any of such diverse sources as the *American Sociological Review*, Shakespeare, the Bible, or Herodotus. The contributions of the humanities and social sciences to the practice and study of politics are what political science always has been and remains concerned with, entirely independently of Trivers' peroration.

There is not a shred of evidence in Trivers' essay that would support a choice between his own innuendo that maybe human reciprocal altruism today is genetically favored because of natural selection for reciprocal altruism per se that took place during the hunter-gatherer stage of human evolution, and the alternative hypothesis that altruism (like cannibalism and infanticide) is a byproduct of cultural evolution in the context of specific biotic evolutionary (niche, habitat, ecological) pressures and constraints.[11] Cultural evolution in that sense certainly is premised upon the chicken-and-egg evolution of the neocortex and language; but at that level all human behavior today is "genetically determined."[12] Indeed,

[11]Donald T. Campbell (1972: 34) has wryly observed that "the moralizing in the Old Testament against onanism, homosexuality, and the temptation to sacrifice one's firstborn son. . .must be directed against socially produced dispositions, since these tendencies would be genetically self-eliminating."

[12]Trivers (1971: 48) claims that "it is reasonable to assume that it has been an important factor in recent human evolution and that the underlying emotional dispositions affecting altruistic behavior have important genetic components." No doubt arousal depends upon human physiological functions that have evolved

Trivers (1971: 48) concedes that "there is no direct evidence regarding the degree of reciprocal altruism practiced during human evolution or its genetic basis today."

In the absence of such empirical evidence it seems prudent to continue to prefer the cultural hypothesis, for which there is a great deal of supporting evidence. Moreover, there are overriding technical grounds for rejecting reciprocal altruism as an admissible extension of inclusive fitness theory. As Eshel (1972: 50) has pointed out, "social compensation may be granted in the form of . . . reciprocation . . . or discrimination against the selfish. Though these mechanisms undoubtedly account for many seemingly altruistic behavior patterns . . . patterns fully explained by social compensation are not truly altruistic by our definition, since *they do not reduce fitness*" (emphasis added).

Wilson (1978: 153) admits that "the form and intensity of altruistic acts are to a large extent culturally determined"; but he argues (1978: 156) that their motivational base remains genetic, because

> "soft-core" altruism. . .is ultimately selfish. The "altruist" expects reciprocation from society for himself and his closest relatives. His good behavior is calculating, often in a wholly conscious way, and his maneuvers are orchestrated by the excruciatingly intricate sanctions and demands of society. The capacity for soft-core altruism can be expected to have evolved primarily by selection of individuals and to be deeply influenced by the vagaries of cultural evolution. Its psychological vehicles are lying, pretense, and deceit, including self-deceit, because the actor is most convincing who believes that his performance is real. . . . [I]n human beings soft-core altruism has been carried to elaborate extremes. Reciprocation among distantly related or unrelated individuals is the key to human society.

And Campbell (1972: 31–32) seems to agree, stating that "what I argue for in the present paper is an ambivalence between socially induced altruistic bravery and a genetically induced selfish cowardice. . . . The conclusion seems to be inevitable that man can achieve his social-insect-like degree of complex social interdependence only through his social and cultural evolution, through the historical selection and cumulation of educational systems, intergroup sanctions, supernatural (superpersonal, superfamiliar) purposes, etc."

On the broader question of cultural evolution in general, Wilson (178: 34) accepts the conventional position that

> we can be fairly certain that most of the genetic evolution of human social behavior occurred over the five million years prior to civilization, when the species consisted of sparse, relatively immobile populations of hunter-gatherers. On the other hand, by far the greater part of cultural evolution has occurred since the origin of agriculture and cities approximately 10,000 years ago. Although genetic evolution of some kind continued during this latter, historical sprint, it cannot have fashioned more than a tiny fraction of the traits in human nature.

genetically; but Stanley Schacter (1964; cf. Schacter and Singer, 1962) and others have shown that the *meanings* associated with most human emotions (perceived arousal) are almost entirely learned (epigenetically determined).

> Otherwise surviving hunter-gatherer people would differ genetically to a significant degree from people in advanced industrial nations, but this is demonstrably not the case.

But then he adds that "it follows that human sociobiology can be most directly tested in studies of hunter-gatherer societies," which patently does *not* follow, for two reasons. If there is not significant genetic difference between hunter-gatherers and the rest of the contemporary species, then it should not matter which people sociobiology draws as samples for its tests. Alternatively, as recent ethnographic reports of San, New Guinea highlanders, and Philippine mountain Pygmies all make clear, there are no more hunter-gatherers living in a state of nature, uncontaminated by the press of civilization. Of course they should be studied; but not under the assumption that they are surrogates for mainstream conspecifics as those people lived ten thousand years ago.

What makes sense is that there was genetic evolution of the human brain, transactionally with the evolution of human language, and that this is what made possible the evolution of human culture. But that genetic evolution certainly was substantially completed several hundred thousand, not ten thousand years ago. And the culture of human altruism is certainly very much older than the agricultural revolution. Trivers agrees in his speculation (1971: 54) that "one may wonder to what extent the importance of altruism in human evolution set up a selection pressure for psychological and cognitive powers which partly contributed to the large increase in hominid brain size during the Pleistocene." He evidently was thinking in terms of genetic altruism; but it may well have been cultural in substantial degree.

Wright (1978: 454–55) postulated a combination of kin and interdeme selection as the genetic basis that made human culture possible.

> The evolutionary advance since the beginnings of agriculture, and of the cities which it made possible, occurred only after some 96% of that after the origin of the species *Homo sapiens* began and 99% of that after the origin of the genus *Homo*. If any appreciable advance has occurred since, it has probably consisted more in the world-wide diffusion of the level attained by the most advanced peoples than in further progress of the latter. This is the last phase of the shifting balance process. . . . The mode of evolution of culture is analogous to that of the genetic system. Invention is the analog of mutation. Diffusion of culture is the analog of gene flow. Cultural variation is continually subject to selection on the basis of utility. There is random cultural drift, exemplified by the breaking up of languages into dialects. Finally, the most favorable condition for cultural advance is local isolation, providing the basis for simultaneous trial and error among many variants and the diffusion of the more successful ones in analogy with the shifting balance process in biological evolution. . . . The state of the culture has been to a considerable extent an index of the rank of populations genetically in the distinctive human line of evolutionary advance, and reciprocally the demands of culture have been the primary selective agent in this advance in its later stages. Aspects of culture are continually being borrowed, but whether such borrowings are effectively integrated into the existent culture to form new peaks (as more conspicuously in the recent period in Japan), or are adopted only superficially and to the detriment of the previous culture, is also an index of genetic capability.

SOFT THEORY

Wilson's distinguished colleague Richard C. Lewontin (1974: 28–29) has told us (in his Jessup Lectures, dedicated to Theodosius Dobzhansky and entitled *The Genetic Basis of Evolutionary Change*) that

> it is a common myth of science that scientists collect evidence about some issue and then by logic and "intuition" form what seems to them the most reasonable interpretation of the facts. As more facts accumulate, the logical and "intuitive" value of different interpretations changes and finally a consensus is reached about the truth of the matter. But this textbook myth has no congruence with reality. Long before there is any direct evidence, scientific workers have brought to the issue deep-seated prejudices; the more important the issue and the more ambiguous the evidence, the more important are the prejudices, and the greater the likelihood that two diametrically opposed and irreconcilable schools will appear. Even when seemingly introconvertible evidence appears to decide the matter, the conflict is not necessarily resolved, for a slight redefinition of the issues results in a continuation of the struggle. It is part of the dialectic of science that the apparent solution of a problem usually reveals that we have not asked the right question in the first place, or that a much more difficult and intractable problem lies just below the surface that has been so triumphantly cleared away. And in the process of redefinition of the issues, the old parties remain, sometimes under new rubrics, but always with old points of view. This must be the case because schools of thought about unresolved problems do not derive from idiosyncratic intuitions but from deep ideological biases reflecting social and intellectual world views. *A priori* assumptions about the truth of particular unresolved questions are simply special cases of general prejudices.
>
> Attitudes about the kind and amount of genetic variation in populations, like all attitudes about unresolved scientific issues, reflect and are consistent with the intellectual histories of their proponents. . . . A scientist's present view of difficult questions is chiefly influenced by the history of his intellectual and ideological development up to the present moment.

Such refreshing and perceptive candor is very reinforcing of the biases about the process of scientific discovery that most younger social scientists now will have been socialized to accept. But more generally, we ought to expect that even the most culturally empathetic biologists will tend to accept genetic evolutionary explanations to a greater extent than the most biologically oriented persons who have been trained as social scientists; the same applies vice versa with regard to cultural evolutionary explanations. Indeed, such ideological differences undoubtedly cut within (as well as between) demes of scientists, so that a critical difference in the perspectives of biologists toward hardcore theory, altruism, and cultural evolution hinges upon whether the animals with which they have had their predominant research experience are social insects or mammals. Five of the hardcore theorists (Alexander, Dawkins, Hamilton, Trivers, and Wilson) work with insects, and Williams with fish; Sewall Wright, on the other hand, worked almost exclusively with mammals. (Unfortunately for the hypothesis, Lewontin also works with insects.)

The unconscious bias shows up in countless remarks, as exemplified by George Williams' (1971: 8) assertion—in the introduction to a book on *group* selection—that "a territory, by definition, is an area that one individual defends against intrusion by other members of its species." Of course, whether this is true depends upon who is making the definition of territoriality. Williams' statement certainly applies to sticklebacks and to many other species of fish and also of birds. But his assertion does not apply to most species of primates, which live in groups and defend group territories, to the extent that they defend any territory; and that must be distinguished, of course, from the "personal space" that may accompany a primate wherever it is. Williams' definition most certainly is a half-truth that failed to include human warfare, which we shall discuss further below. Similarly Williams remarks (1966: 31) in a different book that "a population cannot retreat to a marginal habitat to avoid being killed off by competition"—and here he *is* speaking of mammals (housemice). But he clearly is not thinking of humans, who *were* doing that during the Pleistocene as well as at present.

Ideological bias shows up also in the penchant for teleological explanation that permeates Williams' writing, as exemplified by the selections that follow. "The goal of the fox," he tells us (1966: 68), "is to contribute as heavily as possible to the next generation of a fox population." And as with foxes, so with sexes: "The ultimate goal for both [males and females] is maximal genetic representation in the same population" (1975: 124). Indeed (1966: 69), "duck genes have chosen a life spent largely above the surface of the sea," whereas "tuna genes . . . have cast their fortune in submerged habitats." More generally (1966: 70), "the succession of somatic machinery and selected niches are tools and tactics for *the strategy of the genes*" (emphasis added). In the following phrase he may have intended to say "to be increased," but he in fact assures us (1966: 196) that "it is possible for the donor *gene to increase its* frequency in the population" (emphasis added). Having spoken earlier (1966: 44) of "the real goal of development," he asks (1966: 108): "Do these processes show *an effective design* for maximizing the number of descendants of an individual" [emphasis added]? The clear implication throughout seems to be that Nature moves in mysterious ways, its wonders to perform.

Although in context there is every indication that he is *not* talking about humans, Williams (1975: 138) ventured the observation that "at every moment in its game of life the masculine sex is playing for higher stakes. Its possible winnings, either in immediate reproduction or in an ultimate empire of wives and kin, are greater. So are its possibilities for immediate bankruptcy (death) or permanent insolvency from involuntary but unavoidable celibacy." We can take this as an example of patent anthropomorphism (see ch. 7, above); and the hardcore theory is replete with anthropomorphic metaphors such as "altruism," "selfishness," "cheating," "moralistic aggression," "parental investment," and "kin" selection. In the English language from which they are taken, every one of these concepts implies ethical choice, which, so far as seems likely and so far as is presently known, only humans are capable of making, or else (as in the case of the latter two) they are peculiarly human intergenerational relationships among genetically related animals. It is a mistake to describe the interspecies predation of soldier ants as "altruism," when we believe that they have no choice but to respond as they do to chemical cues. Similarly, dogs and cats do not have mothers, fathers, sisters, or brothers—except in the minds of their human overlords, who frequently (at least in modern urban milieus) compound the confusion by attempting to cast themselves in the surrogate role of parent to their pets. Dogs aren't brothers to people because to be a brother implies at least as much culturally as it does genetically; and whatever the affective and subservient behaviors that some domesticated canids have been (artificially) selected to perform, a dog can share only in part physically, and very little cognitively, those aspects of human culture that define the reciprocal obligations of brothers to each other.

A recent sociobiological best seller is entitled *The Selfish Gene* and merchandised in such a way as to suggest that the subject under discussion is human genes (or at least, one of them). Indeed, the title is wittingly deceptive, because the author is a quite sophisticated zoologist who understands only too well how extremely improbable it is that any set of behaviors as complex as "selfishness" could have evolved under the control of any single gene, and certainly not among humans. In an interview in July 1978, Richard Dawkins explicitly denied that in that book he had intended to emphasize human selfishness or the sociobiology of human inclusive fitness. Many readers, or at least purchasers, of the book must have expected otherwise; even in England books about the population genetics of beetles rarely become best sellers. But if *The Selfish Gene* is not about humans, then surely Dawkins is hoist on the alternative petard of the indicated dilemma: he must be anthropomorphizing genetic control of the behavior of nonhuman animals.

It is unfortunate that more sociobiologists are not as well versed in metaphors and similes as they are in analogies and homologies, because if they were they would be more circumspect in their use of language (Hubbard, 1981: 214). Familiarity with the writings of, for example, Martin Landau (1972), a political theorist who is familiar and has been much concerned with evolutionary theory, sensitizes a reader to the perils of simplistic metaphorizing. Certainly it is simplistic to seek to bridge the gap between human behavior and that of other animals by trying to understand the other animals in terms—in *the* terms—of what are uniquely and idiosyncratically human abilities and relationships. Hardcore sociobiology is going to have to get beyond the Uncle Remus stage before very many social scientists are going to take it seriously, to say nothing of yielding themselves to its purportedly masterful and prepossessing embrace.

What is fundamentally wrong with the hardcore sociobiological approach to human behavior is its pious predilection that human social behavior is motivated by, and explicable in terms of, the competition among individual humans to maximize the contribution of each to a reified metaphor: that of "the gene pool." Most of the better known spokesmen for hardcore sociobiology have asserted that human social behavior not only can, but *must* be, explained in terms of inclusive fitness. Either in its moderate or strong form, the postulate is pious because it rests upon the act of faith that a nineteenth-century idea—Darwinian selection—

must hold for all behavior of all animals, thereby assimilating the social *and the cultural* behavior of *Homo sapiens*, the species that has maximal genetic plasticity, to that of others (e.g., termites) under much tighter genetic control. We should have learned, or should learn, that even in such physical sciences as astronomy and nuclear physics, theory appears invariant only during the stages of substantial ignorance about the phenomena under investigation. Contemporary, modern, recent, and medieval history alike—to say nothing of ancient—are replete with instances of heathens being ripped off by missionaries (and their secular acolytes); and however we might hope that zoologists will achieve the dispassion to recognize the possibility that they may be backing the wrong horse (by worshipping orthodox Darwinian gradualistic selection as the exclusive deus ex machina for biotic change; and cf. ch. 12, below), at least those of us who remain agnostics or even skeptics have an obligation to point out the fundamentally religious nature of the transactions implied by Wilson's call for the conversion of the social sciences (see ch. 8, above).

Many commentators have proposed that genetically supported primate social structure, at least among such hominids as humans and neanderthals during the past hundred thousand years, has been preadapted to hierarchical relationships to such an extent that a hero-god is an indispensable figure to serve as leader of larger, as well as the largest, population aggregations (Tiger and Fox, 1971; H. T. Odum, 1971: ch. 8; and cf. Willhoite, 1977; also ch. 8, above). Sometimes the hero-god (Mahatma Gandhi, Hitler, Mao) appears in the flesh to lead his worshippers; but the supernatural (or at least sanctified) isomorphs (Christ, Jove, Buddha) exemplify one of the many advantages that cultural images enjoy in comparison with their biological phenotypes. Indeed, inclusive fitness—with its inexorable logic of the watering down of *any* individual's possible genetic contribution, with each passing generation—proffers a pale ideal for immortality,[13] in contrast to the possibilities for growth offered by cultural immortality. But if humans are preadapted to worship, such an attitude must apply to scientists also, their socialization toward objectivity in more finite matters notwithstanding.

A scientific paradigm waxes or wanes in its authority and imperialism as a consequence of sociocultural processes that we understand and can document empirically; but to the extent that those sociocultural processes are influenced by

[13]Williams (1966: 24) says that "genes are potentially immortal." More realistically, Alexander remarked, at the Animal Behavior Society's sociobiology panel on June 20, 1978, that genes do *not* survive forever—just considerably longer than populations and environments. But consider: inclusive fitness assumes a genetic value of 1 for each individual human, as a base for purposes of comparison. For each child the progenitor is assigned a score of ½; for each full sibling and each grandchild, ¼; for full cousins and great-grandchildren, ⅛ each; and so on. At the end of only nine successor generations— a bare moment of two or three hundred years in evolutionary time—after the progenitor's own phenotypic death, he or she may have hundreds of progeny; but his or her genetic contribution to any individual member of that ninth generation, a great-to-the-seventh-power-grandchild, is less than .001, one-thousandth of its genetic constitution. And the progenitor's genetic influence is, of course, correspondingly attenuated. Immortality seems to be the wrong word to describe such genetic dissipation.

underlying multiplex genetic determinants, both the putative genes and the interaction processes linking them with the involved sociocultural processes remain unknown and hence unamenable to empirical discussion at this time. The initial acceptance of any paradigm no doubt typically is based upon what at the time appears, to the relevant population of practicing scientists, to be the clear preponderance of the evidence—or at least, of the potential evidence (as in the instance, for example, of Einsteinian relativity in twentieth-century astronomy and physics). But those scientists were members of an ancestral cohort (in terms of the evolution of scientific theory), dwelling in an era past and limited to the information and insights then available. As Kuhn (1962) demonstrated more than a generation ago, the persistence of the paradigm comes increasingly to depend for its acceptance, by the rank and file of successor scientific cohorts, upon faith rather than reason—indeed, upon faith so strong and so passionate that it increases in fervor in direct proportion to the increasing failure of the paradigm to find support in the accumulation of new empirical information. The prophets of a new paradigm, which threatens to displace the old orthodoxy, are first ignored and then castigated (by the many stalwarts of the ancient regime) as fools, heretics, and knaves, for reasons that are readily explained by contemporary social science, or by Shakespeare, or by any member of several generations of philosophers dwelling in Athens some two and one-half millennia ago. Of course this is an evolutionary theory of paradigms; and it would be gratuitous to retell such a familiar story (cf. ch. 19, below), especially in this somewhat involuted context, were it not for the zeal with which social scientists' failure to embrace modern evolutionary theory (but see Corning, 1983) is impugned, and the complacency with which biologist commitment to neo-Darwinism is taken for granted today, in zoological writing generally and hardcore sociobiological theory in particular.

Lewontin (1974: 4) has questioned the inarticulate major premise that underlies the hardcore theory: that Darwinian natural selection is the only way, as well as the truth and the light, to explain genetic evolution.

> For [some leading evolutionary geneticists] natural selection is vital in the divergence between isolated populations while for [others] natural selection is always primarily a cleansing agent, rejecting "inharmonious" gene combinations, and not necessarily the causative agent in the initial divergence between incipient species. Nor is this conflict of viewpoints yet resolved. During the last few years there has been a flowering of interest in evolution by purely random processes in which natural selection plays no role at all. Kimura and Ohta suggest [in 1973], for example, that *most* of the genetic divergence between species that is observable at the molecular level is nonselective, or, as proponents of this view term it, "non-Darwinian". . . . If the empirical fact should be that most of the genetic change in species formation is indeed of this non-Darwinian sort, then where is the revolution that Darwin made?
>
> The answer is that the essential nature of the Darwinian revolution was neither the introduction of evolutionism as a world view (since historically this is not the case) nor the emphasis on natural selection as the main motive force in evolution (since empirically that may

not be the case), but rather the replacement of a metaphysical view of variation among organisms by a materialistic view. . . . *For Darwin, evolution was* the conversion of the variation among individuals within an interbreeding group into *variation between groups in space and time.* (emphasis added; cf. Caplan, 1978b)

Surely no person as open-minded and creative as Charles Darwin could possibly be an orthodox Darwinian today, just as Karl Marx surely would not want to be a Marxist now.

A much better balanced and certainly in terms of relevance to human behavior a much more useful perspective of evolutionary change is presented in the writings of the late Sewall Wright, (d. 1988), a founder of the field of population genetics, whose own doctoral research at Harvard was completed before American participation in World War I. Wright remained a leading contributor to the mathematical theory of the genetics of evolutionary change for more than sixty years; and in the fourth volume of his treatise (1978: 462, 463, table on 461; also 1977, vol. 3: 439, 453–55, 468, 560–62) he remarks:

If mutation pressures and random drift are rarely if ever controlling factors, we are left with only the two kinds of natural selection. Fisher and his followers, including recently Williams (1966), would also remove selective diffusion from consideration, leaving only mass selection, by invoking the principle of parsimony. . . . [Fisher would] be correct if there were only one mode of change of gene frequency, . . . selection in favor of specific modifiers of the heterozygotes. [But] degree of dominance is only one aspect of a complex interaction system (as I had found it to be in extensive experiments with guinea pigs [1916, 1917, 1927, and later]). . . . The occurrence of some sort of mass selection is no doubt more nearly universal than the occurrence of a population structure that favors selective diffusion and thus is to be adopted as the only factor under the parsimony principle. Yet where there is a suitable population structure, selective diffusion may be enormously more effective than mass selection. Where a wide stochastic variation is occurring simultaneously at thousands of nearby neutral loci and more or less independently in many demes . . . a virtually infinite field of potential variability from possible interaction effects is available without change of conditions. There is no such tendency towards exhaustion of additive genetic variance as under mass selection. The cost to the reproductive excess on which selection depends is also minimal. In general, the concept of shifting balance leads to very different conclusions than the parsimony principle.

In further explicit contradiction of Williams, Wright summarizes the Wynne-Edwards thesis and then (1978: 53) remarks:

Elaborate behavior patterns. . .serve to establish a peck order or an allotment of territories in an orderly way, with a minimum of dangerous combat. . . . Under favorable conditions such behavior patterns tend to maximize reproduction by all individuals, but under unfavorable conditions they tend to allot adequate resources only to the fitter individuals at the expense of others. Such patterns are much more easily accounted for by intergroup selection than by individual or familial selection. Since they exist, it would seem probable that intergroup selection is responsible. . . . [L]ocal populations that acquire modes of behavior that obviate overproduction of individuals before actual starvation or permanent destruction of food resources occur, would tend to replace neighboring populations in which individual selection pushes reproductive capacity to the limit under all conditions. It appears probable

> then that reproductivity under favorable conditions tends to be pushed as far as possible by individual (Fisher, 1930) and familial selection (Hamilton, 1964), but that intergroup selection (Wynne-Edwards, 1962) imposes behavior patterns that obviate disastrous overproduction under unfavorable conditions.

In explicit regard to human evolution, Wright observes (1978: 452–53; 1978, vol. 3: 471–73) that

> during the 99% of species history while humans and their forebears of the genus *Homo* remained hunter-gatherers, no doubt there was a steady pressure of mass selection in favor of intelligence, but change of gene frequency according to the net effects of individual genes is a process that is not directed toward what matters most, the effects of interacting systems of genes, and it is subject to the severe cost imposed by selective replacement at one locus on replacement at other loci, especially severe in a species with the relatively low reproductive capacity of primitive man. . . . Simultaneous sampling drift at thousands of sufficiently neutral loci provides different material in innumerable localities without appreciable cost, material that can give the basis for effective interdeme selection. . . . The actual process of interdeme selection may take on different forms. At one extreme, the local appearance of a superior genetic system is followed by expansion of its territory accompanied by complete elimination of its neighbors until it occupies the entire range of the species. At the other extreme there is merely . . . diffusion from the superior center. Neighboring populations are graded up until they reach the point (the crossing of a saddle in the surface of selective values) at which mass selection carries them autonomously to the new selective peak, or perhaps beyond, if they contribute something that improves on the latter. The location of the population with the highest selective peak may shift from place to place in the course of time, as a group of neighboring populations step each other up to heights well above the general level.

As for altruism, Wright (1978: 454) states:

> The heroic virtues, including willingness to sacrifice one's own life for the good of the tribe, are traits that can hardly be developed (insofar as they have a genetic basis) by purely individual selection. They may to some extent arise as a by-product of familial selection . . . [among] close relatives with heredities strongly correlated with that of an individual who gives his own life to save them. . . . The importance of this sort of intergroup selection in evolution has been emphasized . . . by Hamilton. The increase in frequency of traits deleterious on the average to their processors but beneficial to the deme may also, however, be increased by interdeme selection . . . if the benefit to the deme sufficiently outweighs the damage to the individual.

Nowhere in the hardcore sociobiological literature of the past decade and a half do any of its protagonists undertake to explain *why* humans should be consciously motivated to attempt individually and competitively to maximize their respective personal genetic contributions. The stereotyped rationale from evolutionary biological theory is, as we know, based almost entirely on logical deduction and not upon a systematic corpus of either ethological or experimental empirical observations. It goes as follows. Some animals breed more successfully than others of their species. The more successful breeders make relatively larger contributions to "the species gene pool" than do their less successful competitors. Whatever combinations and frequencies of genes are embodied by the successful breeders will tend to become more common, and the frequencies of the genes of the less successful phenotypes will lose out. This does not presume any particular motivation on the part of individual animals, whether successful or not; genes that get reproduced (in any particular environment or set of environments) will become more common while genes found in poor breeders will tend to disappear. So it is all a matter of stochastic processes. Or almost. Because it presumes also that the genes are *controlling* the behavior of the animals concerned. Genes for light-colored moths, once highly adaptive when these insects were roosting on the trunks of preindustrialism birch trees, became displaced by the formerly rare genes for dark coloration once those same tree trunks became chronically coated with soot. As the trees gradually darkened, light-colored moths became relatively easier, and darker moths harder, for predator birds to perceive. To the extent that humans engage in equivalently stereotyped (i.e., nonchoice) behavior, which is equally highly correlated with human morality, then the same arguments might apply—particularly if humans generally had no more cognitive discretion over behavioral choices than moths are deemed to exercise.

Unfortunately for inclusive fitness theory, however, the most remarkable characteristic of our species, together with the brain structure and language behavior with which it is so conspicuously interrelated, is the plasticity of human behavior (Sagan, 1977: 3). Even a wide range of infravisceral systems and functions, many of which used to be classified as "autonomic" and not subject to conscious control or influence, now are increasingly coming under the ambit of cognitive theory as teachable and learnable behaviors. A generation of human evolutionary theorists has proclaimed that it is precisely this variability in behavior, this potentially extraordinarily broad range of adaptability to both natural and social ecologies, that the genes that distinguish our species have supported for half a million years or more. Within that range of choice, at least in genetic terms, is whether or not to have children; when to have them; whether to nurture children begotten by others in preference to or in addition to one's own; whether to share resources with friends or with relatives or with complete strangers.

The metaphor of the "gene pool" is one that, through stereotyped overusage, has become hackneyed and hence misleading. The metaphor invokes the image of some pristine ocean, a maternal symbol by means of which all surviving genes at any particular time are aggregated (and in some sense, summated). But the genes with which we are concerned here exist only in living organisms; the genes nurtured in laboratories for experimental purposes typically appertain to plants or to much simpler animals (at least, socially) than the social insects or the mammals with which sociobiological theory concerns itself (cf. ch. 11, below). Hence genes are in fact aggregated only in units consisting of the organism characteristic of the species of interest, and then infrasomatically. The only conceivable way to "pool" the genes of any species would be to aggregate the entire species population, an entirely unnatural action that would certainly be lethal to the species (and therefore, to all of the genes concerned) even if it could be performed. A gene pool is a purely imaginary hypothetical construct, one that bears no empirical relationship to the structure or the functioning of life in the real world. What is more, it is most dubious that any animal has ever lived that has intended either to make, or to fail to make, a contribution of its genes to the species gene pool—with the possible exception of a handful of hardcore sociobiological theorists. So far as we have any evidence (Griffin, 1978), no nonhuman animal has any idea why it copulates, to what end, or in many instances with whom. Indeed, at least the first two-thirds of that statement could be asserted about most humans, until a very recent moment in evolutionary time. Again, the substantial divorce between copulation and progenitation for the human species, as a consequence of the substitution of chronic for periodic and acute sexual availability among species females, is an idiosyncratic species characteristic; but it is another hallmark of behavioral plasticity that confounds prediction based on simplistic theories of genetic determinism. It is one thing to measure, in terms of ecological variables, the fitness of !Kung females as a function of their birth spacing, as Blurton Jones and R. Sibly (1978) have done using Devore and Lee's data; but it would be quite a more difficult matter to attempt a replication among human females living only a few hundred miles away on the other side of the Kalahari.

Campbell (1972: 30) has pointed out that, "in Trivers' discussion of human reciprocal altruism, he makes use of learned individual tendencies as well as genetic ones, but he fails to give explicit consideration to the social evolution of reward and punishment customs and tends to consider all personality and behavior dispositions as genetically inherited. While he uses the mathematical models of evolutionary genetics, these are not developed for many of his most crucial speculations on human altruism." Trivers' trite example (1971: 35) is of "one human being saving another, who is not closely related and is about to drown," to which he applies the cost-benefit analysis lingo of inclusive fitness, provoking West Eberhard (1975: 81) to carry his reasoning to its logical conclusion: "Theoretically, the most willing lifeguard should be a physically fit eunuch or postreproductive individual (who has nothing to lose in terms of personal fitness), and with few living relatives (little to lose in terms of future gains to inclusive fitness through aid to close kin). A good beneficiary is one with high reproductive value . . . such as a pregnant low-income Catholic teenager who is about to produce her first child." The winter of 1979 confirmed what we already know: that several urban adults could stand on a bridge over the Chicago River and watch a child drown, with no more compulsion to intervene personally than if they were spectators at yet another urban arson. These people were being entertained; they were not whipping out their pocket calculators for guidance in any decision whether or not the odds of ultimate social reciprocity favored intervention. Living in Chicago in the seventies, they had learned how poor these odds would be, and they didn't need to compute. Thirty years ago or earlier, a group of Chicagoans in a similar situation would have cooperated in trying to save the child. It is disingenuous to argue that the change is due to changes in the Chicago gene pool during that interval—although admittedly, considerable colonization (both in and out) did take place during that period. Even for those Americans who did stay put, a new generation was born and came of age subsequent to World War II; and we must keep in mind Williams' admonition (1975: 153) that "in a heterogeneous and fluctuating environment, each new generation may be regarded as colonists entering new environments."

There is biological reformulation of the question of diminishing altruism within a population, alternative to the sociocultural hypothesis suggested above. Clearly the issue is not confined to Chicago, or American megalopolises; it is a social disease—if altruism is good—that is endemic in industrial urbanization (see Green, 1986). To place the biological hypothesis in context, we can note George Williams' observation (1966: 71) that "the phenomenon of fitness can be seen at all epigenetic levels, from genic interactions to the ecological niche," or indeed, between ecosystems—which is highly relevant to contemporary endeavors to redefine the human niche (Dunbar, 1960). So we must now shift gears and transfer our thinking all the way from Trivers' postulated organic competition (which would be midpoint on Williams' scale) to biotic evolution, where competition is interspecific. The hardcore theorists tend to think of all kinds of fitness as a set of games being played against the nature that they typically reify; but our assumption is that "Nature" is not trying to do anything in particular. The interspecific competition game is one at which humans have done extremely well, especially during the past fifty to thirty thousand years, during which we eliminated and subsumed our only surviving hominid competitors, the Neanderthals, and virtually all of the other large mammals of the Northern Hemisphere (Day, 1981; Greenberg, 1982; Martin and Klein, 1984; and Nitecki, 1984), a job that we are now in the process of completing for the Southern Hemisphere, while similarly turning our attention to the mammals of the sea. But in the process of getting rid of our competitors, improving our genetic fitness as a species (see ch. 11, below), and exploiting both plant and physical resources at a logarithmic pace (Green, 1986), we have also imposed important (and for ourselves, highly maladaptive) constraints upon the range of variation left open to our only remaining antagonist— "Nature"—and hence to the only biosphere presently available to us (cf. ch. 12, below).

The mere mention of modern medicine, in the context of the acceleration of species population growth during this century, makes ludicrous the hypothesis that "natural selection" is shaping the future genetic composition of the human species, by the "mechanism" of having those *individuals* who are "best adapted" to their environment in the present generation leaving relatively more progeny, and thereby having the greater influence upon the possible genetic combinations of future generations. Genetic competition today is waged among very much larger population aggregates than any individual (including his kin), and indeed between very much larger ones than the demes evidently envisaged by Wilson. The consistent trend of the past century has been for birth rates to drop to or below zero growth in maturing industrialized societies, while the countries that remain most traditional and most rural maintain the high population growth rates. If there is global genetic competition for resources, it is taking place between Europe, North America, and Japan, on the one hand, and the rest of the world on the other hand. Hardcore sociobiology makes no mention of such problems, and for the very good reason that it has nothing to offer toward either their comprehension or their solution. The reasons why, in general, members of the deme conglomerate of the industrialized West (including, for present purposes, Japan) do *not* undertake to maximize their individual (or collective) inclusive fitness—just as the reasons why most other humans outside the West (or, better said, the "North") *do* try to leave as many progeny as they can afford to raise—have very little to do with genetic determinism; they are instead the byproduct of human cultural differences.

There is a hypothesis—naive in evolutionary and genetic theory alike—that there is some sort of linear and unidirectional process toward greater complexity and perfection ("up from the ape"; e.g., Engels, 1876; Day, 1920) and that therefore the demes of industrialized societies are better adapted than those of the remaining agriculturalists of the Third World, or that either type of deme is better adapted than the remnant hunter-gatherer bands whose existence, though "nasty, brutish, and short," has now been anthropologized. That sort of liberal optimism, which finds its soulmate in a social science environmentalism that postulates no limits to human perfectibility beyond the imagination and creativeness of the perfectors, is increasingly contradicted by the facts of life of the twentieth century. Indeed, the converse hypothesis is much better supported: that hunting and gathering was a better adaptation for humans than agriculture, because the minimal demands made upon the environment by hunting and gathering tended to assure the persistence of the human niche for an indefinitely long future. Agriculture, on the other hand, in barely ten thousand years has exerted and continues to exert an accumulative, consistent, and malevolent effect upon habitats that were for millions of years critical to the definition of the human niche. And industrialism, as we all know only too well, has so accelerated the pace of habitat depletion that the species niche probably already has been irreversibly destroyed by processes already well advanced (see Green, 1986).

In this regard Williams (1975: 147) has remarked that "extinction occurs not because an organism loses its adaptation to an ecological niche, but because its niche becomes untenable" (1975: 154). "Sexual selection," he adds (1975: 154), "facilitates evolution indirectly by making extinction less likely." Nevertheless, "a population may even obliterate its own niche by becoming better adapted to it. All that is required is that increasing adaptation have a progressively adverse effect on total resources." As needed resources shrink, so does the population dependent upon them, until the denouement is resolved when "extinction occurs because there is no corrective feedback between dangerously low population size and the forces of evolution. The last pair of passenger pigeons to nest successfully had no way of knowing that they ought to take desperate actions. Approaching extinction evokes no emergency measures, but rather '. . . the species doomed to extinction, innocently unconscious of its lack of "fitness," continues happily to perform its traditional rites'" (Williams, 1975: 158–60).

When this will happen to humans—and evolutionary biologists generally are certain that it must—is not yet certain. Nor can we be sure that the remnant humans, on the verge of species extinction, will be any more self-conscious about the matter than the passenger pigeons were, or than cetaceans are today. But if it is true that the principal social problem confronting our species today is the loss of altruism that had previously evolved—for whatever combination of cultural factors transactionally with genetic ones—in a niche that as then defined has now *also* been lost, then the principal objection to Trivers' theory of reciprocal altruism is that it is trivial because it diverts attention to a side show and away from what ought to be our central concern. Even if he were correct (and I am convinced that he is *not*) about the relative importance of genetic causation of human altruism, the human groups to which his theory properly applies are those of the hunter-gatherer epoch of more than ten thousand years ago, when people lived in face-to-face groups small enough so that all could know each other as individuals. Of course, his work would be less popular and not seem as "relevant" if it had been proffered—as I am arguing it should have been—as a contribution to human prehistory, rather than as advice about how and why people behave as they do today. Trivers' optimism about how our genes are going to keep churning out altruism now is a dangerous nostrum.

Population genetics has come up with a hypothesis that is helpful in elucidating the problem of altruism, by focusing upon its disappearance in the modern world, as Eshel (1972: 275) has explained.

> Though it may be assumed that the development of a new altruistic trait requires a considerable span of time (see, for example, Simpson, 1945), the extinction of such a trait under a newly imposed condition of high mobility may occur at a higher rate (see a comment on Simpson's work by Wright, 1945). And a rapid increase in demographic mobility is a factor no doubt exclusive to human evolution. Thus, from a theoretical point of view, the quantitative understanding of the effects imposed by this factor on the selection of altruistic traits may add to our knowledge of human evolution.
>
> Trivers (1971; and see also Campbell, 1966; Lee and Devore, 1968) already has suggested that the demographic conditions that prevailed in pre-Neolithic human populations were likely to favor the evolution of altruistic patterns. These patterns may in turn have been prerequisites for the subsequent development and maintenance of human societies. Yet because of the tremendous increase in human mobility during only the last ten thousand years or so, it may be that natural selection no longer favors such altruistic traits, however favorable for human society some of them may still be.
>
> A question repeatedly raised by ethologically oriented authors (Lorenz, 1963; Morris, 1967; and others) is to what extent the relatively short period of human civilization has been sufficient for man to adapt biologically to his new environment. Attempts have been made to explain the malfunctioning of modern society on the basis of the still-persisting "australopithecine" drives of man. But perhaps a more crucial question is whether human evolution is in fact headed toward a desirable social adaptation, i.e., whether present fertility selection

necessarily favors traits that are beneficial to human society as presently constituted. The theoretical conclusion of this study suggests that the opposite may be true. It may be that some intrinsic human drives, altruistic in nature, that are fundamental for the establishment of any human civilization, could possible evolve only under precivilization demographic conditions. And it is possible that just those selection forces imposed by civilization itself act to reduce the frequency of these fundamental drives within human population, thus leading it into the course of misadaptation.

Hamilton, whose earlier work on inclusive fitness inspired Trivers, has in fact already done what I have suggested Trivers should have done, by focusing attention upon the evolutionary genetics of cooperative social behavior in hunting-band humans. It is notable that Hamilton himself (1975: 150) cites Eshel with approval and states as his own conclusion that "civilization probably slowly reduces its altruism of all kinds, including the kinds needed for cultural creativity." It is also notable that in his remarks at the sociobiology panel at the Animal Behavior Society's 1978 annual meeting, Hamilton took pains to pay homage to Sewall Wright, whom he claimed as his own intellectual ancestor.

HARDBALL POLITICS

Hardcore sociobiological theory has little to offer to the solution of the problem of hardball politics, or those of its softball template (political science). Individualism and associated notions of innate human selfishness have formed a major component of political thought for at least three millennia, about as far back as the written evidence extends. Genetic theories of individual natural selection, such as Hamilton's concept of inclusive fitness, have nothing per se to say about political behavior today because they are addressed to what is for political behavior and political science an inappropriate presocial level, characterized by Williams as "cybernetic abstraction." Kin selection is a step in the right direction, however, because it views human behavior as social behavior and as necessarily concerned with the group (population) context within which it is hypothesized to occur.

Nepotism

Nepotism is important in politics, and not only in the social structure of hunter-gatherer or primitive agricultural peoples. It remains a critical factor in political leadership, elite structure, and followership throughout the Third World today, and in much of the Second World as well.

The trend in all of the industrialized countries has been explicitly contrary to extended-family politics, for cultural reasons that have been well discussed by Max Weber (among many others). For Japan, China, the Soviet Union, Europe, and North America (north of Mexico, that is), the surviving monarchies have been reduced to what are conventionally viewed by political scientists as vestigial political functions; Prince Charles' decision to sit with the Cabinet and the apparent involvement of Prince Bernhard as liaison with various international corporations are recent exceptions that remind us of the rule from which they deviate. The reason for reexamining the political role of royal families is that political integration, of the subpopulations of consociational democracies, may be a much more important function than is recognized by the Bentlian, Dahlian, or Eastonian theories of interest-group conflict and organic-system components with which post-World War II political science has been preoccupied.

From the point of view of public policy, antinepotism strictures have been a byproduct during the past decade of the genuflections toward egalitarianism in sex and, to a much lesser extent, in age; but certainly the theory of kin selection has implications for such questions. For example, the extension of kin selection to parent-offspring conflict and parental investment has an obvious bearing upon political socialization and consequently upon political attitudes and participation.

Domestication

Genetic engineering has implications of a very different sort for political science. Local governments and the federal United States government alike have been involved for several years in competition with each other for the regulation of genetic research, N.S.F. guidelines for research in reconstituted DNA constituting one widely publicized case in point. A related policy arena of considerable interest and broad implications concerns the regulation of sperm banks. From a technical point of view, a single (though not necessarily unmarried) male donor could, given appropriate care and distribution of his largess, inseminate every fecund female in the entire country (see Frank, 1947)! Now there, as Humpty Dumpty might have said, is genetic immortality for you! Given the diploid character of the species, this hardly amounts to cloning; and yet it probably proffers the closest empirical analogy to generalized kin selection that even human ingenuity is likely to devise. It promises also to provide the route that would make the hardcore theoretical analogies between humans and certain of the social insects much more meaningful. Under the circumstances envisaged, genetic altruism ought to become a widespread consequence; but then kin selection would certainly constitute a major problem of politics because of its massive entailments.

To show how rapidly significant results could be expected, consider Sewall Wright's discussion (1977: 533; especially chs. 3 and 16) of how bulls do it, or rather one extraordinarily "fit" bull.

> Favourite, born in 1793, shows a random correlation of 0.55 with random Shorthorns of 1920 . . . [while] the breed (in which over 170,000 had been registered in Britain, over a million in the United States, and over 200,000 in Canada by 1921) was radically transformed in type by sires tracing to a single-herd—that of Cruikshank, whose leading bull, Champion of Eng-

land, born in 1859, shows a relation to the breed as a whole of 0.26 for 1850 but 0.46 by 1920.

It is true that Wright does point out (elsewhere in his treatise) that from the standpoint of biotic evolution, humans are a feral and not a domesticated species. But he was clearly thinking in terms of the past, not in terms of a future in which paternal surrogates would rival digital computers in their power of control over human behavior.

Colonization

An intuitively plausible hypothesis that can be investigated is based on Williams' suggestion (above) that each new generation can be viewed as a group of colonists establishing themselves in terra incognita. Population geneticists have developed various models (Levins, 1970; Boorman and Levitt, 1972; Wilson, 1975; 107–17) to predict the adaptation of colonists to novel niches that they exploit; and although this theory, which of course builds upon the work of Wright discussed above, was developed to analyze lateral extension of populations, moving through time together, it can also be employed, as Williams implies, to examine change in populations in relatively fixed places, with the habitats varying according to time rather than place. Or both time and place might be dynamics of equal importance, as seems to be true of internal United States demographic shifts over the period of the past half century. The dependent variable would be change in political cultures and subcultures (Schubert, 1980b). One emphasis in political gerontology (Cutler, 1977) has been upon differences in the political culture of generation cohorts, as exemplified by studies which show how lasting the effect of the political culture of the time of their socialization is upon the subsequent beliefs and behaviors of political elites (see ch. 15, below). Thus the colonist hypothesis from evolutionary population genetics might contribute importantly to research in political change.

War

A promising contributor to the analysis of both public policy and political behavior is Wright's theory of interdeme selection, which as we have seen is opposed by hardcore (individualistic) sociobiological theory as represented in the writings of Williams and to a lesser extent Wilson. Our premise here must be, therefore, that hardcore sociobiological theory has made an unwitting contribution to political analysis, having through its opposition directed attention to a component of more orthodox population genetics theory that does promise to be useful in political analysis.

Physical and social anthropologists agree that hominid and human hunter-gatherers have been living in population groups the size of Wright's postulated demes—

from fifty to five hundred individuals—for a very long time, possibly from up to four million years ago until the transition of circa twelve thousand years BP. Even after the shift to agriculture, most persons continued to live in or near villages of the same size, notwithstanding the great political importance of the rise of a few cities, and subsequently city-states, beginning about six millennia ago. Even as recently as a hundred years ago in the United States, a majority of Americans continued to live in demes of five hundred or less. In the middle of nineteenth century there were only about eighty urban areas in the United States with populations of more than a thousand; and of these less than half were more than ten thousand. In terms of evolutionary time, humans have been primarily compacted in megalopolises for only about one-hundredth of 1 percent (i.e., 10^{-4}) of their specific genetic experience (Pettersson, 1978).

A frequently suggested hypothesis is that at least until twelve thousand years ago, neighboring demes were in genetic as well as phenotypic competition with each other for the resources necessary to individual and group existence: principally food, and within that category primarily animal protein (Ross, 1978). Genetic punishment for too close inbreeding was so direct and so highly probable that cultural norms to reinforce that learning doubtless were invented independently in many times and places, and in any event diffused rapidly (Bischof, 1975; Demarest, 1977), although the universality of their observance was another matter then as now. The complement of incest norms is of course the systematic exchange of females between demes, as Tiger and Fox (1971) and many others have proposed. Catastrophes of a variety of sorts, including but by no means restricted to those stemming directly from climate, always have been an important cause of the extinction or absorption of particular demes. Much of the time contiguous demes—and not necessarily any more "territorial" than the Lapps were fifty years ago—were in direct competition for food; the exchange of females between groups was probably more important as a medium for the reduction than for the incitation of inter-deme aggression. Consequently, chronic (although by no means continuing) aggression between and among groups has been adaptive for human demes for at least half a million years, and not merely during the few thousand years of recent history for which written and/or archaeological evidence demonstrates indisputably that this has been true.

Two equivalent but apparently independently constructed analyses of primitive warfare were published in 1975, both by population geneticists. One was by William Hamilton, who, notwithstanding his own relative youth, became the godfather of hardcore sociobiological theory. He observed (1975: 147–49) that

> effective birth control [practices, including cannibalism] cut warfare at its demographic root. Unfortunately, it is possible that in doing so they also cut . . . the selection for intelligence.
>
> The regards of the victors in warfare [include] tools, livestock, stores of food, luxury goods to be seized, and even a possibility for the victors to impose themselves for a long period as a parasitical upper class. . . . [I]t has to be remembered that to raise mean fitness in a group either new territory or outside mates have to be obtained somehow. The occurrence of quasi-

warlike group interactions in various higher primates . . . strongly suggests that something like warfare may have become adaptive far down in the hominid stock. . . . If the male war party has been adaptive for as long as is surmised here, it is hardly surprising that a similar grouping often reappears spontaneously even in circumstances where its present adaptive value is low or negative, as in the modern teenage gangs.[14]

It has been argued that warfare must be a pathological development in man, continually countered by natural selection, and . . . must always endanger the survival of a species [but] I see no likelihood for it as regards fighting of individuals or of groups up to the level of small nations. . . . [F]or the species as a whole, and in the short term, war is detrimental from the biological demographic point of view, but . . . detriment to the species does not mean that a genetical proclivity will not spread. . . . The gross inefficiency of warfare may be just what is necessary, or at least an alternative to birth control and infanticide, in order to spare a population's less resilient resources from dangerous exploitation. Maybe if the mammoth-hunters had attacked each other more and the mammoths less they could be mammoth-hunters still. . . . Many examples in the living world [today] show that a population can be very successful in spite of a surprising diversion of time and energy into aggressive displays, squabbling, and outright fits. The examples range from bumble bees to European nations. [Nevertheless,] it is hard in the modern world to see warfare as a stabilizing influence for man. . . . Pastoralists tend to be particularly warlike and the histories of civilization are punctuated by their inroads. . . . [But] incursions of barbaric pastoralists seem to do civilizations less harm in the long run than one might expect . . . [because] certain genes or traditions of the pastoralists revitalize the conquered people with an ingredient of progress [*sic*] which tends to die out in a large panmictic population. . . . I have in mind altruism which is perhaps better described as self-sacrificial daring.

Durham (1976b: 406–7) has proposed similarly that in primitive warfare where competition is between relatively small, only distantly related hunting-band demes, an economical way to enhance the genetic fitness of one group is to remove and subsequently display the heads of members of the competing group. As he points out, eating one's (outgroup) enemy is an effective means of eliminating competition for scarce protein, while at the same time directly augmenting one's own supply. His empirical example is not inconsistent with the prediction of inclusive fitness theory; but it is very far from a demonstration that the religious duty to eat one's enemy—however common this may have been among humans of neolithic times—results or resulted from genetic control and not from cognitive choice and learning. We certainly should expect cannibalism to be selected *against*, in genetic evolution, under circumstances such as Durham describes, where reciprocity is not only probable but culturally obligatory and reinforced by kin selection.

Thus we are confronted with two problems. On the one hand, it is essential to distinguish between the genetic and cultural factors that promote human war; yet neither Hamilton nor Durham—or so far as I know, anyone else—has attempted to make such a distinction. And even after that has been done successfully, it will

[14]This is an odd caveat for Hamilton to interpose. Much of the sociological data support the opposite finding: that teenage gang warfare is highly adaptive for most of its individual participants, whatever its impact upon third parties (who are part of the environment, from the perspective of the youthful warriors). For an ethological analysis of juvenile delinquency in the Netherlands, see van Dijk (1977).

be necessary to push considerably beyond the present level of metaphorical theory when neolithic is compared to nuclear (or even to napalm) warfare. The very facts about both demes and their environments, used to support the explanation (whether genetic or cultural) of primitive war, are conspicuously missing from the empirical structure of modern warfare. The hunting band model may fit part of the data for the tactics and behavior of infantry platoons and, possibly, even companies isolated in jungle fighting); but the hypothesis needs a great deal of working over to make it even intuitively plausible to be of help in modeling the possible triangular warfare between the United States, the Soviet Union, and the People's Republic of China. To be useful in political science—to say nothing of the practice of politics—sociobiological theory is in this regard going to have to proffer a causal theory that explains and predicts warfare in, say, the Middle East and Southeast Asia, during the 1960s and 1970s and 1980s and 1990s. And in view of many of its advance notices, hardcore theory ought to be demonstrated to do that at least as well as the presently available, admittedly woefully inadequate, theories produced so far by social scientists. Among the relevant questions to be answered are: Is war adaptive for the very much larger demes (countries) that now constitute the competing "groups" in international politics? Is it adaptive for demes that compete with (and often, now across) individual countries? At the level of biotic evolution, is war adaptive for the human species?

Unreciprocal Altruism

There is a related hypothesis, cited with approval by such leading population geneticists as Eshel and Hamilton, that recent cultural selection has failed to halt the decline in cultural altruism beginning with the rise of civilization, and, what is more, that it is likely to continue to fail at a progressive rate. If that is true, then clearly there are tremendous implications for public policy, politics, and political science. It will be recalled that the hypothesis assumes that cultural altruism was developed over an epoch to be measured in hundreds of thousands of years and probably antedates the differentiation of our present species; yet it has been only about ten thousand years since the breakdown of the face-to-face interdependence of the hunting band resulted in a parallel breakdown in the rationality of reciprocal altruism. The denouement came with the shift from pastoralism to agriculture (and hence to cities, and eventually to the more efficient modes of stored-energy exploitation that characterizes the past two centuries). The phenomenon of mass anomie is a familiar one (Riesman, Glazer, and Denny, 1951), and so is the diagnosis that the underlying cause is the consequence of some regrettable aberrations in contemporary customs (which a new method of teaching reading or of measuring intelligence might tend to alleviate). As a novel element, the Eshel-Hamilton hypothesis suggests that the basic cause is a product of cultural change so slow that it must be measured—like genetic change although not to the same degree—on a scale of geological time. And that suggests that the likelihood must be rated as nil that any feasible kind of cultural change can succeed in reversing the trend toward diminishing altruism in general. Surely the environmental conditions that supported the human niche during the pleistocene are gone, long gone, even if it were otherwise possible to cut the species population down to a fraction of 1 percent of its present size, and even if both the industrial and the agricultural ways of life could somehow be foregone. Obviously, worldwide (i.e., nuclear) warfare might produce such a consequence, but it is even likelier that its result would be straightforward and direct extinction for the species (cf. ch. 12, below). This is a pessimistic conclusion, but if it is correct, it implies that much greater emphasis should be placed upon public policies consciously designed, culturally of course, to attempt to rebuild human altruistic behavior at least on a selective basis. Conceivably, an immensely greater self-conscious effort, if that could be organized on something approaching a global basis, might bring about some short-run increases in altruism. There is an even smaller chance that once such improvements in social altruism were established ("culturally fixed") in some critical mass of populations, then Wright's process of sampling drift would begin to operate through inter-deme selection because the higher-altruism groups prove to be better competitors than the ones of lower altruism. It is pleasing to contemplate such an ironic turning of the tables on inclusive fitness.

10
Cultural

MENTIFICTIONAL RULES

It is disheartening to find so much that is wrong in the statement of a thesis that is so fundamentally right. Lumsden and Wilson (L&W) certainly merit praise for having argued so forcefully the proposition that the relationship between human genes and human culture is both dynamic and recursive; and more particularly, for their emphasis upon the brain as the bridge between genes and culture, for our species. My concern with this aspect of L&W's model and presentation is not that they have gone too far, but rather that they are nowhere near radical enough. Their characteristically linear mode of thinking is exemplified by their suggested "three-step improvement" upon simplistic gene-culture interactionalism: "from genes to epigenesis, from epigenesis to individual behavior, and from individual behavior to culture" (Lumsden and Wilson, 1981: 343). A far better and also much more realistic approach would posit transactional relationships among physiological systems (including the cellular level at which gene-cytoplasmic effects are

---

"Mentifictional Rules" was originally published as "Epigenesis: The Newer Synthesis," *Behavioral and Brain Sciences*, 5: 24–25 (March 1982), © 1982, reprinted with permission of Cambridge University Press; "Fireflies and Foxfire" was originally published as "Promethean Fireflies and Foxfire: Reflections on the Permutation of Coevolutionary Theory," *Politics and the Life Sciences*, 2: 219–23 (Feb. 1984), reprinted with permission of the publisher; and "Rationalism and Reality" is a revision of my comment "Rationalism and Reality," which was published in the *Journal of Social and Biological Structures*, 10: 277–81 (1987), reprinted with permission of the publisher.

manifested in amino acid production), neurological systems (including the brain), and the components of human culture (as both internalized by the brain and externalized by the brain in its effects upon human physiology).

Throughout the book, however, the authors' reach exceeds their grasp, as in the first sentence (p. ix), where they preempt as their own "the first attempt to trace development all the way from genes through the mind to culture," when they claim (p. 230) that their chapter 5 presents the "first concrete[1] model of the coupling between genetic and cultural evolution," and that (p. 256) "gene-culture theory . . . goes far beyond conventional notions." But such huffing claims constitute a less than generous admission of these authors' indebtedness to the prior insights,[2] experimental research, and mathematical formulations of the late Conrad Hal Waddington who, not fifty generations but a full generation ago, formulated the key concepts of epigenesis, developmental canalization, and genetic assimilation upon which L&W's thesis depends.

The major arguments in the early chapters of the book exemplify culture as archaeological artifacts, whereas the later chapters, in which more sophisticated versions of human culture are sometimes mentioned, deal with culture in contemporary terms on a merely verbal level—and the latter is an activity that social scientists and humanists probably can do just as well as physicists and entomologists. Frequent mention, for example, is made of the relatively specific epigenetic "rule" against incest, which is said to be illustrated by the culturally established brother and sister relationships experienced by children brought up in kibbutzim, who manifest a striking aversion for heterosexual intercourse or marriage among each other. This is said to be the consequence of "an automatic sexual inhibition between persons who lived intimately together . . . to the age of six" (p. 86). There certainly is cultural (i.e., literary) evidence to the contrary, involving dizygotic heterosexual twins (Mann 1938); but given Wilson's longstanding interest and expertise in haploid breeding systems, which were also the subject of Hamilton's (1964a; 1964b) primogenitive articles on inclusive fitness, the preoccupation of this book with brother-sister incest (e.g., pp. 147–58) seems odd. The evidence from human culture, contemporary as well as historical, suggests a much higher probability of parent-offspring incest (and especially, of course, father-daughter), where the coefficient of relatedness is just as high as between full siblings.[3] This implies just as high a risk of negative genetic consequences. One would therefore expect that it should provide an example, on both scores, of greater interest and utility to these authors than the alternative that they proffer.

[1]Describing the human brain and nervous system as "concrete" is a less than felicitous way to suggest its most important and (in the context of the quotation) transcendent function in human development and evolution.

[2]Waddington himself would not necessarily have taken an Olympian view of Wilson's cultural assimilation; (see Waddington 1975, p. viii and ch. 10).

[3]"A single generation of inbreeding between parent and offspring or between full sibs produces an F [Wright's coefficient of inbreeding] of 1/4" (Lerner, 1968: 263).

There are some patent asymmetries, notwithstanding L&W's emphasis upon rules, in their application of them to genetic, as distinguished from cultural, data. Binary classification, for example, is quite rational—one is tempted to say "eu-functional"—for the authors themselves to use when they "wish to emphasize the power and flexibility of two-'culturgen' models, whose utility parallels the two-allele case in theoretical population genetics" (p. 274; cf. the cell-color dichotomies discussed at p. 48); but when social scientists do something equivalent, then "there is a nonrational proneness to use two-part classifications in treating socially important arrays, such as in-group versus out-group . . . and so forth" (p. 95).

Clarence Day (1920) provided some intriguing speculations about what humans would be like behaviorally if they had evolved as superfelids instead of as super-primates (cf. Marais 1969); and in their scenario for what genetic determinism really would imply for humans, L&W (p. 331) describe a society with which at least a specialist in the behavior of social insects ought to feel right at home (cf. Marais, 1937). Such a robot civilization is portrayed as archetypical in contrast to the tabula rasa notion of human consciousness which L&W attribute to most social scientists; and this leaves their own "gene-culture" interaction theory to occupy all of the middle ground between the "pure genetic" and the "pure cultural" approaches (p. 99). In practice, however, virtually all of their mathematical modeling (as distinguished from graphic displays of more qualitatively based ideas) clusters around the genetic pole of their continuum, where they examine binary choice between cultural units, each of which is assumed to be monotonically controlled by a single and corresponding monogene. And the further they get from perception, the more frequently (and vociferously) they justify their inability to say anything explicit about genetic effects upon learned behavior as resulting from the underdevelopment of population or behavior genetics in relation to other fields such as developmental neurobiology or developmental psychology. But the latter complaints also cut the ground from under their sometimes guarded (e.g., pp. 300–301) but usually hyperbolic (pp. 295–96) claims on behalf of their so-called thousand-year rule: that human culture can have positive feedback upon its own genetic and epigenetic bases within as few as fifty human generations. The models from which this deduction is made are built in terms of the (for humans) tremendously oversimplified assumptions of dichotomous choice, and large and randomly mating populations "in order to exploit the deterministic equations of population genetics" (p. 266). Nothing in the models takes into account the empirical parameters mentioned elsewhere (pp. 197–200) of 100,000 human genes, mostly acting epistatically with both polygenic and pleiotropic effects, of which only 1,200 have been identified and barely 210 have been mapped. Given those parameters and their absence, in any case, from the gene-culture coevolutionary model, it is a difficult feat even in imagination to contemplate how the "thousand-year rule" can be tested with data on humans, during any future in which any of the latter are likely to be reading the L&W book.

The book includes its fair share of vacuous remarks, beginning quite early on with the invocation (p. 6) of "sheer drive" as the explanation why chimpanzees, who lack it, cannot conceptualize language like humans. The authors subsequently speak (p. 330) of the genus *Homo* as having overcome "the resistance to advanced cognitive evolution by the cosmic good fortune of being in the right place at the right time." But this kind of poetic remark could be applied equally to dinosaurs, army ants, sharks, or indeed to any other, temporarily well adapted species; certainly the remark explains nothing, or at least nothing new. Soon thereafter readers are reassured (p. 345) that "the inseverable linkage between genes and culture does not also chain mankind to an animal level." Of course, the authors mean to imply "of mind"; but their statement remains a fatuous one for a biologist to make. Similarly, at the very beginning of the book (p. 2) and throughout it, the authors speak of "the epigenetic rules feeding on" this and that; in this initial instance (and also at pp. 272 and 349) it is on "information derived from culture and physical environment." One suspects that in writing this, the authors have not taken the time to think through the implications of what some sensitive readers might consider to be a disgusting alimentary metaphor of information processing, the logical output of which is, of course, a great deal of excrement in one form or another.

An unhappy feature of this book is found in the authors' penchant for gobbledygook: pretentious and dysfunctional neologisms. We can briefly note "heterarchy" (p. 108) and "social contagion" (p. 113) with all of its pathological undertones, notwithstanding the ready availability of such accepted ethological concepts as "mood convection" which offers the additional advantage of already having been introduced into the research literature of political and social science (e.g., Masters 1976: 225–26). L&W's designation of perception as "primary epigenetic rules," and of learning as "secondary epigenetic rules" (p. 36), is a more glaring example of taking feckless (if not reckless) liberties with the English language. Perception is a well-established and active field in physiological psychology; and so are the parallel fields of learning (in psychology), socialization (in sociology and political science), and acculturation (in social and cultural anthropology). Calling perception "primary epigenetic rules" and learning "secondary epigenetic rules" certainly makes no contribution to present empirical knowledge. These proposals are likewise of most dubious value as theoretical contributions, particularly in the use of the concept "rules" in relation to the effect of rules upon human behavior; and this is a matter about which political scientists, like lawyers, can be presumed to have some specialized professional knowledge (including rules applied to the behavior of research scientists; Schubert 1967a; 1967b; 1975). L&W's repeated emphasis upon epigenetic rules (in lieu of the Waddingtonian concept of epigenetic *processes*) would, if taken seriously, be likely to encourage a giant step backward toward premature closure, especially if the actual network relationships in the brain turn out to be more open, more variable, and more rapidly changing

(Davidson and Davidson 1980; Geist 1978; Pribram 1979, 1980; Roederer 1978) than the circumstances defined or associated with "rule-ordered" modes of canalizing behavior.

L&W's misuse of "reification" constitutes another example of semantic obfuscation. For reasons that they nowhere divulge, they decided to call conceptualization reification. Social scientists, including psychologists, use "reification" pretty much as it is defined in *Webster's New International Dictionary* (2d ed.) to mean hypostatization, or the regarding of an abstraction or mental construction *as though it were* a material thing, discrete and objective. L&W chose—in a patently arbitrary manner—to describe as reification "the operations of the human mind [that] incorporate (1) the production of concepts and (2) the continuously shifting reclassification of the world" (p. 5), a usage in which they persist throughout the book, with unfortunate consequences for their communication with literate readers. They refer to a metaphor in which "the archipelago-society can be treated as one insular example; . . . as though it were a single space open for occupation by competing culturgens" (p. 306). Now *that* is reification in the usual sense (although L&W do not, of course, call it that); indeed, the authors instead reveal that "language is the means whereby the culturgens are labeled" (p. 253). Another good example of reification in the conventional sense is found in their explanation of the especially powerful role that mentifacts play in culturgen packing. Mentifacts "are the nearly pure creations of the mind, the reveries, fictions, and myths that have little connection with reality but take on a vigorous life of their own and can be transmitted from one generation to the next" (p. 316). There, as Humpty Dumpty said, is glory for you.

L&W's understanding of Marxism (pp. 354–56) appears to be informed exclusively in terms of Western culture, but there are lessons in this regard to be learned from the East as well. Like the late (and until recently great) Mao Tsetung, L&W are enthralled with the idea of making a Great Leap Forward, in the authors' case in cultural understanding of the biological basis for human social behavior; and this leads them to deplore the "balkanization" of the social sciences, which labor under a "crippling state of affairs" (p. 345). Like Mao, they would have been better advised to have taken a different page from the Little Red Book, seeking instead to "Let a Thousand Flowers Bloom."

FIREFLIES AND FOXFIRE

Lumsden and Wilson write like superscientists who describe their purported subject (which is the function of the human brain) as though psychobiology were a blank slate created "xenidrinic" (Lumsden and Wilson, 1983: 56) for them to write upon. It is true that coauthor Lumsden, in a contemporaneous article for a refereed scientific journal (Lumsden, 1983) discusses the same subject with considerably more concern for the use of the relevant research literature of both psychobiology and social science; but the tenor and tone of the second Lumsden and Wilson book (1983) are very different from that of Lumsden's article. The junior author of the book (who is the senior author in real life) has not even by a whit relinquished his self-proclaimed and grandiose ambition to catalyze the cannibalization of the social sciences by what he calls "sociobiology"; however, perhaps due to the forthright rebuffs attracted by his frontal assaults during most of the past decade, he employs here a somewhat more indirect—though by no means subdued—approach. The present book is the fourth in sequence in what is best understood as a tetralogy, all published by the press of the university that employs coauthor Wilson. That tetrad consists of two pairs, each in the pattern of a "scientific" book followed by a "popular" book designed to "explain" the scientific technicalities to a lay audience. Thus, Wilson's *Sociobiology: The New Synthesis* (1975) is to his *On Human Nature* (1978), as L&W's *Genes, Mind, and Culture: The Coevolutionary Process* (1981) is to their *Promethean Fire: Reflections on the Origin of Mind* (1983). So here, in *Promethean Fire*, we deal with the popular version of *Genes, Mind, and Culture*. But, the comparison breaks down in one important respect; the writing of *On Human Nature* appears to have been motivated, in considerable degree, by criticisms that had been made of the concluding chapter of *Sociobiology*, whereas *Promethean Fire* never confronts and actually distorts (as we shall see below) the published professional critique of *Genes, Mind, and Culture*.

Because I have already reviewed *Genes, Mind, and Culture* (Schubert, 1982), I should much prefer to restrict my comments here to whatever new and different Lumsden and Wilson say in *Promethean Fire*; but their disdain generally for the scientific criticism of the former (see Lumsden and Wilson, 1982b; and below) makes that impossible. They begin their response to some two dozen comments upon their own precis (Lumsden and Wilson, 1982a) of *Genes, Mind, and Culture* in the leading journal of current research and theory, *The Behavioral and Brain Sciences*, with the remark, "The reviewers do not identify any systems of explanation that are competitive to the theory of gene-culture coevolution under discussion. Schubert implies that it has been done before, but does not say where or by whom" (Lumsden and Wilson, 1982b: 31). What I said then (Schubert, 1982: 24) has been quoted here on page 157 ("Throughout the book . . . L&W's thesis depends").

What I meant then, and believe now, is that in *Genes, Mind, and Culture*, the authors proffer a highly speculative, nonempirical, mathematical exegesis upon a theory of gene-culture coevolution—epigenesis—that virtually all living evolutionary theorists today deem to represent Waddington's major contribution, as represented in such books of his as *The Strategy of the Genes* (1957), *The Ethical Animal* (1960), and *The Evolution of an Evolutionist* (1975; cf. also ch. 14, below). To be blunt, I thought and think that L&W have embellished Waddington's theory, without giving him proper credit, in an insignificant way—unlike such alternative and in my opinion considerably more significant (if heuristic) contributions to epigenetic theory as those of Cavalli-Sforza and Feldman (1981) or Pulliam and Dunford (1980).

Lumsden and Wilson ignore in their response what I pointed out earlier (p. 159, above: "emphasis passim upon epigenetic *rules* . . . Pribram, 1979, 1980; Roederer, 1978"). Yet it is true that L&W at least acknowledge Waddington's existence by citing *Genes, Mind, and Culture*, and they even concede awareness of "the imagery invoked by Waddington (1957) and other biologists *in their original conception* of the epigenetic field" (1981, emphasis added). In *Promethean Fire*, however, Waddington is simply not mentioned. Having usurped his concept of epigenesis for their own purposes in *Genes, Mind, and Culture*, they now speak of it throughout as though it were an idea that originated—no doubt by Promethean fox fire—in their own minds. Their ignoring of Waddington leads them to treat epigenesis only metaphorically in *Promethean Fire*, notwithstanding their mathematical gloss—a methodological approach that stands in sharp contradistinction to the use of mathematics to develop biological theory in, for example, the late Robert MacArthur's *Geographical Ecology* (1972; and cf. MacArthur and Wilson, 1967), to say nothing of Waddington's own work.

Promethean Fire has other major faults. Most conspicuously, it is weakest of all in its apparent awareness and use of the scientific literature in which, given its purported subject, it ought to be *strongest*—that of psychobiology and neurobiology. This is exemplified by the authors' failure to discuss laterality and sex brain differences in the evolution of genes in relation to culture. In addition, we can consider their denigratory treatment of the role of emotion in conscious as well as unconscious thinking. They speak (p. 3) of the "less focused flurries of cell activity" that "do not, for the moment at least, enter its [the mind's] mainstream." Such a simplistic and primitive statement of the affective interface with effective thinking cries out for some indication of awareness of the research literature exemplified by the contributions of MacLean (1973), Edelman and Mountcastle (1978), Kent (1981), Gray (1982a, 1982b), and Panksepp (1982). Both Sagan (1977) and Geist (1978) are among the many recent works that do a very much better job than L&W of relating the discussion of gene-culture coevolution to its substrates in the biology of the human brain. Equally conspicuous is their failure, notwithstanding their mathematical pretensions, to make any serious attempt to link up their concerns with "epigenetic rules" to the extraordinary, dynamic development throughout the past generation of the use and study of artificial intelligence (Dreyfus, 1972)—a mode of cultural evolution that manifestly they ought to integrate into their (as they present them) idiosyncratic notions about bootstrapping the pace of genetic evolution through cultural feedback, as in their so-called thousand-year rule (and cf. Schubert, 1983a, for a more extended discussion of the importance of artificial intelligence to future epigenetic phylogenesis of the human mind).

Of lesser importance but equally conspicuous is L&W's penchant for dealing romantically with contemporary hunting and gathering peoples and their continuing preoccupation with the notion that cultural rules against sibling incest provide the proof par excellence for their sociobiological pudding. *Promethean Fire* (e.g., at pp. 132–36), like *Genes, Mind, and Culture*, is replete with this particular theme; and characteristically, the authors ignore in *Promethean Fire* what was brought to their attention (Schubert, 1982: 24) about *Genes, Mind, and Culture* (p. 157, above: "The evidence from human culture. . .the alternative that they proffer").

As a resident of Hawaii, I can point out to them that their expectation (pp. 64–65) that "even if a society could somehow begin anew with brother-sister incest as the norm, it would probably develop a cultural antagonism toward the practice in a generation or two" certainly is not true of many Polynesian peoples into very recent history, where brother-sister incest was the basis for the reproduction of the *alii* (ruling class). For L&W, sibling incest is an ethereal red herring, which they attempt to use as a negative example of the proof of the rectitude of hardcore sociobiological theory (as distinguished from sociobiological theory: see ch. 9, above) by aligning themselves on this issue on the side of the angels (p. 176). This gets rather ridiculous, because of the essential irrelevance of sociobiological theory (whether as kin selection or otherwise) to their nominal subject in *Promethean Fire*. Even so, one would think that authors who proclaim so vociferously their desire to co-opt the social sciences would make more of an effort to become better aware of the social scientific response to their version of sociobiological theory (e.g., White, 1981b; Baldwin and Baldwin, 1981; and cf. *Promethean Fire*, p. 192, n. 44).

Lumsden and Wilson's handling of the political dimension of the response to their work is odd. They are undisturbed by the scientific critique of *Genes, Mind, and Culture*, mentioning quite blandly (p. 206, n. 121) that "critiques and defenses of the culturgen concept are given by many of the twenty-three authors who review [*Genes, Mind, and Culture*]." This certainly puts the scientific response to the book in the most favorable possible light from the authors' point of view. By my own reading and count of the peer commentary to which they refer, the score is: five defenses (by Fagen, Masters, Shepfer, van den Berghe, and Williams), two neutral (Charlesworth and van Gulick), and sixteen negatively critical (Barash, Caplan, Ghiselin, Gruber, Hallpike, Hartl, Johnson, Kovack, Loftus, Markl, Maynard Smith, Plutchik, Rosenberg, Schubert, Slobodkin, and Wholwill). By my reckoning, the overall scientific response, judged by the evidence that L&W themselves invoke is about 80 percent *non*-supportive of their culturgen approach. But they are considerably more exercised about the critique of human sociobiology that they characterize as that of "the American radical left," including a letter signed by Ruth Hubbard and others of Wilson's Harvard biology colleagues, which was published in *The New York Review of Books* (Beckwith, 1975). Evidently L&W perceived that popular response to constitute a much greater *political* threat than the views of scientific colleagues published in a scientific journal of much smaller and more specialized circulation. Given such political sensitivity on their (or at least, on Wilson's) part, it is strange that they failed to recognize any political scientists—and there were in fact two, both clearly identified as such—among the commentators in *Behavioral and Brain Sciences*, a group that L&W describe as including "geneticists, psychologists, anthropologists, sociologists, and philosophers" (p. 191, n. 38).

Promethean Fire contains many errors en passant, of both fact and interpretation, regarding which I am going to be highly selective, noting only a couple from the first few pages, plus one other. In view of their announcement (p. vi) that "we have been careful to distinguish fact and true theory from mere speculation," it seems fair to take a close look at page 2. It is claimed there that "the modern synthesis of evolutionary theory, which joined genetics to the remainder of modern biology during roughly the half century from 1930 to 1980, was not stretched to include psychology or any significant part of the social sciences." That claim is an unmitigated misstatement of fact, certainly as concerns not merely psychology but also anthropology and political science (see Hinde, 1970; Greenwood and Stini, 1977; and Somit, 1976).

A few pages later (p. 7) they wonder: "Armed with scientific insight we can ask again with rising hopes: what is humankind, what created us, and what is our purpose in the world?" Armed with scientific insight into the premises of evolutionary geneticists, such as Jacques Monod in his *Chance and Necessity* (1971), one would never raise such a transcendental question. To Monod, evidently as distinguished from L&W, chance meant chance—dumb luck—and not necessity, theological or otherwise (cf. Day, 1920).

Not least, they assert (p. 83) that "the human genes prescribe the epigenetic rules, which channel behavior toward the characteristic human forms of thought." This statement is a contradiction in terms and no doubt reflects the authors' overenthusiastic habit of employing epigenesis in a metaphorical rather than a mathematical sense. Genes prescribe genetic, not epigenetic, rules; epigenetic rules are the consequence of interactions between genes and environments, including cultural environments (Waddington, 1957).

There remains, however, at least one point on which I am in unqualified agreement with L&W, that "the brain . . . is the ultimate object of scientific inquiry" (p. 167). Irrespective of their speculations about the origins of the human "mind," the authors are on sound ground in emphasizing the basic importance of placing the human brain at the center of our agenda for future research, in the social and the biological sciences alike.

RATIONALISM AND REALITY

In this paper (Lumsden and Wilson, 1985) restate an argument that they previously have published in a pair of books and two articles, as augmented by five articles (including one in this journal) authored by the nominal senior author of this iteration. My comment here will focus on four questions. The first of these reflects a relatively novel (for L&W) emphasis in their present paper: this I discuss in the section on "Reality" immediately below. The other questions are not novel at all, constituting recurrent issues raised by L&W, regarding which they have demonstrated thus far no ability to learn from previous critique. These are discussed in the sections on "Emotion," "Epigenesis," and "Evidence."

Reality

Not only do "fruit" and "fish" not exist in the "real" world (L&W 1985: 344, par. 1); neither do the "objects" that these authors hypostatize (cf. Peterson, 1983a). What L&W fail to comprehend, or at least to accept, is that "reality" is *created* by precedent brain events such as perception and cognition and that language is, for humans, an integral tool for the *making* of cognitive reality—which for phylogenetic reasons is different from reality, in confrontation with the same (otherwise) environmental context, for insects, birds, dogs, and cats. L&W postulate an objective, external world that exists independently of eyes to see, ears to hear, fingers to touch, and a brain to process neural and hormonal information; and notwithstanding all of their talk about the importance of evolutionary theory, their *Weltanschauung*—to speak only in terms of Western culture—is antecedent to both Darwin and Bohr and therefore innocent of the observer paradox (cf. Schubert, 1983b: 105–10).

L&W claim that if the human brain could operate in a more holistic and less linear mode, then "the world would be perceived and classified more exactly *as it is*, true to the *real* continua and boundaries of nature. . ." (p. 346: par. 1; emphasis added). They speak of the "*objective* existence" of "an *actual* nucleotide sequence" (348: par. 1; emphasis added); but then they proceed to repeat the orthodoxy of contemporary genetics theory, which hypothesizes "a continuous genetic text lacking obvious breaks or pauses" (*sic*: definitionally!). Nevertheless, two paragraphs later they refer to "genes as generative *units*" (emphasis added) that are operated on by "epigenetic rules"; and this leads them directly to "the question of greatest immediate interest [which is] the nature of the generative unit, . . . the culturgen" (see Slobodkin, 1982). That conclusion is clearly a non sequitur and, one would think, obviously so for authors who purport to be as dedicated to logic and rationality as L&W; if genes are, as they premise, continuous nucleotide sequences, then neither genes nor their (many-times-removed) cultural byproducts are best conceptualized as *any* kind of empirical, objective unit. In the context of this article, the authors' reference en passant (p. 347: par. 4) to "the *actual* mental processes transmitted from one brain to the next" (emphasis added) is a statement that must represent either a Freudian slip of transcendentalism (cf. Schubert, 1984c: 222, col. 1, par. 3) or else a subconscious affirmation of ESP.

As Graubard (1985: 113) remarked, a human "is a culture- building animal because culture is the product of creativity, a unique characteristic of [humans]." And L&W's topic sentence for "Innate preferences" once did speak of "people . . . *creating* memory"; but in an apparent change of mind, the author correcting the galleys substituted "experiencing" for "creating"—which I think was most unfortunate, because they had it right the first time. (Maybe L&W became lulled by an overdose of reading experimental psychology, at the cost, as I argue below, of not reading virtually any psychobiology.) The human brain is an active and positive source of reality; and substituting "experiencing" for "creating" changes the whole enterprise to a passive, object-centered, negative portrayal of human thought. Thinking people are *subjects*, not objects whose only fate is to experience the slings and arrows of outrageous fortune. As the Bard elsewhere put the same point, the fault lies not in their stars.

Emotion

L&W almost completely ignore the importance of emotion in relation to any kind of human cognition *or* perception (as several peer commentators pointed out to them four years ago: e.g., Plutchik, 1980, 1982); the result is an extraordinarily sterile model of the human brain. L&W also leave it entirely up to culture to supply the holistic grist for their epigenetic rule mills (i.e., *if* the human brain could operate in a more holistic and less linear mode; see above) with the consequence that their preoccupation is with what amounts to a strictly left-hemisphere model of the brain of a right-handed human male: serial, linear, and rational to the max (cf. Morgan and Corballis, 1978; and Bradshaw and Nettleton, 1981). They postulate a human brain that is an *ef*fective machine, untroubled by and innocent of the *af*fective transactions that characterize (and for the best of evolutionary reasons) mammalian thinking as a process (MacLean, 1973, Edelman and Mountcastle, 1978; Pribram, 1979, 1980a; Kent, 1981; Gray, 1982a, 1982b, Panksepp, 1982).

One apparent reason for their difficulty is their decision, in venturing to explore some previously unfamiliar (to them) interdisciplinary boundaries, to settle on (and for) cognitive psychology instead of psychobiology and neurobiology. But this is a problem that I tried (Schubert, 1984c: 222) to bring to their attention two years earlier.

No one familiar with psycho/neurobiology could imagine a discussion of thinking and memory that ignores emotion, which probably is *the* most important endogenous factor affecting either the storage or retrieval of information in the human brain.

Epigenesis

This is a concept that L&W continue to abuse; as Charlesworth (1982: 9) states, their "view of epigenesis is fundamentally one of genetic determinism" (cf. Markl,

1982); in regard to L&W's simplistic view of genetics, see Hartl (1982); page 159, above ("epigenetic *rules* . . . epigenetic *processes*"); page 164, above ("the human genes prescribe . . . genetic, not epigenetic, rules"); and page 161, above ("L&W have embellished Waddington's theory . . . in an insignificant way"). Waddington's theory of epigenesis is reviewed in detail in chapter 14 (below), where its importance to contemporary studies of human development also is discussed.

Evidence

There are at least two major problems concerning the use of scientific evidence by L&W. First they ignore the most obviously relevant sources of research bearing upon what they claim to be central to their theory. I am speaking here about research in *biology*, the disciplinary orientation in which these authors are presumably (and certainly avowedly, in Wilson's case) most expert; I am *not* talking about their use of relevant research in social science and/or the humanities, upon which I take no position and make no comment, for present purposes. Second they pay no attention to, and make no use of, what has by now become a substantial previous critique of the ideas that they present here; the *only* critics whom they cite and make any apparent use of are supportive and friendly critics.

If L&W believe (356: par. 2) that "the uniqueness of the human mind is due to the specific cognitive mechanisms that . . . are genetically based, that is, they are grounded in programs of neural development," then it is not unreasonable to expect that they will evince some minimal familiarity with the *technical* research literature in psychobiology and neurobiology (e.g., Ebbesson, 1984). *The Behavioral and Brain Sciences* is a journal that even a social (political) scientist such as I have read regularly since it began publication almost a decade ago; but L&W's only citation to it is to their own precis of their first book (L&W, 1982); nor do they cite anything in any other brain science journal.

These authors have proven to be, and continue to appear, impervious to unfavorable criticism of their ideas. Their second book, *Promethean Fire*, never confronts and actually distorts (as the discussion that followed demonstrated) the published professional critique of *Genes, Mind, and Culture*" (Schubert, 1984c: 220).

A very straightforward and explicit example is provided by their claim (L&W, 355: par. 3) that "all societies have imposed a taboo against brother-sister incest. Only a very few, such as the ancient Egyptians and Bunyoro of Uganda, permitted this form of pairing." In regard to their first book, in 1982, I observed (p. 157, above), with what I thought then and think now was admirable restraint, that parent-offspring incest is in fact common in many human cultures, past and present; and it presents a much higher risk of negative phenotypic consequences empirically than does the full-sibling incest that obsesses L&W so unduly. L&W ignored that. So in 1984 I endeavored to enlighten them (p. 162, above) about the entrenched reliance upon full-sibling incest for the reproduction of the ruling class in aboriginal Polynesian societies, contradicting L&W's expectation that such a practice would "probably develop a cultural antagonism toward the practice in a generation or two. And it isn't just I whom they ignore; as noted earlier (p. 163, above), 80 percent of the twenty-three peer commentators on their target article for *BBS* were negatively critical of L&W's culturgen concept and theory of epigenesis, but L&W are much less disturbed by the scientific than by the *political* response to their work. They castigate Wilson's Harvard colleagues, evolutionary theorist Richard Lewontin and feminist biologist Ruth Hubbard, as radical leftists—which can hardly be stretched to subsume Sir Peter Medawar, the author (1981) of a highly publicized review of their first book. Evidently they don't bother to read disapproving reviews of their work. That would explain why they seem to learn so little from their critics.

11
Eugenic

At first glance, it might seem manifestly appropriate that biology should inform public policy concerning bioethics, in any serious discussion of the subject of eugenics as applied to humans. In that regard, we would be remiss not to recall that for Darwin, his own studies in the domestication of animals were at least as important as were Galapagos finches in his thinking and formulation of the theory of evolution, and not least as it applies to the descent of our own subspecies (Darwin, 1871).

Human domestication of other animals has been extensively studied during the century and more since Darwin, as exemplified by the encyclopedic compendia of Zeuner (1963) . . . and Ucko and Dimbleby (1969) and by Paul Shepard's (1978) provocative discussion of how animals think (cf. Griffin, 1984) in relation to how thinking about animals remains critical to the human species adaptation. Indeed, it has been studied throughout at least the past hundred thousand years for which artifactual evidence has been found—and probably much longer in the minds and lives of our hominid forebears (for a current summary and commentary on domestication, see ch. 8, above).

We certainly should be, and no doubt are, dizzy with success in contemplation of the great strides and vast progress that we have made—and in the rather miniscule geologic moment of barely ten thousand years—in transforming such wild products of blind natural selection (Monod, 1971) as jungle fowl, peccaries, and leopards (which I select almost at random from among the hundreds of examples

---

Originally published as "On the Domestication of Eagles: Designer Genes as Kentucky-Fried Dysgenics," *Politics and the Life Sciences*, 3: 155–56 (Feb. 1985). Reprinted with permission of the publisher.

available) into our artificially selected and preferred factory chickens, hogs, and pussy cats. Given such dramatic progress in the application of the science of eugenics to hundreds of other species of animals than ourselves (and let us not forget, thousands of species of plants also), it is mind-boggling to consider what we are likely to produce in the perfection of ourselves, once we begin to apply to *Homo sapiens sapiens* (and, as one is tempted to say, with a vengeance) the accumulated wisdom of our experience in domesticating other life forms—an experience that stretches over ten times the duration of Lumsden and Wilson's "thousand-year rule" (cf. the reviews of their *Promethean Fire* (1984), *Politics and the Life Sciences*, 2: 213–24).

Candor compels the admission that, even among evolutionary biologists, substantial disagreement obtains concerning the extent to which contemporary humans already constitute a partially domesticated species (see the discussion of "Domestication" in ch. 9, above). The thrust of the argument of the proponents of the self-domestication view is that an entailed consequence of our forebears' shift from the gatherer-hunter to pastoral and agricultural adaptation(s) was the epigenetic effect that, in domesticating other plants and animals, we—however inadvertently and unwittingly—could not avoid beginning to domesticate ourselves also. From that perspective, of course, human eugenics constitute only the planned, rational escalation of a developmental trend that already is well underway.

Nevertheless, from a biological perspective, such optimistic scenarios for further progress in human development must be tempered with at least passing mention of the traditional and of course conservative credos of the leading theorists of modern evolutionary biology (e.g., Williams, 1966; Lewontin, 1974; and Gould, 1977a). These paragons of the dominant paradigm in contemporary biological science stand fast in their faith in unfettered natural selection as the best way to preserve species gene pools sufficiently heterogeneous to optimize genetic protection against the constraints of the no doubt vastly changed environments that surviving humans will confront in even proximate decades (cf. Yanarella, 1984)—as compared to the dysgenic alternative of putting all of our (chicken) eggs in one basket (see Nitecki, 1984; Martin and Klein, 1984; Dror, 1984; and ch. 12, below). But it is *we*, after all, who are political scientists and the experts in public policy making (cf. Lasswell, 1963); what do evolutionary biologists know, anyhow, about how to balance the national budget?

12
Catastrophic

EVOLUTIONARY CHANGE

Gradualism

The synthetic theory of evolution is preeminently one of gradualism: that genetic change is normal but that it takes place at a glacial pace. As Eldredge and Tattersall (1982: 37, 39; Eldredge and Gould, 1972: 94–96; and Eldredge, 1985a: ch. 4, 1985b: 196–201) have summarized: (see also Fisher, 1930; Haldane, 1932; Dobzhansky, 1937; Huxley, 1942; Simpson, 1944, 1953):

> The prevailing view of the evolutionary process is called the "synthesis" because it integrated, in the 1930s and 1940s, the seemingly disparate data of genetics, systematics. . ., and paleontology into a single, coherent theory. . . . Then population genetics . . . developed a mathematical theory showing how frequencies of alternative forms of genes (alleles) could change within populations through time, given certain mutation rates and intensities of natural selection . . . [Thus] natural selection, working on a groundmass of genetic variation, changes gene frequencies each generation. Mutations are the ultimate source of the variation, but it is selection, working to perfect adaptations or to keep a population in step with changing times, that is the real agent of generation-by-generation change. Over long periods of time—thousands, millions of years—such minute step-by-step change will have large effects. So large, in effect, that all the evolutionary patterns seen by systematists working on living plants and animals, as well as the fossils seen by paleontologists, are nothing more than the results of these small-scale processes summed up over geologic time.

This chapter is abridged from "Catastrophe Theory, Evolutionary Extinction, and Revolutionary Politics," *Journal of Social and Biological Structures*, 12 (1989). Reprinted with permission of the publisher.

Nonconformism

There are three principal components of a new evolutionary paradigm—which may already have displaced gradualism by the time these remarks get into print. All three had their beginnings fifty years or so ago, when neo-Darwinism really *was* new and just gaining acceptance as the dominant evolutionary paradigm. The first of these is Sewall Wright's (1940) small population and genetic drift model (Stanley, 1979: 24, 165–68). The second is the Waddington/Geist model of epigenetic/dispersal change. And the third is punctuationism/catastrophism.

I have recently discussed the evolutionary ideas of Wright (1968, 1977, 1978, in Schubert, 1981d: 202, 208–10, and ch. 9, above); of Geist (1978, in Schubert, 1981c, and ch. 13, below); and of Waddington (1957, 1960, 1975, in Schubert, 1985b, and ch. 14, below); but I have not previously discussed those of punctuationism/catastrophism. Therefore I deal here only briefly with the first three, but at greater length with the latter.

Wright suggested that genetic change takes place very rapidly and with dramatic effects in colonizing and other small populations detached from their main species habitats. He also proposed a theory of adaptive landscapes and genetic peaks that supported an interpretation of group rather than individual selection in human populations. Waddington emphasized the importance of behavioral response to environmental stress as a device for reorganizing the *expressive effects* of genes, which in turn bring positive feedback to bear upon possible behavioral response to further and frequently different environmental stress. Similarly, phenotypic structures vary in development following the transactional interplay between behavior and environment. Waddington emphasized that *populations* of organisms evolve; and Geist, agreeing with Waddington, added that the population selection Waddington heralded occurs only under conditions of resource abundance, and therefore for small colonizing populations of what Geist calls the "dispersal phenotype" that such animals exhibit. "Therefore," he says, "new adaptations, diagnostic features, which differentiate the colonizing from the parental population, can arise only during dispersal" (Geist, 1978: 122).

Punctuationism

Punctuationism begins with the work of Ernst Mayr (1954), as expanded and didactically stated by Eldredge and Gould (1972; for discussion see Stanley, 1981: 77–78), who asserted that new species arise by the splitting of lineages; new species develop rapidly; a small sub-population of the ancestral form gives rise to the new species; and the new species originates in a very small part of the ancestral species' geographic extent—in an isolated area of the periphery of the range.

Now, fifteen years later, many additional contributions to punctuationism are available, including Stanley (1979, 1981, 1986), Eldredge (1985a), Gould (1984), and Green (1986: 163–83), and more briefly Goldsmith (1985: 148–49), Gould

(1985: 242), Elliott (1986: 38), and with particular application to hominid evolution, Eldredge and Tattersall (1982) and Stanley (1986).

Stanley (1979: 183; cf. Eldredge and Tattersall, 1982: 64–65) contrasts *micro*-evolution (infra species) with *macro*evolution (which takes place between species) as follows (for his extended discussion of how and why macroevolution is different, see pp. 63, 99–100, 141–42, 211–12; and Green, 1986: 168):

Macro	*Micro*
1. phylogenetic drift	genetic drift
2. directed speciation	mutation pressure
3. species selection	natural selection

Stanley remarks (1979: 209) that "evidence for the punctuational model becomes a double-edged sword, undercutting two of [gradualism's] premises. In the first place, it presents a strong empirical case that natural selection within established species is somehow stifled, forcing us to look beyond phyletic evolution to account for large-scale transition. In the second place, it shows that quantum speciation [see Stanley, 1979: 179; Eldredge, 1985b: 79–80; and Goldsmith, 1985: 148–49] is a real phenomenon and a source of great variability, thus opening the way, on a theoretical plane, for rapid macroevolution via species selection." Furthermore (Stanley, 1981: 198), "genetic recombination . . . is crucial to the process of quantum speciation: new body plans and behavior patterns must develop rapidly, with genetic rearrangements, and repairings playing an important role . . . divergent speciation would be difficult without genetic recombination."

Stanley emphasizes (1981: 181) that "in the punctuational scheme speciation is seen as the focus of evolutionary change. . ., [although] at any time, the direction of the next event of speciation will be heavily dependent upon unpredictable [environmental] and genetic accidents." However (Stanley, 1981: 94), "no single species ever becomes very highly diversified . . . diversification proceeds by [adaptive radiation] from already established species [in] the punctuational scheme of evolution." But (Stanley, 1979: 221) "sex can have a profound effect only within small interbreeding populations of the sort that are involved in speciations." And not least (Stanley, 1981: 5), "the punctuational view implies . . . that evolution is often ineffective at perfecting the adaptation of animals and plants: there is no real ecological balance of nature; that most large-scale evolutionary trends are not produced by the gradual reshaping of established species but are the net result of many rapid steps of evolution, not all of which have moved in the same direction." The negative consequence of this for paleobiology is that most such change "has taken place so rapidly and in such confined geographic areas that it is simply not documented by [the] fossil record" (Stanley, 1981: 5).

All three of the leading protagonists of punctuationism have indicated (Gould, 1985: 198; Stanley, 1979, 1981, 1986; and Eldredge, in Eldredge and Tattersall, 1982, *passim*) their conviction that punctuational evolutionary theory best explains hominid evolution. Stanley, for example, states (1979: 63) that "phyletic evolution in the Hominidae (human family) has apparently been even slower than that for other groups of the Mammalia, implying a punctuational pattern in the ancestry of *Homo sapiens*"; and (1981: 206) "the fossil record shows no evidence of human evolution in Western Europe during the entire forty thousand years of our species existence [cf. E. Morgan, 1984; and Gould, 1985: ch. 12]. . . . [and earlier] *Homo erectus* survived for upwards of a million years without altering." Indeed, Eldredge and Tattersall demonstrate (1982: the front and rear inside cover chart of the last four million years of hominid speciation) and discuss at considerable length (chs. 5 and 6, pp. 67–159) that four hominid species, three of Australopithecines and one of *Homo*, each survived for a million years or more, and each without significant evolutionary change (within-species selection) taking place. These include: A. *afarensis* (4.0–3.1 mya [million years ago]), A. *africanus* (3.0–2.0 mya), A. *boisei* (2.2–1.2 mya), and H. *erectus* (1.6–0.4 mya). *Africanus* was apparently the successor to *Aferensis*, about 3 mya; but almost a million years later when *Boisei* appeared, soon threafter there were two other species, one Australopithecine (*robustus*) and the other *Homo* (*habilis*), all three of which co-existed (although probably not, and certainly not necessarily, in the same places at the same times) in Africa for more than half a million years (from 1.9 to about 1.4 mya).

Both A. *aferensis* and A. *africanus* were gracile (i.e., not heavy set) as were also two of the successor species of the *Homo* genus: *habilis*, and *sapiens*; but both of the other Australopithecines (*boisei* and *robustus*) were robust, and so were all of the other *Homos* (*erectus*, *arachaic sapiens*, and *neanderthal*). The lack of directional progress in evolution is illustrated by the fact that H. *erectus*, a quite robust species, succeeded H. *habilis* and therefore preceded (with the intervention of *archaic sapiens* for half a million years) the reappearance of gracility in modern Homo Sapiens forty thousand years ago.

Eldredge and Tattersall comment (1982: 140–42) that "the period of 2 million to 1 million years ago does provide us with the first absolutely unequivocal evidence that we have the co-existence of at least three separate hominid lineages, and also with a nice example of stasis in the hominid fossil record. For sites all over Africa have yielded evidence not only of gracile hominids but also of robust ones." As an example of quantum speciation, they note that

> *boisei* springs forth fully fledged and hyper-robust at its first appearance in the geological record and continues unchanged as far as one can tell for the million years until its last occurrences. Its South African contemporary, *robustus*, is more conservative than *boisei*, and it is even possible that the latter species was derived from an early [but thus far undiscovered] population of *robustus*. . . . The placement of *Homo erectus* at 1.6 mya was at the beginning of the Pleistocene when the current ice age accelerated with a worldwide climatic impact, and the environmental fragmentation which this must have involved presumably provided plenty of opportunity for the isolation of populations on which speciation so largely depends.

Catastrophism

According to Gould (1985: 242) there are

> two distinct levels of explanation—punctuated equilibrium for normal times, and the *different* [emphasis added] effects produced by separate processes of mass extinction. Whatever accumulates by punctuated equilibrium (or by other process) in normal times can be broken up, dismantled, reset, and dispersed by mass extinction. If punctuated equilibrium upset traditional expectations. . ., mass extinction is far worse. Organisms cannot track or anticipate the environmentalist triggers of mass extinction. No matter how well they adapt to environmental ranges of normal times, they must take their chances in catastrophic moments . . . [mass] extinctions can demolish more than 90% of all species, [so] groups [are being lost] forever by pure dumb luck among a few clinging survivors designed for a [prior] world.

To exemplify his point that catastrophic mass extinction has a key influence upon the history of life on earth, Gould (1985: 408–9) invokes that best studied catastrophe, the Cretaceous, to generalize upon the significance of catastrophism.

> The great world-wide Cretaceous extinction . . . snuffed out about half the species of shallow water marine invertebrates . . . [as well as the dinosaurs that] had ruled terrestrial environments for 100 million years and would probably reign today if they had survived the debacle. Mammals . . . spent their first 100 million years as small creatures inhabiting the nooks and crannies of a dinosaur's world. If the death of dinosaurs had not provided their great opportunity, mammals would still be small and insignificant creatures. We would not be here, and no consciously intelligent life would grace our earth. Evidence gathered since 1980 . . . indicates that the impact of an extraterrestrial body triggered this extinction. What could be more unpredictable or unexpected than comets or asteroids striking the earth literally out of the blue. Yet without such impact, our earth would lack consciously intelligent life. Many great extinctions . . . have set basic patterns in the history of life, imparting an essential randomness to our evolutionary pageant.

POLITICAL CHANGE

Gradual Progress

Evolutionary gradualism is a model for dictatorships, oligarchies, and constitutional democracies alike, thereby cutting across and applying to all of the traditional categories in terms of which political scientists conventionally distinguish among basic types of polities. From a gradualist perspective all are alike because, once established, the focus of political analysis is upon the maintenance of their stability, with slow, moderate, and gradual change postulated as the expected (and desired) norm.

Gradualist theory insists (Fisher, 1930; Hamilton, 1964; Williams, 1966) that natural selection operates on *individual* animals, not on social groups of them (or at least, not on larger groups than closely related "kin"), and most certainly not on entire species, genera, families, or orders. The mechanism postulated by orthodox gradualism is genetic competition between and among animals of the same species; and the aggregation of gene pool changes for a species, due to the differential breeding success of individuals, *constitutes* natural selection. Mutation functions at a minute pace to favor fortuitously occasional individuals while disadvantaging most others.

Gradualism glorifies stasis and justifies it on the basis of genetic competition among individual humans; and the only use for such an ethic as applied to twentieth-century industrial and postindustrial politics (Green, 1986) is ideological, as a purported justification for conservative exploitation of the masses and for a failure to take concerted political action— which requires considerably more synergy than the residue from a host of individual human breeding choices—to avoid or avert "Apocalypse Tomorrow." Surely the chronic problems of niche degradation cannot be resolved by gradualistic solutions. Certainly, too, the questions of nuclear power and weaponry cannot wait ten years, or twenty years, or (like our national debt) for our children and grandchildren to solve, or for the future to heal: they demand revolutionary political change right now, in the immediate present. Biological gradualist evolutionary theory as a model for human politics is a blueprint for the doomsday of the human species, by reinforcing the politics of certain failure to change either enough or in time. It is not surprising that nineteenth-century Darwinian theory (in the guise of Social Darwinism) was rejected as a model for social, economic, and political change, and that the twentieth-century neo-Darwinian modern synthetic theories of evolutionary gradualism and sociobiology have had virtually no attraction for political scientists or use in serious modeling of politics (but for the rare exception, see Kaufman, 1985; and Kort, 1983 and 1986).

Eldredge and Tattersall remark (1982: 161–62) that "the prevailing view among historians . . . remains an expectation of slow, steady, gradual change—in a word, progress. Historians . . . have suffered from the same sort of Victorian myths that have afflicted [gradualist] evolutionary biologists." This contrasts sharply with Eldredge and Tattersall's own impression "that human history over the past 6,000 years . . . reveals a continuation of the same basic pattern [as that of the prehistoric record of both genetic *and* cultural punctuationism] of rapid innovation followed by far longer periods of little or no change." Eldredge and Tattersall (1982: 162) quote sociologist Kenneth Bock (1980) in explicit support of their own stated position; and they note that Bock "goes on to argue even more strenuously against the idea that there is [any] inevitability underlying social change—[either] biologically based or [otherwise]." Bock reports that "far from being unilinear, graded, progressive, and inevitable, social change is a rare phenomenon, the result of specific events. When change happens, it happens quickly. For the most part, history is nonchange unless and until something happens."

The relation between biological and cultural evolution is then discussed by Eldredge and Tattersall (1982: 177–79, 181):

> the traditional response of cultural evolutionists to biologists who, like sociobiologists, seek to reduce cultural evolution to the terms of biological evolution, justifiably emphasizes inheritance. . . . But [the punctuationist] idea that once a way of being is invented it will persist until change is forced on the system is hardly a metatheory of evolution. . . . [Punctuationism] sees species as basic units in evolution: species are the ancestors and descendants of the evolutionary process. The pattern of change in the fossil record strongly suggests that most change in the evolutionary process is related to the origins of new species. . . . That patterns of genetic and cultural differentiation are not coextensive should be enough . . . to deter biological determinists (like sociobiologists) who seek to reduce cultural evolution to biological terms. . . . Sociobiology is an astonishing anachronism.

And they add that "geographic patterns—especially isolation—can foster change and consequent stability in both evolutionary systems. Why the change need be rapid . . . is [perhaps] 'sink or swim' [at least] for biological systems."

Eldredge and Tattersall (1982: 172) attribute the illusion of progress to "misperception of pattern" rather than to "social ideology"; but this is a distinction without a difference. As their editor Schopf said in his introduction to the pilot Eldredge and Gould paper on punctuationism (1972: 83), "the larger and more important lesson" of their work is not punctuationism, but rather their demonstration that "*a priori* theorems often determine the results of 'empirical' studies before the first shred of evidence is collected . . . [the] idea that theory dictates what one sees." About half a century earlier, according to quantum physicist Werner Heisenberg (1977: 5), Einstein had remarked to him that "it is the theory which decides what can be observed." This means that it is always culture that corrupts biology, rather than vice versa (as in the notorious but mythical "biological determinism" associated with the Social Darwinism of yesteryear, or with today's counterpart in feminist fears of biological corruption and contamination of their goal of social, economic, and political equality). The psychobiology of the brain makes it possible for a culture—any culture—to be expressed; but the culture supplies the content of what is thought, and biology provides the process of thinking. Progress and punctuationism are equally cultural; and neither is biological.

Punctuating Political Stasis

Clearly gradualist evolutionary theory has little relationship to international relations among polities, because competition there is at the equivalent biological level of species, not of individual animals; and gradualism denies that species are selected other than through competition among their individual members. From the perspective of philosophy of science, the ascribed posture of political international relations, as "explained" by evolutionary gradualism, is methodological individualism with a vengeance.

Punctuationism rejects the monotheistic individualism that characterizes exclusive reliance on gradualism as the mechanism for change. Punctuationism thus takes a very different stance than gradualism towards progress. As Gould (1985: 444) puts it, "no internal dynamic drives life forward. If environments did not change, evolution might well grind to a virtual halt. . . . [The] primary 'struggles' [of animals] are with changing climates, geologies, and geographies, not with each other. Competition . . . [is] a sporadic and local interaction . . . not . . . [life's] driving force." Stanley (1979: 203) also has questioned gradualism's monomaniacal preoccupation with competition as the universal deus ex machina of natural selection. He makes the point that vulnerability to predation must also be considered. "Disease may have caused the extinction of many species . . . [but] this special form of predation is seldom detectable in the fossil record."

Punctuationism emphasizes species selection as the principal mechanism for speciation and evolutionary genetic change, which are postulated to occur very rapidly in small populations detached from the main species population and in habitats that demand rapid phenotypic adjustment to dynamically accelerating ecological disturbances. New species result from stress, by invoking from among fortuitously available epigenetic combinations of structures and behaviors some that fit the new demands better than the older species adjustment did or could. Some of the new species and their clades are better adapted than are others; those better adapted may flourish while maladapted ones may rapidly become extinct under the stress of drastic ecological change.

If we are to analyze international political change from the comparative perspectives of punctuationism instead of gradualism, then we must clarify the presumptions of indulging in the metaphor that *polities* can be conceptualized as *species*. An animal species is a population of relatively highly genetically interrelated individuals capable of the pairwise sexual reproduction of fecund offspring of the same average degree of relatedness as the other animals constituting the population. A polity is a population of relatively highly politically and culturally interrelated individuals capable of speaking the same maternal language(s) (see Masters, 1970) as the other humans constituting the populations. An animal species must exploit for sustenance a niche to which it becomes and remains successfully adapted, often for millions of years and with annual generations so that with a million successive population cohorts, the millionth will be very much the same as the hundredth. A polity controls the exploitation of a geographically defined space which provides sustenance for the population during a culturally defined series of regimes (political generations). A polity typically persists until its subject population is displaced, absorbed, or extinguished by a different and successor population of humans who define a new polity. For large mammals the genetically defined biological life span of a cohort may be many decades; for humans it is approximately eighty years (Fries and Crapo, 1981). The longevity of regimes usually varies from a few years to several decades; and polities sometimes persist for centuries.

Many consequences follow from the punctuational assumption that polities are equivalent to different species in competition for the same or overlapping resource bases in the same time/spatial environment. Many species predate upon others, which function as vital resources for the predators; and even sympatric species (of the same genus) may relate to each other in environmental settings in such a way that one competitively excludes (i.e., extinguishes) the other. There are of course many political analogues of biological competitive exclusion, as exemplified by the displacement of most of the Moslem population from Palestine when the post-World War II Israeli polity was established.

Punctuationism can also be used to model domestic politics, through cohort analysis of generational change and differences in political culture, attitudes, and behavior. The American generation that came of age in the early 1950s, as contrasted to the one maturing in the late 1960s, can serve as an example: products of the McCarthy era versus those of the Vietnam era. Evolutionary theorist George Williams has suggested that each new generation be viewed as colonists trying to establish themselves in terra incognita; this is the Wright concept redefined so that habitats vary according to *time* rather than place. By also conceptionalizing each cohort as the cultural analog of a biological species, the metaphor to punctuationism is evident.

The great advantage of catastrophism in political theory is that it proffers an ideology, premised in the biology of our species, that can be used to legitimize the revolutionary solutions demanded by the catastrophic changes already well advanced in our biospheric niche, in our phenotypic selves, and in our chances of having either the time or the capacity to preclude the kind of macroevolutionary megacatastrophe that has happened to our planet many times in the past (but even the most recent was long before hominids evolved). Such "natural" catastrophes are highly likely to happen in a future as relatively remote—in relation to our probable species duration—as the relevant past; and as for the present, a nearby supernova may already have exploded without yet quite having manifested itself to earthlings. Goldsmith (1985: 113) suggests that "if a star within 50 light years of the sun [i.e., about one and three-fourths trillion miles]—one of the thousand or so nearest stars—were to become a supernova, its cosmic-ray flux would be sufficient to produce catastrophic effects." There is no point to worrying about such matters because they are all utterly beyond our control or influence; what we should worry and do something revolutionary about is first the preclusion of nuclear holocaust and second the reclamation of a niche that we can adapt to while still remaining human for yet a while longer (see Yanarella, 1984).

The acceptance of the implications for ourselves of catastrophic evolutionary theory would certainly constitute a major paradigm shift for all of the social sciences (Kuhn, 1962; Schubert, 1983b). Shifting from gradualism to punctuationism and catastrophism might be presumed to have no automatic effect on the manifold of public policy problems that confront us, and both as individuals and collectively imperil our continued existence as an animal species—however some individuals may continue to save their souls. But that assumption—that how we think about problems doesn't change their nature—is fundamentally wrong. How we first perceive and then conceive them makes all the difference in what we can (imagine) doing about problems, and indeed even whether we are able to define them as "problems"; not to mention what might seem possible and appropriate for us to do about them. So the paradigm shift is itself the first step to be taken in trying to do something to postpone the extinction of our species.

PART IV

Evolutionary Development

Part 4 focuses upon how and why human development has a major influence upon political behavior. This is a very different subject from what political scientists usually have in mind when they refer to "political development," by which they typically mean the extent to which the societies and economies of less industrialized countries (or portions thereof) can be "assisted" by the leading exemplars of industrialization in North America and Western Europe to "modernize" and progress as industrial satellites of their "benefactors" (Green, 1986). But here our interest lies, instead, in how human animals change in their growth, both individually and collectively as populations, because of the reciprocal influence of environmental constraints and opportunities upon genetic capacities, and vice versa.

For mammals life begins in utero, and the uterine environment of prenatal life is where and when the processes discussed below begin. Genetic constraints tend to be maximal and environmental opportunities minimal (and usually malevolent) for prenatal human existence; reciprocally, environmental constraints—and opportunities—tend to have a greater relative influence than genetic influences for humans once sexual maturity is achieved and for the next several decades. Both genetic and environmental influences are of more equal but extremely dynamic importance during infancy, childhood, adolescence, and subsequent to both the male and female menopauses. Very little research has yet been done to explore how such a complex and continuously changing transactional patterning of the interfacing conjoint influences of genetic with environmental produced information affects growth in the ontogeny of human individuals and populations, as that relates to those facets of human behavior that political scientists deem to be "political" (see ch. 7, above). The three chapters of this section explicate some of the key concepts and theories about human growth that must be understood for political behavior to be linked to human development, if such research is to be done.

To illustrate this in greater detail, I shall rely first upon recent statements by several persons who are specialists in the biology of human development, in particular with regard to how laterality differentially characterizes fetal growth from its earliest stages; how the genome controls normal sexual differences in postnatal development; and how laterality becomes dynamic again in the normal continuing brain development of mature adult humans. Then I shall characterize the most important differences among the three chapters that follow.

Commenting on the articles by Corballis and Morgan (1978) and by Morgan and Corballis (1978) that discuss the evidence for a maturational left-hand gradient in the biological basis for human laterality, Mittwoch explains (1978: 306) that their paper is important for three reasons: it places within the general framework of mammalian asymmetries certain paired structures of the human brain; it proposes a hypothesis to explain why the left hemispheric rate of growth is greater than that of the right; and how that differential growth gradient is attributed to previously observed functional differentiation of the cortical hemispheres. Mittwoch adds, however, that in her opinion

> the growth gradient begins with a left-to-right bias in early human development, [which] subsequently changes over to a right-to-left direction, and that it may finally change over once more to . . . another left-to-right bias. Since different parts of the embryo develop at different times, the asymmetry in these parts will therefore run in different directions. Throughout the entire period during which the body and its parts are laid down, development and differentiation appear first in the head region and then advance tailward. For this reason, many structures in the embryo show a gradation in their development, with progressively advanced stages located at higher levels. . . . Accordingly, by the first month of human embryonic development, the head and trunk regions are well established, whereas the buds of forelimbs and hind limbs do not appear until the beginning of the second month, the former somewhat earlier than the latter. . . . The gonads also begin their development in the second month. We may assume that a change in the direction of the growth gradient from left-to-right toward right-to-left occurs at some stage between that of the major development of the brain and that of the forelimbs and gonads. A second change over toward a left-to-right gradient is suggested by the fact that in human beings the left leg tends to be longer than the right, whereas the opposite relationship holds for arms.

Morgan has pointed out (1978: 326) that "the absence of consistent functional asymmetries in nonhuman primates might be related to the considerably greater time man spends in growth and development . . . [so] that differences depending upon developmental rate will be more marked when development is spread out over time." To this Taylor adds (1978: 319) that

> the function of the Y chromosome was to slow the developmental rate of males. This slower rate leads to a more detailed read-out of information from the genome as well as to longer periods spent in any given immature and vulnerable state [and see Gualtieri and Hicks, 1985]. With regard to the cerebral hemispheres, it has been difficult to decide what to regard as "mature" or "developed." While we agree in this respect with [Corballis and Morgan], their inference that visuospatial gnosis is a secondary or second-rate skill is unacceptable; much depends upon it from the first day of life. Further, using the more advanced female as a criterion, it is indisputably the function of the left hemisphere of *boys* that is most laggardly.

Moving further along the human development cycle, the question arises what effect maturation has upon lateralization. Kocel (1980: 299–300) discussed aging in mature adults—but not the effect of aging upon lateralization in *elderly* humans (which would surely involve the reciprocal question of the effect of lateralization upon aging). Her findings are that interhemispheric communication becomes, through continuing reinforcement, more efficient as a consequence of aging and that this leads to greater isomorphism in the two hemispheres because of a relatively sharper *decline* in right-hemisphere abilities (thus making the right less different from the left). Her data indicate that with age comes increasing reliance on *left*-hemisphere strategies (whether causing, exacerbating, or in response to right-hemisphere deterioration).

Chapter 13 is a review essay of a book that presents a biological theory of human health, from the perspective of a field ethologist who has spent considerable time tracking bighorn sheep and mountain goats in the high Canadian Rockies. Those animals are adapted to a habitat that is periglacial much of the year, which is obviously suggestive of the environmental conditions confronted by many humans during most of the most recent fifty thousand years. Cro-Magnons had spread through Europe forty to fifty thousand years ago, during the interglacial warming following the retreat of the first Wurm glacial; then the Wurm returned in a second glaciation about thirty thousand years BP, not to retreat until the transition to pastoralism and agriculture began some eleven thousand years ago (see ch. 8, above); a corresponding scenario transpired in North America, where the retreated, advancing, and ultimately retreating ice sheet was called Wisconsin. The ethologist Geist's thesis is that humans both evolved and increased in numbers most rapidly in the most rapidly changing regions at the edge of the retreating (or advancing, as the case may have been) ice sheet, where vegetation was lush and ungulates and smaller game and birds prolific. In warmer and less demanding climates and habitats where environmental stimulation and risks were much lower but cultural homeostasis accompanied genetic homeostasis (cf. Lerner, 1968; Underwood, 1975, 1979), humans were subject to minimal selection and were confined to at best replacement population growth.

Geist defines as a "dispersal phenotype" the increase in body size and social adaptation that follows a species' migration into a habitat previously unoccupied by conspecifics. He remarks that a characteristic of such populations is their neoteny—which is true of species generally, but especially conspicuous in humans for whom extreme neoteny is a defining species specific trait (compared to other animals) for genetic reasons independent of the strongly reinforcing effects of the novel habitats and environments described by Geist. He states (1978: 257) that "neotenization is apparently a consequence of selection for enhanced and plastic body growth during dispersal." And he adds (1978: 259–60) that "neotenization is adaptive to individuals in the dispersing fringe of the population . . . because it increases body size . . . [and] also because it retains the plasticity and adaptability of the juvenile. . . . the more neotenous an individual the better its chances to adapt to environments outside the norms encountered by the [more stable, and nonmigrating] parent populations." Furthermore, such evolutionary regression by breaking down genetic behavioral mechanisms evolved by the parent species "opens up new avenues for selection to shape new behaviors" (Geist, 1978: 318). As a specific example of the effects of migrationally enhanced population neoteny, Geist speculates (1978: 252–53) that as humans pushed beyond the relatively favorable climates and environments of Europe and the eastern Mediterranean, beyond central Asia into Siberia and via the Bering landbridge (from time to time) into North America, the harshness of those habitats during the Upper Paleolithic evolved "people better adapted to cold and with the largest brain sizes . . . [which are] a product of neoteny; the more neotenous the population the better the changes for maximal growth of body and organs. This interpretation explains the gradient across Eurasia into America of people with increasingly neotenous features . . . such as a very large brain and a reduction in secondary sexual characteristics, as well as a reduction in sexual dimorphism."

Chapter 14 discusses both the biological and the political theories of one of this century's best known and most creative evolutionary theorists: C. H. Waddington, who was a lifelong critic of and dissident from orthodox neo-Darwinism. The conventional view postulates, as we have seen (part 2, "Political Evolution"), a sharp dichotomy between genes (which through sexual reproduction transmit information from one generation of a species to the next) and culture (which must be transmitted exogenetically). Waddington's developmental theory of epigenesis was, instead, that genetic expression is highly dependent upon variation in cultural stimulation and that transactions between an animal and its habitat during its development shape its ontogeny in a manner unpredictable and inexplicable by either its genotype or its habitat, considered independently. Growth, structure, and behavior alike are themselves both products and causes of the unique life experience of each animal in relationship to the sophisticated transactional processes through which its genome and habitat interface. But Waddington's epistemology was not limited to developmental biology; he was an advocate also for a moral philosophy which he mistakenly believed to be the inevitable consequence of his biological premises; and in addition he was a talented and creative proponent of dancing and critic of modern visual arts. The chapter discusses all of these facets of his career and ideology.

Chapter 15 discusses the relationship between human aging and ideological continuity and change. The chapter undertakes to design and carry out an empirical test of two competing hypotheses, both common (and unreconciled) in the research literature of political science and other social science disciplines. The first is a cultural hypothesis: that adult humans continue to believe in the ideology into which they were socialized as children and adolescents, which are (definitionally) the formative years for determining their views as adults. The hypothesis postulates that humans are born into cohorts of age-mates, whose views reflect the climate of opinion into which they are born and raised. The second is a biological hypothesis. As they age, humans (like other mammals, and indeed like most other forms of life) begin (in general) to senesce approximately as soon as they achieve sexual maturity with adolescence. In old age, as a function of both their biological development and their cultural experience and learning, their increasing senescence (in general)—*not* their *senility*, as the chapter takes pains to point out and explain—results in what Veblen termed the "trained incapacity" to behave in many respects in the same ways (and at the same speed) characteristic of their adult years until "middle age." The learned adaptation to decreasing behavioral competence is conservatism in decisionmaking and motor behavior, which for many persons projects into social, economic, and political conservatism as well.

The empirical test takes advantage of available data on the voting behavior of United States Supreme Court justices, using chronological age as an index of biological senescence and historical cultural analysis to categorize the ideological thrust of socialization eras into which justices of diverse ages were born and raised. These data indicate no support for the cultural hypothesis, and only very modest support for the biological hypothesis. The chapter closes with a critique suggesting how the design of replicatory research might profitably improve upon this chapter's pilot venture into the adduction of empirical negation of either or both hypotheses. The chapter also discusses in some detail the more general questions of the relationships between human biocultural and biosocial development, and of persistence and change in political ideology as a function of aging.

13
Periglacial

Valerius Geist is a free and creative thinker, whose most important ideas stem directly from his own idiosyncratic experience as a field ethologist tracking bighorn sheep and mountain goats through the lofty heights of the Canadian Rockies throughout the preceding two decades. Geist has also worked with moose and other large ungulates, so his previous interests and experience clearly have focused on herbivores; nevertheless, if a comparison were to be made with him, the prototype that leaps to mind is George Schaller, whose work has centered on carnivores but who (like Geist) has been concerned for a long time with both environmental effects on behavior and human evolution; Schaller's more recent experiences in the high Himalayas (like his earlier work with gorillas) mark him too as a mountain man.

The core of Geist's book (1978) is concerned with the opportunity and impact of periglacial environments on mammalian adaptation, behavior, morphology and selection; and as a special case of that more general theme, it is concerned with human evolution and demography during the Upper Pleistocene, with the last Interglacial and the Wurm ice age extending from approximately a hundred thousand years ago to the beginnings of the Recent era—variously some ten to thirty thousand years ago depending upon the locale. It was during that epoch that the Neanderthals filled their spectacular but precarious niche as the hominid specialists who were top carnivores of the gigantic ungulate species that flourished close to the cutting edge of the great European ice sheet; at the same time, the

---

This chapter is a revision of "Glaciers, Neoteny, and Epigenesis." *Journal of Social and Biological Structures*, 4: 287–96 (1981c). Reprinted with permission of the publisher.

Caucasoid *Homo erectus sapiens* were adapting to the forests and river valleys further to the south, in Europe and North Africa. But the thrust of the book is concerned with the gross hiatus between the periglacial environments, to which humans remain best adapted genetically, and the artificial environments created by themselves to which they have become, at increasing peril to both their own species and the rest of the biosphere, morphologically and behaviorally adapted; it also deals with the implications of the latter adaptation for the possibility of health for humans as a species and of healthful environmental design by them. In addition, the book includes three chapters of special interest to social scientists: "The Biology of Art, Pride—and Materialism" (ch. 5), "The Biology of Inequality" (ch. 6), and part of chapter 13 on the biology of music, dancing, laughter, language, and games ("On the Evolution of Modern Man").

In any truly individualistic book that is also (like this one) highly opinionated, it is inappropriate for a reviewer not to tolerate considerable latitude in the hyperbole of the author's style of self-expression, and to be equally magnanimous when confronted with the author's occasional lapses on substantive issues. Social scientist readers may, for example, question Geist's observation that "the upper classes" (as he calls them) are "characterized by relatively large bodies, large brains, dolichocephaly, and superior intellectual competence"—at least on the evidence from Great Britain (p. 430).

This discussion does not purport to follow the formal structure of the book itself, but instead is organized in terms of the principal themes that recur in the author's development of his subject. These include: evolution, genetics, behavior, environment, humans, and health.

EVOLUTION

Geist's model specifies that "new human adaptations arise at the geographic fringes of a species where populations are in contact with adverse environments and where populations may be trapped in pockets of temporarily favorable landscape. Once a new form evolves, it disperses away from the origin of its evolution and may move into more physically benign habitats where it displaces more primitive predecessors. The driving force of human evolution is seen to be the pulses of glaciation, and human evolution ought to be most rapid where it contacts the temperate and periglacial zones" (pp. 269–70). One of his key propositions, however, is that "natural selection and evolution are not synonymous . . . [and] evolution appears to be a rare occurrence, [while] most populations live in environments that produce no noticeable change in the gene pool" (p. 21). Geist distinguishes also what he terms "the phylogeny fallacy," that "the adaptations of today result from phylogenetic determinism directly derived from the adaptations of yesterday"; indeed, "anthropology, in embracing primate studies, has already been guilty of it. . . . Thus, evolutionary theory dictates that similarity is due to convergent selection in similar ecological professions (niches), not necessarily due to genetic similarity. The origin of a system is thus irrelevant to an understanding of the function of a system. . . [and hence] the study of aggression in primates will tell us very little about human aggression; in fact, aggression as practice[d] in carnivores and in ungulates teaches us considerably more" (p. 71).

"[T]he evolutionary progression in various mammalian lineages . . . has been from the forest to the savannah and then to the steppe, from the warm and wet habitats to the hot and dry or cold and dry ones" (p. 221). And "during the ice ages, mammalian lineages that radiated from the warm climates to the cold ones tended to increase in body size. . . . The very same trend is evident for our own genus, *Homo*" (p. 256). Moreover, "large body size appears to be the prerequisite for evolutionary innovation" (p. 11).

"[A]ntipredator strategy powerfully shapes the body proportions, social behavior, and weapons of [a] species" (p. 11) and "social systems are a function of a species' adaptive strategies" (p. 398).

"[P]rimitive species last the longest" (p. 177); and 'mouse' and 'rat' are the most successful life forms and are the most likely to re-evolve in future geologic periods. In contrast to the common life forms which are clearly recognizable evolutionary successes are the unique life forms, of which we are one. These are life forms whose success is suspect, since they fill a unique ecological niche and, given past history, are not likely to last long. Successful life forms reappear in evolution, unique life forms can be expected to be short-lived experiments of nature. Since we are a unique life form, this has implications for our strategy of survival" (p. 184).

"Evolutionary regression by way of neotenization appears to break down genetic behavioral mechanisms evolved by the parent species and opens up new avenues for selection to shape new behaviors" (p. 319).

> Neoteny was a probable consequence of selection for adaptability and large body size during geographic dispersal. That is, the diversity of tasks to be performed plus the need for novel solutions to new problems in heterogeneous environments, selected for individuals with retarded or extended maturational processes, and thereby a maximum of juvenilelike plasticity in shaping behavior and body structure. It is the process of retardation of maturation that creates almost inadvertently the neotenous features of humans if the benefits of maturational retardation outweigh the benefits of a morphology adaptive in parent populations. Thus, the further the species penetrated into unexploited uninhabited terrain, the more neotenous it became, that is, the more juvenile features characterized the sexually mature adult form . . . [whereas] the most primitive or least neotenous race of human beings would be found in the geographic area of origin of modern man. . . . [Hence] the gradient across Eurasia into America of people with increasingly neotenous features. (pp. 352–53)

"Therefore, where hominids colonized last we can expect them to be most neotenous" (p. 352), which implies that if humans recapture the fading dream of colonizing space, then the egg-shaped head and unarticled speech of Doctor Huer of the Buck Rogers comic strip of the Thirties proffers a much more appropriate caricature of future space colonists, than does the motley crew of archtypes populating contemporary sci-fi flicks.

GENETICS

To a greater extent than other animals, "humans can develop—and lose—specialized physical (body) adaptations—a magnificent attribute, since it permits our bodies to adjust to seasonal or local environmental demands" (p. 23). This plasticity of humans assumes particular importance, in the light of Conrad Hal Waddington's theory of epigenetic selection, which Geist accepts as an indispensable correction to "the inadequate neo-Darwinian paradigm [which] is still king" (p. 116). The theory assumes that "genes appear to have at their disposal alternative strategies of development which they switch on or suppress depending on the messages from the individual's environment. . . . these [are] epigenetic mechanisms." Waddington himself "was concerned with how characters appear, become genetically enhanced, and finally, fixed in the developmental process of a species." But Geist's "emphasis here is quite different." Geist is "concerned with how genes in the developmental process adjust an individual to the environmental demands" (p. 116).

> The concept of epigenetic mechanisms implies that in large long-lived organisms evolution is a rare event and that it tends to proceed rapidly when it does occur; evolution does not proceed continuously, but in steps, with long periods of genetic stability in between. . . . What initiates selection for a given morphological feature? It is the behavior of individuals. Behavior is the cheapest means of adjustment. Specific activities result in physiological and morphological change of organs during heavy use, [due] to phenotypic adjustment. If all individuals are subject to this adjustment due to a severe environmental contingency, then the differences between individuals in that adjustment are largely due to differences in genetics. Now the genetic variance is large. Natural selection rapidly acts on the genetic composition of the population, increasing the frequency of individuals genetically most capable of achieving the given adjustment. This much is Waddington's. . . . We add the following: New adjustments which lead to morphological change are expensive. They cannot therefore, arise under maintenance conditions, but only under conditions of resource abundance. This condition is found during colonization when the dispersal phenotype exists. (p. 122)

But Geist later states that "*Waddington* . . . show[ed] that major evolutionary changes [can] occur only during dispersal, when resources [are] abundant" (p. 403; emphasis added); evidently Geist is not of one mind—or at least, recollection—on this subject.

Nevertheless, "once . . . organisms find themselves in environments within the range of adaptability by epigenetic mechanisms, selection on genes comes to a halt . . . until the population decreases in size or colonizes [a] new habitat. . . . [T]he differences observed between species within genera, tribes, or even families could hardly be due to the direct action of genes, but must result from different degrees of expression of very similar genomes. [For example,] genetically the genera *Homo* and *Pan* are apparently as close as sibling species . . . [so that] the differences here lie less with genes than with epigenetic mechanisms" (p. 121).

Geist emphasizes that "behavior . . . is the first level of adjustment [to environmental change], followed by physiology, followed by morphology. Only when these levels of adjustment are exhausted does genetic adaptation set in" (p. 241). Behavior is discussed below, in the next section of this review; at the physiological level, Geist asserts that

> tissues of *high growth priority* are essential to the organism's function under environmental conditions of low energy and nutrient availability and high intraspecific competition; they are highly canalized tissues, or tissues of K-selection. Tissues of *low growth priority* or low canalization are tissues highly functional under conditions of r-selection, such as during dispersal into vacant habitat. Tissues of high growth priority have been subject to selection for phenotype redundancy; their development is so essential under conditions of K-selection, that is, when individuals are competing for scarce resources at carrying capacity, that a multiplicity of diverse genes can be expected to contribute to the strongly canalized development of organs that are most adaptive under conditions of resource shortages. Growth priorities are evolved because the organs of an individual enter into competition for the body's limited resources of energy and nutrients. (p. 137)

But "the physiological mechanism . . . that permits the individual to adjust to unfavorable aspects of its environment . . . may severely damage the individual if exercised. This condition is . . . stress" (p. 129). Natural selection *favors* individuals who cope with stressing conditions by displaying *psychopathic* traits; and it selects *against creative* individuals (p. 396).

At the level of morphology, "bone shape is very much a consequence of the behavior of individuals" (p. 243) and "brain size enlarges phenotypically with use" (p. 245). Indeed, "learning in neurophysiological terms *is* growth, and learning proceeds best when accompanied by adequate physical exercise . . . and a high protein diet" (p. 144); the brain "enlarges in response to specific exercises well learned. . . . [and] domestic animals have brains 25–50% lighter than those of wild counterparts" (p. 368). In humans, "dolichocephaly is the product of the body in time and space. . . . Brachycephaly is the product of . . . a strong emphasis on verbal skills and knowledge. . . . In pioneering populations . . . [the "phenotype hypothesis"] predicts dolichocephaly; once the population has settled . . . at relatively high population density, there ought to be a shift toward brachycephaly" (p. 388).

The phenotype characteristic of r-selection is "a special and rare type of phenotype, one capable of dealing with contingencies, when exposed to unexploited environments. This [is] termed a 'dispersal phenotype'" (p. 403). And "the dispersal phenotype is a mechanism by which genes maximize their survival

> under conditions of uncertainty—which do, after all, prevail when individuals disperse into habitats unoccupied by their species (p. 354). On the other hand, under conditions of resource scarcity we . . . have selection for a phenotype adapted to cope with the consequences of a high density of conspecifics . . . the "maintenance" phenotype, or K-phenotype. Re-

> sources are reduced primarily by intraspecific competition; a high density of conspecifics favors the spread of parasites and pathogens; there may be more social partners to deal with; predators may pay more attention to the population. All these factors increase the cost of maintenance to the individual. Hence, the phenotype, under conditions of resource scarcity, can invest less in growth and reproduction, and must invest more in diverse mechanisms of individual maintenance. Therefore, it is smaller in body size. . . . It must be lethargic and must not readily enter into social interactions. . . . It must invest a relatively large amount of resources in antibody development to cope with parasites and pathogens. We expect it to stick very closely to the law of least effort. (p. 119)

Moreover, "selection in populations at carrying capacity [is] characterized by individuals with maintenance phenotypes" (p. 301).

Geist notes that "resources shortage . . . is the normal condition of populations" (p. 120); and under such conditions, "phenotypic development is reduced, the expression of the genes is no longer maximized, and the development of any one characteristic reflects less the genetic potential than the environmental constraints . . . [but] the population even during K-selection, holds onto its new genome . . . [and] once carrying capacity is reached, the population freezes in its evolutionary advancement" (p. 262). Also, "under maintenance conditions, . . . selection reinforces existing adaptations via the evolution of growth priorities. . . . This type of evolutionary environment *minimizes* health!" (p. 403n.). On the other hand, it is also true that "excessive resources for reproduction and growth can lead to fat unhealthy bodies and organs and reduced life expectancy" (p. 141).

BEHAVIOR

According to Zipf's Law, "individuals ought to minimize expenditures on maintenance in order to maximize reproductive fitness, thereby sparing maximum resources for reproduction"; "moreover, one of the overriding principles discovered by bioenergetics is that animals use energy in their physical system, not in accordance with their uniqueness as a species, but in accordance with their mass, surface area, temperature, etc., that is, in accordance with their physical parameters" (pp. 2–3). However questionable the application of Zipf's Law to humans—and this is my caveat, not Geist's—the stated bioenergetics principle certainly does apply. A corollary of Zipf's Law is that "the simpler the signal, the less costly it is of energy, the more frequently it will be used . . . [and hence] behavior patterns very costly of energy, such as clashing, butting etc., tend to be used less than their corresponding threat patterns, which are in turn less frequently used than the simpler patterns such as the horn displays" (pp. 50–51). Field ethological observations of humans (Swiss supreme court justices) support this latter proposition (Schubert, 1982a), not literally (I hasten to add) but rather figuratively, in that the least energy-expensive communicative behavior (e.g., oral speech in one's maternal language) was most common; while at the other end of the energy continuum, perambulation was least common.

In classical ethological terms, stimulus contrast evokes changes in animal behavior because it "bodes no good and it is [therefore] adaptive [for the animal] to respond with arousal, that is, to be prepared for the eventuality of flight" (p. 59). This is because "signals come to the attention [of an animal] by rising above the common sounds, motions, and postures simply by being relatively rare, and novelty . . . create[s] arousal" (p. 55). Hence it is common to observe "overt hostility triggered against individuals that act 'abnormally'" (p. 82)—among humans just as among animals generally. Likewise, "under kin selection, displays enhancing overt aggression would be common, since demes fight demes as groups" (p. 103).

Geist defines "aggression as social behavior that displaces individuals and/or bars their access to scarce resources, as well as those actions that protect the individual from bodily harm. This . . . makes [aggression] part of *agonistic behavior*" (p. 69). But "the study of aggression in primates will tell us very little about human aggression; in fact, aggression as practice[d] in carnivores and in ungulates teaches us considerably more" (p. 71). This is because of "familiarity with groups of species in which excellent morphological and behavioral defences prote[c]t the fighters" (p. 84). Recent studies "of large mammals . . . have shown clearly that death through interspecific combat is by no means rare, and that injury during such combat is reduced less by inhibitions against using weapons than by skillful uses of defence strategies, behaviors, and morphological adaptations" (p. 73). But "only humans, by virtue of cultural weapons and defences, can escape prompt retaliation either by killing the opponent outright or, if not by killing outright, by ducking behind some protective shield" (p. 77). Geist does not say so, but contemporary American policies toward juvenile delinquency no doubt exemplify such a shield, as may be implicit in his remarks that "only the instant death of the victim, or its inability to retaliate, or some means of escaping retaliation, will permit damaging aggression to flourish" (p. 77); he adds that "clearly competition in adulthood, combat, violence, callousness, and bloody ritual ought to come 'naturally' to individuals raised in societies that survive by aggression and that have large peer groups" (p. 395).

"Since sexual behavior," he remarks, "can normally only be performed after successful competition against conspecifics of equal or near-equal status . . . it appears logical that an individual prepared to copulate must also be prepared to do combat. Thus neural mechanisms controlling sexual and agonistic behavior may have to be equally susceptible to any given level of arousal" (p. 85). Recent research in human neurophysiology has underscored the propinquity of brain centers that are critically involved in both sexual and aggressive behavior.

ENVIRONMENT

"Unless we pay attention to the environment that nurtures us as that unique life form," underscores Geist, "we shall, over time, lose the very genetic base that makes us humans. Therefore, our concern with guarding our genome ought not be expressed so much in eugenics as unenvironmental design" (p. 173). But "there cannot be for humans *an* environment they are genetically adapted to. Rather, we expect them to cope with a great diversity of environments" (p. 401).

> [T]he populations of central Asia evolved under harsher conditions than did those in the periglacial environments of southwestern and southeastern Europe, and changed toward greater neoteny, reduced sexual dimorphism, and cultural institutions that enhanced the social role of the female. This was adaptation to a harsher environment that extracted heavy work for both male and female. . . . In Europe, during deglaciation the rich fauna of the periglacial egosystem had vanished and with it the primacy of big game hunting. . . . [M]uch of the forest area became uninhabited during the Mesolithic, and population levels did not rise again until the advent of agriculture. . . . The archeological evidence from the Mesolithic . . . indicates a concurrent decline of population size, population quality, and culture, while violence and cannibalism rose as environmental quality deteriorated. . . . [Indeed,] homicide is associated with a marginal existence. (pp. 380–81)

The same principles apply, of course, to animals generally, so that "environments very unfavorable, such as those created purposely under laboratory conditions, are so far removed from the norms encountered by the respective experimental species that severe malfunctions are their response. Under field conditions individuals would probably have departed in search of new living places long before conditions deteriorated to the extent to which we can permit them to deteriorate in laboratories" (pp. 136–37). Geist does not suggest the analogy, but social scientists will immediately think of the impact upon inmates of such institutional environments as those of prisons, orphanages, poor farms, mental hospitals and other "asylums," and during the past generation in the United States the public schools: human laboratory animals become similarly stressed and distressed.

We are only too self-conscious about the evils, for human health and humane behavior, of industrialization (see Green, 1986); but Geist shows that the agricultural revolution entailed its fair share of detriments to betterment in the adaptation of the species. "A hunter who must be mobile cannot amass possessions, since they are a burden to him; but with a sedentary existence possessions can be hoarded and become symbols of prestige, prowess, and power" (p. 390). There is a trade-off involved, however, because with the materialistic surplus comes "extreme vulnerability that generates suspicion, hostility, intolerance, and [overt male] aggression toward others" (p. 390). There are other psychological consequences as well: the survival practices of "hunters in cold, hostile climates" should select for "introverted personalities;" while in agricultural societies, natural selection should have favored extraversion (pp. 393, 395). But for either a sparsely populated hunting band or a settled agricultural village, strangers would rarely be encountered; and Geist's analysis suggests a consequent biological basis for ethnocentrism: "Rare events in nature are, on principle, dangerous" (p. 43).

HUMANS

"The home of the man we can consider human," according to Geist, "is not Africa with its warm climates but the rich, diverse, and demanding periglacial environments at the fringe of glaciers in Eurasia" (p. 188). Although "it is commonly assumed that the interglacials were 'better' periods for human habitation of the glacial refugia than the glacial periods[,] I believe this assumption to be in error. Although the interglacial climate is milder, it brings about tree growth which mitigates against ungulates developing as high a biomass as on the steppe" (p. 210 Moreover, "the time between *H. sapiens'* appearance and his occupation of the periglacial environment exceeds 150,000 years" (p. 348). But "the case for hunting by the *early* hominids has been greatly overplayed, and the idea that *mid*-Pleistocene men hunted cooperatively in groups appears to me a case of wishful thinking" (p. 226; emphasis added). When the hunting band did evolve during the Upper Pleistocene, its average size "consisted of some 5 men" (p. 375), which leads Geist to query whether "strong supportive bonding between 5 to 7 individuals, and somewhat weaker long-term bonding within a group of 20 to 25 persons, [is] of benefit to man in modern times?" (p. 378).

It is not news, of course, that among humans there has been a "long evolutionary history of social bonding through sex" (p. 328). Geist adds that "under the conditions postulated for early *Homo* there would be powerful selection for the female advertising her worth. . . . [However,] in modern man there would be less selection for secondary sexual signals in the female, hence the waning of steatopygous buttocks, chromatic labia minora, and sexual dimorphism of the face" (p. 328–29). As a consequence of social bonding through sex, "the most primitive of human families is the monogamous or nuclear family; all other family forms are deviations from it" (p. 330). "[I]t is with the late Paleolithic people that the three-generation family arise[s]" (p. 323). No doubt this was because "tolerance and generosity can only be expected when resources are reasonably abundant. . . . The extended family could arise only under conditions of stored food surpluses" (pp. 345–47). However, the ongoing "progression from extended to nuclear to single-parent family increases the workload of child rearing on the adults, increases worries and frustrations, reduces social contact, and increases the chance of suboptimum development of the children" (p. 417).

In dramatic contrast to the assertions of hardcore sociobiologists, Geist proffers as one of his major conclusions that "our species has for a long time, probably 100,000 years or more, structured its adaptive strategy by *cultural* means in such a fashion that it maximizes the individual's physical, intellectual, and social development at the expense of population size. . . . Under cold and periglacial conditions *Homo sapiens* experienced continual r-selection that he brought about artificially under K-conditions" (p. 354; emphasis added). "No evolution toward human characteristics could have been conceivable," he adds, "had we been subject to biological mechanisms of population control" (p. 375).

Among humans neoteny also takes the form of relict behaviors, which are "traces of juvenile mimicry . . . found in human submissive behavior, such as crying and childish behavior of persons pleading in despair. . . . Precisely because of the superb ability of humans to discriminate, it is likely that all our submissive gestures are biological relicts of the past and a motley collection of gestures from the sexual and infantile realms" (pp. 349–50). Also typically—though not uniquely—human are such behaviors as language, art, and music; of these Geist comments "that human language is based in part on a genetic program whose unfolding is environmentally determined; thus, grammar has a biochemical structure in the CNS, and we can safely predict that it is formed from some precursor common to subhuman [*sic*] animals. . . . It is in the planning and execution of complex communal seasonal hunts, rather than in confrontation hunting, that we probably find the origin of complex symbolic communication, namely language" (pp. 61, 312). Like language, "art is apparently based on mechanisms of perception that human beings share with other large mammals" (p. 97).

> Music does indeed arouse, . . . [but] rhythmic, slow, and "familiar" music produces the opposite physiological responses; it relaxes and can lead to drowsiness and sleep. . . . Our bodies can follow music physiologically to a remarkable extent. . . . The original evolutionary reason for music and dancing is . . . to maximize physical fitness. Merriment, laughter, and dancing are . . . biological adaptations of humans in the best sense of Darwinian fitness . . . [and] males— who did the hunting originally—should be more easily stimulated by music, song, and laughter than females. (pp. 340, 342)

There is a side cost, however: "Creativity can be maximized only at the expense of cooperation and group harmony" (p. 373).

Geist's "theory of phenotypic response to environment state[s] that health [is] maximized when the diagnostic features of a species [a]re maximized"; and for humans, these include "our vocal and visual mimicry, humor and laughter, superb self-control [and] self-discipline, altruism, [and] our very intellectual competence" (p. xiv). Geist reports a list of human diagnostic features (at pp. 412–13), ranging from "(1) Upright posture" to "(29) Menarch." Such features "are developed maximally under conditions of resource abundance" (p. 22). "Therefore, new adaptations, diagnostic features, which differentiate the colonizing from the parental population, can arise only during dispersal" (p. 122).

HEALTH

"[A]nimals are healthy," Geist remarks, "only under exceptional circumstances" (p. xii). However, "the evolutionary environment under conditions of abundance of resources *is the same as the one maximizing health*" (p. 22; emphasis added). Stated otherwise, "health must be at a maximum when the phenotypic expression of a species' diagnostic features is maximized. The environment that does this is the evolutionary environment of a species" (p. 412). Various contributors to good health are mentioned throughout, for example "that intense, intimate, strong social bonding of long duration is most conducive to good health" (p. 378). But of course these maxims are subject to qualification through the interactive effects of the strongest in both physiological and social terms; and so when Geist purports to illustrate "the sensitivity of human beings to environmental factors" by noting "an average progressive decrease in body size with birth order" (p. 144), it is equally pertinent to remark that there is an inverse correlation between fetal and neonatal health and maternal age (at least among sexually mature human females). Without statistical or experimental controls and relevant empirical data, therefore, it is more plausible to attribute later but smaller babies to maternal aging rather than to environmental factors.

Geist summarizes "a model life style that maximizes health . . . [including] a great amount of diverse, skillful, physical activity, of intense learning of knowledge and skills, of complex interactions with nonpeers, of long-lasting intense social bonds, of developing mastery over a broad range of difficult tasks and a high level of discipline over one's intellect and emotions, and also a life filled with humor, good fellowship, a thorough exercise of bodily pleasures and a diet both abundant and of high quality." To maximize health, he concludes, "is to maximize humanity" (p. 402).

There is a final question about the future implicit in much that Valerius Geist has written in his book, although he never quite confronts it directly. What he does make clear is that our species will have become well settled—possibly as soon as the end of the 1980s—into a globally extensive K-selection maintenance phenotype. We have conquered this earth only too well (if not too wisely); but even should there be new worlds to explore, where in the world—where in *this* world—can we ever get the energy necessary to colonize them?

Surely we have enough neoteny; and epigenesis should assure that the human phenotype follows wherever behavior takes the lead. Our problem is that we no longer can command abundant resources, in relation to the survival needs of our species' population (cf. Cohen, 1977). Perhaps we are running out of glaciers, too.

14
Epigenetic

Conrad Hal Waddington, to whom I shall frequently refer by his nickname, "Wad" (Robertson, 1977: 612) in the interest of brevity (if not euphony), was an archetypal figure in twentieth-century biology. He was a creative, independent individualist who swam consistently against the tide of neo-Darwinism during the last four decades of his long and productive life—to the considerable benefit of those of us who are students of biological and social processes. How and why considerable is the subject of this review essay.

For Wad's own statement of his ideas I shall rely primarily (although by no means exclusively) on what he says in his three best known books (1957) [SG], (1960) [EA], and (1975) [EE]. I shall refer to these books by the indicated acronyms, with page citations as suitable; but because each of them was widely reviewed at the time of its respective publication, I shall make no attempt to review his books as such. I have selected them as representative of his evolutionary thought as he expressed it in a score of books and at least ten times that many reports of experimental research and other briefer scientific writings (Robertson, 1977: 614–22).

The focus here will be on four interrelated themes that, in my judgment, characterize Waddington as a scientist and as a somewhat peripherally located Englishman. Note that I do *not* say "a peripherally located scientist or biologist," since the University of Edinburgh ranked among the top three British Universities in scientific teaching and research throughout the middle decades of this century. My social scientist's point here is that it is located in Scotland rather than England—

This chapter is a revision of "Epigenetic Evolutionary Theory: Waddington in Retrospect," *Journal of Social and Biological Structures*, 8: 233–53 (1985b). Reprinted by permission of the publisher.

and most importantly, not in Cambridge. As Robertson (1977: 578–79) observes, "His position in Cambridge (1933–1945) was always slightly insecure."

First I shall discuss Waddingtons's critique of neo- Darwinism, in his own terms but also in relation to constructional biology and other critiques of what remains today the mainstream channel of evolutionary theory, in "modern biology" (as it has been called for more than sixty years). Then I shall discuss his more general philosophy-of-science stance, as the basis for an understanding of his epistemology. Both of these sections of the review are introductory to the remaining two sections, which develop and discuss his major contributions to evolutionary theory but also his (as he claimed) biologically based theory of social and political ethics.[1] The indicated structure of the essay will involve an alternation between consideration of Waddington's biological and his human-social theories and opinions; concerning Wad's corresponding but earlier double life at Cambridge, see Robertson (1977: 578). This constitutes a fair representation of Wad's interests and self-perceived priorities; but my own judgment, as I shall detail below, is that his contributions to these diverse fields of study—evolutionary theory, evolutionary philosophy, and sociopolitical theory—were by no means of equal importance.

A brief methodological caveat also is necessary. Wad was a writer who chewed his cabbage much more than twice; and thus his major concepts of epigenetic development were initially stated in well-developed form as early as 1940 (EE: 219), and then reiterated (and I do mean "re-") continuously throughout the corpus of his subsequent writings, often with direct phenocopies of the original (or of a slightly improved) statement. Thus, his three books of greatest relevance here are mostly based on aggregations of his previously published articles; but given the nominal difference in their respective subjects, there is less direct overlap between SG and EA, while there is considerably more overlap (and necessarily so, given its theme) between EE and both of the two earlier books. One might say that each of the books functions as a related genotype, combining chromosomal chapters that link sociogenetic themes: my function is that of an exogenous environment that stresses Wad's genotypes and assimilates their sociogenetic content into the phenotypic reconstruction of Waddington's theory that appears below.

EVOLUTIONARY PARADIGMS
NEO-DARWINISM MENDELISM RECONSTRUCTED

As a graduate student at Cambridge during the dozen years from 1926 to 1938, Waddington turned from geology to two years of research in paleontology, after

[1]I shall *not* undertake to appraise his well-developed and longstanding interest in and relationship to modern art and artists—including both of his wives, of whom the first "had strong artistic leanings" and the second was an architect—but it was substantial, and included a labor of love (1969) in the form of a "magnificently illustrated comparative study of science and painting in the twentieth century" (Robertson, 1977: 608). See also Robertson (1977: *passim*) for a discussion and appraisal of Waddington's psychological and social involvement in arts and artists. Robertson concluded (1977: 614, emphasis added; and cf. EE: 1) his official Royal Society obituary of Waddington with the rhetorical query: "Is it prophetic that *he* . . . has described *Wittgenstein* as a poet forced by chance to behave as though he were a philosopher?"

which he shifted to genetics and experimental embryology—a field in which he continued to work for another thirty years. His first article explicitly focusing on evolutionary theory was published in 1941; in his own words, "in the late thirties I began developing the Whiteheadian notion that the process of becoming (say) a nerve cell should be regarded as the result of the activities of a large number of genes, which interact together to form a unified 'concrescence' [or 'creode' as he subsequently came to call it]" (SG: 9). He had already begun to speak of "epigenetics" by 1940 (*Organisers and Genes*) and of "canalisation" and genetic assimilation by 1942; but further work along these lines was postponed for the next half dozen years because of his naval service activities during World War II. So his position as a critic of the then (and still now) dominant neo-Darwinian paradigm of evolutionary theory was already defined by the early 1940s, but his point of view did not become more widely publicized and recognized until a decade later, apparently because of the intervention of World War II.

Waddington described (EE: v) this century's prevailing orthodoxy of "modern biology" in terms of three principal components, of which the first two were already well established by the time he was doing his undergraduate work (in other fields of study) at Cambridge in the mid-1920s:

> 1. Variation between individual organisms is due to changes in discrete units (genes) which do not "blend." (Paradigm: *individual* genes in *individual* organisms. Bateson, Morgan.)
> 2. Evolution is to be considered in terms of changes in frequencies of *individual* genes in *populations* of organisms. (Haldane, Fisher, "neo-Darwinism.")
> 3. Evolution is concerned with *populations* of genes (gene pools) in *populations* of organisms. (Sewall Wright, Dobzhansky.)

Wad points out that the second paradigm was not superseded by the third until the late 1930s and that since then, he had found himself "impelled to envisage evolution as dependent on processes which *affect phenotypes*, and have only *secondary* repercussions on frequencies of genotypes." He remarks (EE: vi):

> This is perhaps a more profound change of paradigm than the other, since it demands radical alterations in some of the most deeply ingrained biological dogmas:
> 1. It points out that, although an "acquired character" developed by an "individual" is not inherited by its individual offspring, a character acquired by a *population* subject to selection will tend to be inherited by the offspring population, if it is useful.
> 2. It argues that genotypes, which influence behavior, thus have an effect on the nature of the selective pressures on the phenotypes to which they give rise.
> 3. It introduces into the theory an inescapable indeterminism quite different in nature from quantum intermediacy, but almost as intractable, since identical phenotypes may have different genotypes, and identical genotypes may give rise to different phenotypes.

Very much to the contrary, said Waddington (EE: 197–98):

> Classical neo-Darwinian theory may be summarized as a system which involves: (a) random gene mutation, treated as a repetitive process so that each mutational change can be assigned a definite frequency. (b) Selection by "Malthusian parameters," i.e., effective reproduction rates. (c) An environment which is treated as uniform, that is, it can be neglected. (d) The

phenotype has no importance other than as the channel by which selection gets at the genotype.

This system is theoretically a closed one, which does not lead to continued evolution, but at best to a passage leading to a state of equilibrium. The possibility of continued evolution required the postulation of one or more of the following additional points: (1) A continued change in the environment, arising independently of the existence of the organisms within it. (2) An initially heterogeneous environment, whose heterogeneity is continually increased by the fact that different populations evolve into adaptations to the initial sub-environments; i.e., the organisms adapted to other environments form part—a changing part—of the environment of the organisms in the subenvironment under consideration. (3) The existence of epigenetic organization of the phenotype, when it occurs in a later-evolved organism from what it had been at an earlier stage. (4) The possibility of the occurrence, at later stages, of types of gene mutation which were theoretically impossible at earlier stages (e.g., the occurrence of an intra-genic duplication would then make possible types of mutation which previously could not occur).

It seems almost certain that all these factors have in fact played a part in evolution as it has actually occurred. The questions then to be asked are: how have they operated so as to result in the appearance of only a restricted number of different types of organisms? Why are there no organisms which simultaneously exploit the full potentialities of myosin (active movement) and chlorophyll (photosynthesis)? Why are there no vertebrates as thoroughly hermaphroditic as some molluscs? In other words, *once you have got a theory which makes continued evolution possible the* major problem is to provide a general theory of phenotypes *. . . [and] the crucial issue is . . . the epigenetic organization of the processes by which the genotype becomes developed into the phenotype.*

And "it has to give up the claim," he adds (EE: 236), "to be the sole and sufficient explanation of all evolutionary phenomena." (For his corresponding critique of neo-Mendelian theory, see EE: 44–45.) He sums up his critique by stating (EE: 280), "In my opinion the conventional Neo-Darwinist theories of Haldane and Fisher (and to a lesser extent, Sewall Wright) are inadequate *both* because they leave out the importance of behavior in influencing the nature of selective forces, *and* because they attach coefficients of selective value directly to genes, whereas really they belong primarily to phenotypes and only secondarily to genes."

Therefore Waddington became perhaps the earliest major dissenter from and critic of the neo-Darwinian paradigm that has dominated evolutionary theory for the past half century (Williams, 1966; 1971; 1975) and that achieved an apogee of sorts during the later half of that period in the form of sociobiology.[2] Wad was a fellow of the Royal Society for almost forty years, having been elected in 1947; and he certainly was the earliest British biological evolutionary theorist of em-

[2]Sociobiological neo-Darwinism evolutionary theory finds its most vulgar (at least, in the sense of "popular") expression—and certainly that which is maximally antithetical to Waddington's thinking—in Dawkins (1976), which Webster and Goodwin (1982: 42; cf. Schubert, 1981d: 204–6) characterized as "its most absurd and degenerate stage in the concept of the 'selfish gene' in which the entities are freely invented and endowed with whatever properties are required 'to explain' biological phenomena." But Waddington does not appear to have engaged himself with sociobiology. Thus his discussions of fitness ignore Hamilton's work and the concept of *inclusive* fitness (e.g., Wad's comments in 1969 [EE: 252] could have been, but were not, updated for publication (in 1975) as a result of which he looks forward to David Bohm, but backward to his former colleague Wynne-Edwards, in his [the cited] exchange with Maynard Smith).

inence to break ranks from the Fisher-Haldane-Maynard Smith phalanx. Only American evolutionists (such as Wright, Dobzhansky, and Lewontin; and see EE: 231–32) took equally independent and heretical theoretical positions regarding neo-Darwinism/Mendelism and/or Sociobiology (see Schubert, 1981d: 203, 207–16). It certainly was no coincidence that the same three Americans were among those with whom Wad was most closely associated, both professionally and personally (EE: 96–98 [Dobzhansky]; EE: 181–83 [Lewontin]; EE: 233–34 [Wright]); and Robertson (1977: 581) reports that during his year-long visit at Wad's Institute of Genetics at Edinburgh, Sewall Wright had given "a long course in biometrics, [and had] laid the foundations of his four-volume *Evolution and the Genetics of Populations*" (which was published decades later—given its extraordinary scope, complexity, and profundity—during 1968–1978).

But the North American whose ideas about evolutionary theory are perhaps closest to those of Waddington is Valerius Geist (1978), whose ideas were discussed in ch. 13 above. Although Wad was an experimentalist whose major research was done in laboratories on developmental processes of widely varying species of animals, and Geist is a field ethologist whose speciality is tracking sheep and goats in the high Canadian Rockies, nevertheless the two biologists had much in common (see ch. 13, above). They shared a central focus on neoteny and epigenetic processes; iconoclastic views of neo-Darwinian orthodoxy; and by no means least, athletic personal phenotypes. The latter was a prerequisite for anyone with Geist's occupational psychosis; and Wad (whose undergraduate sport was running) was an enthusiast who organized and led the Cambridge Morris Men in performance tours of such available highlands as the Cotswolds. Both also shared a strong interest in art; and at least in Waddington's case this was a lifelong obsession (Robertson, 1977: 577–78); and see especially Waddington, 1969: 118–27, in which he presents his most sophisticated and complete discussion of organization theory in biology, in relation to modern physics—both of them in counterpoint to their equivalences and analogies in twentieth-century painting and scientific photography).

Geist prefaces his book (1978: xvi) with a recognition of "the need to abandon the neo-Darwinian paradigm of natural selection in favor of the Waddingtonian." And Geist (1978: 116, 118, 122) remarks:

> It is C. H. Waddington . . . who had a deep insight into [the] significance [of epigenetic mechanisms] to an understanding of evolution. His thoughts are greatly neglected, and invariably misunderstood when mentioned, in the polemics about human evolution; the inadequate neo-Darwinian paradigm is still king, as illustrated, for instance, in discussions by Wilson (1975), Trivers (1974), Alexander (1974), Durham (1976a), or Ruyle *et al.* (1977).
>
> Waddington paid far more than lip service to the accepted notion that the phenotype is the unit of natural selection; he developed a theory explaining how environmental dictates are ultimately translated into genetic responses by species. He was concerned with how characters appear, become genetically enhanced, and finally, fixed in the developmental process of a species. My emphasis here is quite different. I am concerned with how genes in the developmental process adjust an individual to the environmental demands. . . .

> Although "epigenetic mechanisms," strictly speaking, refers to physiological processes, they are simply one means of adjustment, and there is logically no reason the term could not have a meaning as broad as to include cultural mechanisms. Anything that changes individuals adaptively becomes, then, at least an expression or consequence of epigenetic mechanisms. However, for semantic reasons I shall retain a narrow meaning for the term "epigenetics," and let it stand strictly for the physiological responses, although I shall champion here—as did C. H. Waddington-the view that it is behavior that triggers these epigenetic responses, and that morphology is the consequence of epigenetic responses. . . .
>
> What initiates selection for a given morphological feature? It is the behavior of individuals. Behavior is the cheapest means of adjustment. Specific activities result in physiological and morphological change of organs under heavy use, to phenotypic adjustment. If all individuals are subject to this adjustment due to a severe environmental contingency, then the differences between individuals in that adjustment are largely due to differences in genetics. Now the genetic variance is large. Natural selection rapidly acts on the genetic composition of the population, increasing the frequency of individuals genetically most capable of achieving the given adjustment. This much is Waddington's (1957; 1960; 1975) thinking. We add the following: New adjustments which lead to morphological change are expensive. They cannot, therefore, arise under maintenance conditions, but only during conditions of resource abundance. This condition is found during colonization when the dispersal phenotype exists. Therefore, new adaptations, diagnostic features, which differentiate the colonizing from the parental population, can arise only during dispersal.

Geist (1978: 403, and cf. 22) subsequently confirms that "*using the concepts of C. H. Waddington*, it was shown that major evolutionary change could occur only during dispersal, when resources were abundant" (emphasis added).[3]

Waddington represents the developmental critique of neo-Darwinism; and Geist, the ethological. Certainly more conspicuous, and apparently more successful, has been the direct assault (by several paleobiologists, who built upon the theory of evolutionary speciation proposed by Mayr, 1954; 1963; 1970) upon neo-Darwinism's central tenet of gradualism in evolutionary change. The punctuated equilibria theory of these paleobiologists clearly asserts that speciation (a major unsolved problem for Darwin himself and his orthodox contemporary followers) and other major morphological change in species is the result of behavioral response to catastrophic environmental change (and hence, strong stimulation of organisms; see ch. 12, above). This theory fits like hand in glove with both Geist's theory of behavior epigenetics and Waddington's theory (to be explicated below) of epigenetic genotypic reorganization under strong environmental stress to produce major phenotypic changes in organisms in the absence of prior change in the locus or chemical definition of the structure of individual genes. S. J. Gould is both one of the coproponents of contemporary punctuational theory (Eldredge and Gould, 1972; Gould and Eldredge, 1977; Stanley, 1979, 1981, 1986; Gould, 1983: ch. 17, 18, 20) and also a well-publicized and highly skilled writer on evolutionary theory for both professional and popular audiences (Gould, 1977a; 1977b; 1980a; 1981; 1983). It is therefore the more remarkable that in those five books

[3]I wish to take this opportunity to correct my former statement of this same point (Schubert, 1981c: 291), where I improperly—even if mildly—criticized Geist for what I can now recognize was my own mistake in how I quoted him, for which I apologize.

collectively Gould mentions Geist only once (and then for Geist's earlier ethological monograph on mountain sheep); and likewise he mentions Waddington only once—and then (1977b: 43) as a "gadfly" among evolutionary theorists.

Gould comments (1980a: 186) on: "the great neo-Darwinian synthesis forged during the 1930s and 1940s and continuing today as a reigning, if insecure, orthodoxy. Contemporary neo-Darwinism is often called the 'synthetic theory of evolution' because it united the theories of population genetics with the classical observations of morphology, systematics, embryology, biogeography, and paleontology."

In contrast stands Gould's own (1980a: 184)

> model of *punctuated equilibria*. Lineages change little during most of their history, but events of rapid speciation occasionally punctuate this tranquility. Evolution is the differential survival and deployment of these punctuations. (In describing the speciation of peripheral isolates as very rapid, I speak as a geologist. The process may take hundreds, even thousands of years; you might see nothing if you stared at speciating bees on a tree for your entire lifetime. But a thousand years is a tiny fraction of one percent of the average duration for most fossil invertebrate species—5 to 10 million years. . .)

His model presumes (1980a: 183) that

> all major speciation maintains that splitting takes place rapidly in very small populations. The theory of geographic, or allopatric, speciation is preferred by most evolutionists for most situations (allopatric means "in another place"). A new species can arise when a small segment of the ancestral population is isolated at the periphery of the ancestral range. Large, stable central populations exert a strong homogenizing influence. New and favorable mutations are diluted by the sheer bulk of the population through which they must spread. They may build slowly in frequency, but changing environments usually cancel their selective value long before they reach fixation. Thus, phyletic transformation in large populations should be very rare—as the fossil record proclaims.

To that Gould appends a footnote stating:

> I wrote this essay in 1977. Since then, a major shift of opinion has been sweeping through evolutionary biology. The allopatric orthodoxy has been breaking down and several mechanisms of sympatric speciation have been gaining both legitimacy and examples. (In sympatric speciation, new forms arise within the geographic range of their ancestors.) These sympatric mechanisms are united in their insistence upon the two conditions that Eldredge and I require for our model of the fossil record—*rapid* origin in a *small* population.

And *then* he asserts (1980: 192)—evidently on the basis of his own appraisal (1977a: *passim*) of empirical studies in embryology by other scholars, and entirely independently of any reliance upon (or familiarity with?) Waddington's well-developed and equally well publicized theories on the subject—that "the problem of reconciling evident discontinuity in macroevolution with Darwinism is largely solved by the observation that small changes early in embryology accumulate through growth to yield profound differences among adults"; this quoted remark is *pure* (albeit unacknowledged) Waddingtonianism. But in addition to Wad's major contribution as an embryologist it is relevant to remember (in the present context of how he was perceived by Gould) that Wad (EE: 5) "began to work as a paleontologist, studying the evolution of certain groups of fossils." And Wad chose, as his main interest, cephalopods (which laid down spiral shells), "a group which forces on one's attention the Whiteheadian point that the organisms undergoing the process of evolution are themselves processes."

In another essay focusing upon catastrophic change in a physical environment, Gould (1980a: ch. 19) recounts the process—in which he evidently was a participant observer—by means of which a geologist of independent mind and views was pilloried by the American geological establishment of sixty years ago (at, of all places, The Cosmos Club in Washington, DC) for his effrontery in having proposed on the basis of considerable empirical evidence that he had collected, but in the teeth of the Darwinist gradualist dogma, that the southeastern part of the state of Washington had been sculpted in its present form "by a single, gigantic flood of glacial meltwater" at the end of the first Ice Age (within a matter of *days*!) It took more than half a century for theory to catch up with the facts in this instance; but Gould reports a happy (though ironic) ending to his tale: the Geological Society of America presented its highest award, the Penrose Medal, to the geologist—by then ninety-seven years of age but evidently (Gould, 1980a: 203) fully conscious of what had so long been his due—who had been so universally professionally denounced on grounds of principle, for his temerities as a mere forty-three-year-old. Waddington's day may come yet!

The fourth stance critical of neo-Darwinism is that of constructional biology, as exemplified by Webster and Goodwin (1982: 44), who conclude their discussion of "Organisms as Structures" with the statement that "it is clear that, in general terms, the picture of morphogenetic process presented above is similar to that which Waddington (1957) arrived at on the basis of his analysis of genetics and experimental embryology. Waddington pictured the orderly change in 'competence'—the set of possible transformations available at any given time—and the capacity for self regulation in terms of a spatial metaphor into a specific theory." And they conclude their article (1982: 46) with the observation that

> chance events enter into a structured system, a system which, because it is law governed, results in a "a priori" necessary order. As Waddington (1957) implied, and as we believe we have made more explicit, living organisms are devices which use the contingent "noise" of history as a "motor" to explore the set of structures, perhaps infinitely large, which are possible for them.
>
> A structuralist conception of living organisms with its emphasis on the logical, the universal and the necessary, implies that the organismic domain as a whole has a "form" and is therefore, intelligible (which does not mean predictable) and that the "content"—the diversity of living forms, or at least their essential features—can be accounted for in terms of a relatively small number of generative rules or laws.

They add that they expect this claim will, no doubt, seem absurd to neo-Darwinists.

EVOLUTIONARY EPISTEMOLOGY

Throughout his adult lifetime, Waddington considered himself to be philosophically a follower of Alfred North Whitehead. That in itself is a datum of ambivalent significance: Whitehead was renowned not only as a philosopher but also as a mathematician for his work in abstract algebra and symbolic logic, (and as the co-author, with his student Bertrand Russell, of the *Principia Mathematica*). He departed Cambridge (for Harvard) not long after Wad had matriculated. The ambivalence lies in Waddington's own ambiguous feelings toward math, concerning which he frequently expressed self-deprecatory views; it was an appropriate tool for *others* to employ in operationalizing (not a word that he himself used) his theoretical insights (cf. Webster and Goodwin's concluding remark, above; and Robertson, 1977: 613: "Some of his concepts turned out under use to be rather fuzzy at the edges—perhaps because he was often content with presenting them as analogies and did not develop them into precise theories"). In his "autobiographical note" (EE: 3), Waddington acknowledges the importance of "a large body of much . . . explicitly rationalized thinking; in the first place that of Whitehead, to whose writings I paid much more attention during the last two years of my undergraduate career than I did to the textbooks in the subjects on which I was going to take my exams"; and for his appraisal forty years later, of Whitehead's persistent influence on his thinking, see Waddington (1969: 111, 113–16, 118). Robertson (1977: 603) comments that Wad "was much influenced by Whitehead, whose portrait forms the illustration for Waddington's posthumous article in *Nature*—'What I was reading 50 years ago.' He admitted that Whitehead's later writings were very obscure and, in so far as I can appreciate the latter's doctrines through Waddington's words, they appear as a not very helpful holism. Perhaps the most clear cut is that . . . what we describe are 'events' or 'processes' and any event can only be properly described in relation to every other event in the Universe. The concepts that Waddington himself introduced do in general refer to processes" (cf. EE: 223).

Waddington himself claimed (EE: 4),

> It was, for me, Whitehead who suggested new lines of thought. What was the Whiteheadian metaphysics? I will sketch briefly what were the salient features in my eyes. 1. The raw materials from which we were to do science—or with which we finish the scientific testing of a theory—are "occasions of experience." An occasion of experience has a duration in time (cf. David Bohm, "there are not things, only processes"). 3. An occasion of experience is essentially a unity. . . . 4. The content of any occasion of experience is essentially infinite and undenumerable. . . . 5. The experience, which Whitehead refers to as an "event," has, however, some definite characteristics. . . . Definiteness of the Whiteheadian objects in an event implies that, although the event has some relation to everything else past or present in the universe, these relations are brought together and tied up with one another in some particular and specific way characteristic of that event. (Whitehead was *writing before* quantum mechanics became a dominant influence in our thought, but compare these notions with

the idea that a particle must also be thought of as a wave function extending throughout the whole of space-time.) . . . Later he described the way in which an event here and now incorporates into itself some reference to everything else in the universe as a "prehension" of these relations by the event in accordance with its own "subjective feeling." This is a metaphysics very close to that advocated by David Bohm when he speaks of creativity. (see ch. 19, below)

"I tried," said Waddington (EE: 11), "to put the Whiteheadian outlook to actual use in particular experimental situations."

As an undergraduate Waddington held an Arnold Gerstenberg studentship in philosophy—an award made to encourage natural science students to study philosophy—and wrote a thesis on "The vitalist-mechanist controversy"; simultaneously he held also a studentship in paleontology. He was a friend of Wittgenstein, whom Wad got to know (Robertson, 1977: 603) "when he came to a lecture I gave in my room in Christ's College to the Cambridge Moral Science Club, about some rather general topic on the philosophy of science, which I now forget. Wittgenstein hid on a window seat behind the curtains for most of the lecture, but then emerged to state that I had not been talking philosophy at all, but had talked science. I had at that time no idea who he was . . . [but] he came back to my room the next morning, and . . . we met fairly frequently for a few months, after which I went away . . . and the contact was lost for a time." Then several years later, "he and R.H. Thouless and I used to meet one evening every week, and, spend three or four hours after dinner discussing philosophy in . . . Cambridge." (Webster and Goodwin preface their article [1982: 15] with an introductory quotation from Wittgenstein, about a slippery slope—a metaphor that has figured frequently in the jurisprudence of the United States Supreme Court in recent years.) Waddington thought (Robertson, 1977: 604) that the crux of Wittgenstein's teaching—although this became much more vividly and clearly apparent in Wittgenstein's oral discourse than in such works as his *Philosophical Investigations* (from which Webster and Goodwin gleaned their introductory quotation)—was "a method which explicitly recognizes the importance of a developmental analysis of language. . . . [But] Wittgenstein failed to realize that what he had to convey was essentially a poetic awareness of *the otherness of reality*" (emphasis added).

More generally, Wad's philosophy-of-science stance was characterized by his staunch opposition to Popper over such issues as rationality (see ch. 10, above), the relevance of history to biological development (see Gray, 1983), and holism (EA, p. 56n., and ch. 7), but expressions of support for Kuhn's theory of paradigmatic change in science, and for Piaget's structuralism. Wad spoke disparagingly (EA: 206) about one of Morley Roberts' books (1941; but cf. Roberts, 1938) because of his classification of "nations" as "low-grade invertebrates" and although I agree with Wad that comparisons between human society and animal organisms are not very satisfactory, nevertheless Roberts is probably on the right track in his thinking that *if* such comparisons are to be made, then similes to jellyfish (rather than mammals) come closer to the mark (Schubert, 1976: 190 n. 36).

EVOLUTIONARY EPIGENESIS

Piaget (1970: 49–50) has remarked that

> in embryology the structuralist tendencies that were first given currency by the discovery of "organizers," structural regulations, and regenerations, have now become accentuated through the work of C.H. Waddington, which introduces a notion of "homeorhesis". . . according to which embryological development involves a kinetic equilibration whereby deviations from certain necessary paths of development . . . are compensated for. More important still, Waddington has shown that environment and gene complex interact in the formation of the phenotype, that the phenotype is the gene complex's response to the environment's incitations, and that "selection" operates, not on the gene complex as such, but on these responses. By insisting on this point, Waddington has been able to develop a theory of 'genetic assimilation,' i.e., of the fixation of acquired characteristics. . . . Waddington views the relations between the organism and its environment while being conditioned in it. What this means is that the notion of structure as a self-regulating system should be carried beyond the individual organism, beyond even the population, to encompass the complex of milieu, phenotype, and genetic pool. Obviously, this interpretation of self-regulation is of the first importance for evolutionary theory.
>
> Just as there still are embryologists who remain wedded to an entirely preformational view of ontogenesis and who, accordingly, deny all epigenesis (restored to its plain sense by Waddington), so it has occasionally been maintained of late that the entire evolutionary process is predetermined by the combination established by the constituents of the DNA molecules. . . . Waddington, by reestablishing the role of the environment as setting "problems" to which genotypical variations are a response, gives evolution the dialectical character without which it would be the mere setting out of an eternally predestined plan whose gaps and imperfections are utterly inexplicable.

A characteristic statement of this point, by Waddington himself (EE: 170), is that

> the relation of the behavior of an animal to the evolutionary process is not solely that of a product. Behavior is also one of the factors which determines the magnitude and type of evolutionary pressure to which the animal will be subjected. It is at the same time a producer of evolutionary change as well as a resultant of it, since it is the animal's behavior which to a considerable extent determines that nature of the environment to which it will submit itself and the character of the selective forces with which it will consent to wrestle.

In various of his writings (e.g., EAP: 95; EE: 57) Wad presents the same figure as an explication of the logical structure of an evolutionary system. Four partially intersecting sets ("systems") are arrayed along a linear dimension of historic time, ranging from one generation of genotypes to its next successor generation. In Waddington's terminology, the initial system is the exploitive one (with the animal selecting its niche). Animals choose their own environments (EE: 170, above; and cf. SG: 104); alternatively, Wad speaks (EA: 94) of an "exploitive system" containing the set of possible environments from among which an animal modifies some that constitute its environmental niche: "Animals are usually surrounded by a much wider range of environmental conditions than they are willing to inhabit." And "It is when we turn to consider the evolution of the exploitive system that we find ourselves confronting for the first time the whole mass of data which constitutes the major part of our knowledge of the course of evolution" (EA: 133), as in comparative anatomy, physiology, behavior, and ecology, plus, of course (although Wad did not mention it in this context), ethology.

Next comes the epigenetic system of development, with environmental stresses revealing a subset of (from among the much larger set of genetically possible) phenotypic potentialities. Waddington points out (EE: 259) that "in the twenties, it was remarked that all individuals of given species captured in the wild state. . .look remarkably alike; they were examples of 'the wild type.' But this is because the wild type masks genetic variation: 'individuals that are phenotypically almost identical, looking as alike as two peas, contain wildly different genotypes, each a sample drawn from the population's highly heterogeneous gene pool. The uniformity of the wild-type is a *phenotypic* uniformity, as a result of the canalisation of development, which conceals the heterogeneity of the genotypes and of the epigenetic environments.'" "What we seem to meet in embryology," said Waddington (SG: 17), "are situations in which small initial differences lead to large divergences in later development" (cf. the quotation, above, virtually identical to this in substance, from Gould, 1980a: 192), with end products that are sharply distinct from each other. This is illustrated by embryonic induction in vertebrates, for whom (Robertson, 1977: 589, quoting Waddington) "the very tip of the brain induces the formation of the nose organs; slightly behind this, two outgrowths from the brain, which later develop into the eyes, induce the overlying skin to form the lenses; still further posteriorly, behind-brain induces the ears, and so on." Such processes of embryonic development (i.e., the regionalization of specialized cells; the histogenesis of tissues; and the morphogenesis of organs) are under tight epigenetic control. But that control depends very little upon which genes are located where on which chromosomes; instead there is another type of association among genes that arises during development (EA: 112).

> The formation of the different tissues of the body, such as muscle, nerve, etc., depends on the inter-related functioning of sets of genes. In a certain sense, therefore, each specialized tissue of the body represents the activities of a group of genes (or perhaps better, represents a group of gene activities), the association of the members of the group being in this case not at all a matter of chance, since it is essentially dependent on the way in which the gene activities interlock and interact with one another.

Most importantly, Wad stresses that such epigenetic groupings of genes, groupings that arise from their activities during development, are *not* "reflected in the normal transmission of information between generations of animals by the genetic mechanism."

Embryonic development constitutes endogenous adaptability, whereas Waddington (SG: 151) speaks of exogenous adaptability to signify any other capacity to react to external environmental stresses.

Natural selection is Wad's third system, and environmental stresses provoke behavioral reactions from the animal; in the fourth genetic system, which comes last in his causal sequence, mutation modifies the selected potentialities of the animal for reproduction. Both the natural selection and the genetic systems are discussed below; but Waddington explains (EE: 58) that "we have to think in terms of circular and not merely undirectional causal sequences. At any particular moment in the evolutionary history of an organism, the state of each of the four main subsystems has been partially determined by the action of each of the other subsystems. The intensity of natural selection forces is dependent on the condition of the exploitive system, on the flexibilities and stabilities which have been built into the epigenetic system."

Waddington (EE: 218) defined as *epigenetics* "the branch of biology which studies the causal interactions between genes and their products which bring the phenotype into being." A lifelong neologizer (typically on the basis of classical Greek words: Robertson, 1977: 613), he came to use "epigenetics" to describe also the causal analysis of the processes of development; and he used "epigenetic landscape" (Robertson, 1977: 593) to refer to "a visual analogy of the stable pathways of development," for which he eventually coined the word *chreod*, signifying a necessary path to which a system returns after disturbance. (Actually, he coined it in 1952 as *creode*, which is how he spells it in SG and EA; it appears as *chreod* in EE.) The dynamic movement through a "time-extended course" of phenotypic change (transversing such an analogical landscape) he referred to as "homeorhesis" (EE: 221). But (SG: 122) "evolutionary processes affecting the epigenetic *canalisation* of the individuals in the population will arise in connection with the effects of *both genic and environmental variability*" (emphasis added), with *canalised* denoting internal mechanisms that tend to regulate its development towards an initiated end. "An epigenetic system which is relatively unresponsive to genetic variation must almost inevitably also show considerable stability in the face of environmental variations, since the environmental conditions must usually produce effects on the developing processes similar in kind to those which could be produced by gene alterations" (c.f. SG: 76, 78, 88, 127). Elsewhere (EE: 104; also pp. 255–56) he explains further that

> natural selection will tend to eliminate those alleles which in the normal environment cause the development of abnormal phenotypes; this may be called normalizing selection. A rather different type of natural selection will tend to remove those alleles which render the developing animal sensitive to the potentially disturbing effects of environmental stresses, and will build up genotypes which produce the optimum phenotype even under suboptimal or unusual environmental situations. This type of selection has been referred to as "canalising selection," since it brings it about that the epigenetic systems of the animals in the populations are canalised, in the sense that they have a more or less strong tendency to develop into adults of the favored type.

But it is (SG: 132) "the totality of circumstances physically exterior to the organism which affect[s] its reproductive contribution to the next generation."

> It must be too difficult for natural selection to produce organisms which always respond in a perfectly adjusted adaptive manner to fluctuating environmental circumstances, and that

> *faute de mieux* it tends to fix, by canalisation, a type which is reasonably well adapted to the situation it will most frequently encounter. When this occurs in a population living in an environment which remains relatively unchanged for considerable periods, it is the process which we have called canalisation selection. When it happens to a subpopulation which is carrying out exogenous adaption to a new environment, it converts this into a pseudoexogenous adaptation, and the "acquired character" becomes genetically assimilated. (SG: 168)

"Epigenesis" appears in the seventh edition of the *Penguin Dictionary of Biology* (Abercrombie et al., 1980: 106), but it is very narrowly defined as the "origin of entirely new structure during embryonic development" (cf. "preformation"); "genetic assimilation" also is listed (p. 127) but only to be crossreferenced to "acquired characteristics inheritance of"—which discusses primarily neo-Lamarckism and explains genetic assimilation as imitating Lamarckism *but depending on mutation.* This definition misses or ignores completely the point of Waddington's use of the term, but constitutes no doubt a fair representation of neo-Darwinian/Mendelian stonewalling of the heresies of apostates such as Waddington—who was entirely familiar with that process, as exemplified by his description of the role of William Bateson (the father of Wad's college friend and colleague Gregory) as recounted by Arthur Koestler in *The Case of the Midwife Toad,* which Wad reviewed (EE: 177–81). Wad's own stance towards Lysenko was very pragmatic, although the same cannot be said of Wad's apparent attitude towards his own Soviet counterpart, Schmalhausen (see the exchange of correspondence between Wad and Dobzhansky, in EE: 96–98).

Waddington argued (SG: 176) that "'acquired'" characters may become inherited *without* our having to suppose that the external conditions have been responsible for calling into being the necessary genetic determinants . . . [and there are] strong indications that the genetic basis for the assimilated genotype was in fact present in the initial population, while there is little that positively suggests . . . that the environmental stress has called for the new variation" (emphasis added). His own view (EE: ch. 8–10) was that the relevant change took place *not* in genes per se—hence, there was no need to presume mutational causation (EE: 22)—but rather in the *reorganization* of the transactional effects of preexisting (but not necessarily expressed) genes: such change was in genotypic *organization* rather than in *genes.* Thus (EE: 69) "in all cases in which complete or nearly complete assimilation has been achieved, the process has involved changes at many loci throughout the whole genotype." So (SG: 178) "the process depends on the utilization of genetic variability present in the foundation stock with which [an] experiment begins—and irrespective whether that takes place in the laboratory or in nature due to changing environmental conditions. An alternative way to state this is that environmental stress changes the *expression* of genes before it changes genes themselves." (But, of course, it changes behavior first, as was discussed, above, in regard to Wad's concept of "exploitive systems.") After such genetic reorganization has begun, however, the mutation that occurs will relate to it differently than could have occurred earlier (EE: 56): "Genes newly arising by mutation will operate in an epigenetic system in which the production of such co-ordinated adaptive modification has been made easy." Thereby (EE: 91) "a phenotypic character, which initially is produced only in response to some environmental influence, becomes, through a process of selection taken over by the genotype, so that it is formed even in the absence of the environmental influence which had at first been necessary."

If the effect of environments is most direct on animal behavior (Wad's "exploitive system") and, of course, vice versa, less direct on the animal's physiological responses (to environmental stress), and even less direct on epigenetic systems, then the recanalization of developmental effects through variance in epigenetic responses (i.e., changes in pleiotropic expression) will result in "genetic assimilations." Consequently, changes in multi-gene expression, shifting also through developmental time, will result in genotypic variation: it is the possible and expressed genetic effects that count.

Waddington summarized his theory of epigenesis as follows (SG: 188–90):

> For the ability to develop adaptively in relation to the environment [natural selection] will build up an epigenetic landscape which in its turn guides the phenotypic effects of the mutations available. In the light of this, the conventional statement that the raw materials of evolution are provided by random mutation appears hollow. The changes which occur in the nucleoproteins of the chromosomes may well be indeterminate, but the phenotypic effects of the alleles which have not yet been utilized in evolution cannot adequately be characterised as "random"; they are conditioned by the modelling of the epigenetic landscape into a form which favours those paths of development which lead to end-states adapted to the environment. . . . The fundamental characteristics of the organism . . . are time-extended properties, which can be envisaged as a set of alternative pathways of development, each to some degree . . . a creode towards which the epigenetic processes exhibit homeorhesis. And in this way we can conceive of organic Form, not only as occupying four dimensions instead of only three, but as comprising potentialities as well as what is actually realised in any given individual. . . . Natural selection . . . builds into the epigenetic system tendencies to be easily modifiable in ways which are adaptive. And since animals, at least, have some ability to choose their environment, natural selection has still another dimension [with] which to work.

Waddington was persistent and emphatic (and no doubt, also right) in his insistence that selection only operates on genes in vivo—that is to say, indirectly as a consequence of direct selection effects on the phenotypes of organisms. But (as he likewise asserted: EE: 253, and ch. 17–21 and 26) individual organisms do not evolve; they may live and always eventually die, and hence they are selected. "Selection depends on the ability of *organisms* to leave offspring. It is the phenotype throughout its development which is exposed to the rigors of the environment; which falls prey to disease, the inclemency of natural conditions, or predation; which is successful or unsuccessful in mating, which is fertile or infertile" (SG :65).

It is *populations* of organisms that evolve, although (EE: 208) "populations exist in heterogeneous environments, so that the applied selection criteria are not the same for all individuals." Thus (EE: 280) "entities which undergo evolution are not simply populations of genotypes but are populations of developing . . . organisms." More generally (EA: 89–90):

> natural selective pressures impinge not on the hereditary factors themselves, but on the organisms as they develop from fertilized eggs to reproductive adults. It is only by a piece of shorthand, convenient for mathematical treatments, that indices of selective value are commonly attached to individual genes. In reality we need to bring into the picture not only the genetic system by which hereditary information is passed on from one generation to the next, but the "epigenetic system" by which the information contained in the fertilized egg is translated into the functioning structure of the reproducing individual. As soon as one begins to think about the development of the individuals in an evolving population, one realizes that each organism during its lifetime will respond in some manner to the environmental stresses to which it is submitted, and in a population there is almost certain to be some genetic variation in the intensity and character of these responses. . . . Mutations, which we can think of as random when we are considering nucleoproteins in the chromosomes, will have effects on the phenotype of the organisms which are not necessarily random, but which will be modified by the types of instability which have been built into their epigenetic mechanisms by selection for response to environmental stresses.

Robertson (1977: 613) thought that "as an experimentalist [Waddington] will be remembered mainly for two things—firstly, the work on induction in birds and mammals and, secondly, the experiments on genetic assimilation." But I think it is possible that his most important theoretical contribution is found in his sophisticated concept of time as a parameter of change in biological systems. Wad occasionally mentioned quantum theory in his biological writings, but not the relativity theory[4] which is most obviously relevant to his four-dimensional space-time theory of biological time, which I shall now describe.

Wad posits three scales: (1) population of organisms that change both phenotypically and genetically in the slow time of evolution; (2) individual organisms that develop throughout their lifetimes, initially embryonically from fertilized eggs, and subsequently through exogenous epigenesis as their phenotypes mature and reproduce in phase with a time scale of moderate duration; and (3) the organ systems of all the individual organisms which remain, during the organism's lifetime, in continuing physiological flux but within sharply defined limits of range, in what is (relative to either of the other scales) very fast time. In his own words (SG: 6–7; cf. EE: 218), with emphasis added:

> In the biological picture towards which we are finding our way, the three time systems will have to be kept in mind together. That is the feat which common sense still finds difficult. Even in current biology, most of our theories are still only partly formed because they leave one or other of the time scales out of account . . . [and] all three time-scales are essential

[4]His most extensive and sophisticated discussion of modern physics theory—with which he was evidently quite familiar—came in his book of modern *art*, in which he discusses particularly Einsteinian special relativity, in relation to the equivalence of multiple frames of reference, and cognition (1969: 9, 13, 39, 104). The only place where I found him to discuss quantum mechanics in relation to biology is in the context of their conjoint implications for the work of (and for the understanding of) modern painters (1969: 104–6, 108–9, 113, 116, 118).

for the understanding of a living creature. One might compare an animal with a piece of music. Its short-scale physiology is like the vibrations of the individual notes; its medium-scale life-history is like the melodic phrases into which the notes build themselves; and its long-scale evolution is like the structure of the whole musical composition, in which the melodies are repeated and varied.

His fourth dimension of course is time, the one common to all three of his scales. Thus Waddington proposed a dynamic concept of mutual feedback for genotype and phenotype *within*, and *between* and *among*, individual organisms; and also between and among populations of organisms, of both the same and diverse species. This puts life in a theoretical frame of reference with the momentary "present" for any individual related by development to both its historical past and potential future, and by reproduction, to both the historical past and potential future of the species. Such a striking insight cries out (in my view) for extrospection in terms of an equally sophisticated perspective of time that is informed by the philosophy of modern physics (e.g., Grunbaum, 1973; Reichenbach, 1956; 1958; Salmon, 1980).

EVOLUTIONARY ETHICS

Only a geneticist could appraise the extent to which the experimental empirical data with which Waddington illustrated and purported to support his theory of epigenetic development and variation have been superseded during the sixty years since his research began. But as a social scientist I am qualified to evaluate the quality, the utility, the extent of obsolescence, and the relevance of Waddington's sociopolitical ideas. He presents these as a scientific philosophy of ethics that is logically derived from his scientific, biological, evolutionary theory discussed in the immediately preceding section of this review.

It should be clear from what I have already said that Waddington's political philosophy is strictly normative and nonempirical, because it is based entirely on deductive inferences from another theory (i.e., epigenetic evolutionary theory in biology) and is supported by no systematic observations whatsoever of political behavior. Neither does Wad evince any familiarity with modern political theory, at least in the sense in which that is understood by most political scientists. Consequently, the values projected (to use the psychological concept—cf. Gianos, 1982: 40) in Wad's *Ethical Animal* are patently personal derivatives of his own social class and cultural indoctrination: that of a middle class gentleman, scholar, and by no means least) patriot. I am aware that this essay is a retrospective appraisal of a distinguished and deceased scientist schooled in a discipline other than mine; but Waddington was neither Alfred North Whitehead nor (most assuredly) Bertrand Russell. Robertson (1977: 613) remarks that Wad described himself as "leftist," but that he never was politically active, and that the leftist inclination of his earlier writings probably reflected the influence of his Cambridge habitat and companions. His pro-Soviet remarks in 1941 ("the socialists have made out a strong case that they would be able to organize a better system [than that of the British Empire under Chamberlain]. It is difficult to overlook the considerable increase in wealth in the Soviet Union, the only socialist country, throughout the twenty years during which the capitalist countries suffered an appalling depression and finished up in a war" [1949: 113]) overlook the fact that "socialists" ended up in the same war and for the same reasons as the "capitalists"; no doubt Wad's views here are best compared with the smiling poster of benevolent old "Uncle Joe" that adorned United States army bulletin boards in induction camps at about the same time that he wrote, and hence rest on no worse—nor better—"scientific evidence" than did Wad's own effusion. Since the publication of the third edition (1968) of his *Scientific Attitude* (cf. Robertson, 1977: 613), Wad's social and political views must be described as conservative, certainly by the standards of his own day, and even from the perspectives of contemporary English or American society, both of which have themselves become increasingly conservative during the 1980s.

Waddington portrays humans (EA: 14) as the "complete master[s]" of "technological advance" that "was bound to sweep over the rest of the world." He asserts "that the recent changes in the human condition can be considered as progress. In Western Europe, where the whole process originated and has been going on longest, the word progress is now extremely unfashionable. Anyone who is bold enough to assert that it has occurred or even that the word has a definite meaning, is likely to be dismissed as merely naive and unsophisticated. I hope to show that this accusation is unjustified. One can be quite sophisticated . . . and still believe in progress." Furthermore, Wad believed (EA: 133–34) that "if there are any biologists who, while accepting the notion of evolution, reject that of evolutionary progress, they must [be] doing so merely in order to provide grounds for some future argument. I think that all biologists who have no ulterior ends in view have always from the time of Aristotle, agreed that one can discern a real hierarchy or progression in the forms of the organic world." Consequently (1948: 114), "the important line in politics is between those who judge the value of a society by its efficiency in maintaining itself and by its advance along the whole line of human evolution, *and those who judge it by some other criterion, whether based on mysticism, nostalgia for the past, or motives of personal advantage*" (sic: emphasis added). This is simplistic ideologism; Victorian Englishmen of the upper classes, together with Social Darwinists everywhere, believed in progress too (see Hofstadter, 1944: 31, and *passim*).

By *hierarchy* Waddington meant a dynamically stratified and scientifically justified Great Ladder of Being (in lieu of the static one derived from Christian theology, which Wad of course disavowed), ranging from "such elementary organisms as bacteria and viruses" to mammals, with humans at the top of the structure (and therefore closest to the perfection towards which all living beings are pointed).

Naturally (EA: 100), "the processes of evolution in the subhuman world do not, I feel, have of their own nature any ethical quality whatever." Thus, other beings not only have no ethics; they are also relegated to the *sub*human world, a vision of superiority that he never gave up, because his major contribution to the (for him, posthumous) book that Erich Jantsch and Wad coedited (1976) was a chapter entitled "Evolution in the Sub-Human World." His concept of hierarchy impelled him to several other chauvinisms also—those of age, sex, and social class (EA: 113): "The task of getting new ideas through [a *man's*] skull becomes progressively harder after a certain age"; his sexual bias is demonstrated throughout the publication, as exemplified in the immediately preceding quote and by his invariant linguistic practice of specifying the male gender when his clear reference group is the species, humans, generally, people as distinguished from "lower" animals (Jantsch and Waddington, 1976). Even in attempting to advocate equal opportunity for all, Wad is betrayed by his sexist bias, as exemplified by the concluding sentence to his foreword to the third edition of *The Scientific Attitude* (1968: xxxi): "The social and political system should make it as easy as possible for everyone—male and female, clever and stupid, tough or weak—to develop as far as possible every potentiality that they have in them." Evidently, in his thinking, humans can be divided into two classes: a positive class consisting of tough and clever males; and a negative class of weak and stupid females.

But he also committed reverse age discrimination, remarking (EA: 102; emphasis added) that "in *man*, the processes of teaching by the older members of the population and learning by the younger ones, have been carried to an incomparably higher pitch than is found in any of the prehuman forms of life, where they play only a relatively minor role." This is contradicted by a growing research literature in both experimental animal behavior and field ethology and primatology, testifying to the major importance of play and autogenous group learning processes among juveniles of most primates and many carnivores as well; and enough of this literature had become available during Wad's lifetime (Groos, 1898; Welker, 1956; Hutt, 1966; Mason, 1965) to have cautioned a writer less ideologized on this question. Wad was also very late in recognizing ethology (EA: ch. 12 was at the time of its publication out-of-date ethology and primatology); and his references to human culture vis-à-vis communication among "lower animals" evince no awareness of the then available research on animal culture, including such innovators as the celebrated Japanese macaque Imo, whose exploits (to use Wad's own term) were first observed in 1952. He did, however, anticipate behavioral ecology (EA: 94), which was still embryonic at the time he wrote in 1959.

For examples of Wad's class biases, perhaps the following (from EA: 191–92) will suffice: "The pre-eminently stable British political system has been based on the acceptance of a social hierarchy"; he also rejects the "misinterpretation" that "a belief in the virtues of social organization . . . must be countered by a belief in . . . social freedom." Indeed, in his concluding chapter (EA: 205) he avows that "the major ethical problems of today in the context of individual-to-individual behavior . . . have to be sought in those types of attitude and activity which facilitate or hinder the development of a healthy authority structure."

The achievement of such a goal requires appropriate treatment of the developing child, a task much facilitated by "the authority structures within the mind" (EA: 105); but to comprehend how he had arrived at such a conclusion we must go back to the beginning of his book, where he initiates the discussion of "biological wisdom" (i.e., criteria for evaluating human social ethics). There (EA: 29) he asserts that "the most crucial step in the whole chain of argument is the thesis that the development of a newborn infant into an authority accepter—an entertainer of beliefs—involves the formation within *his* mind of some mental factors which carry authority, and that it is some aspects of these same authority-bearing systems that are responsible for *his* simultaneously moulding into an ethicizing creature" (emphasis added). Furthermore, "the function of ethicizing is to make possible human evolution in the sociogenetic mode." "Authority acceptance" by humans is justified over and over throughout the book as the fundament upon which civilization rests; for example (EA: 151), he stresses that "the functioning of a sociogenetic system depends essentially on the existence of the role of authority-acceptor. In *man* the formation of this role is brought about by the processes which involve the internalization of authority. And some aspects of this internalized authority have the character for which we give the name ethical. Thus 'going in for ethics,' or 'ethicizing,' is for *man* an integral part of the role of the taught or the authority-acceptor, without the existence of which his cultural sociogenetic evolutionary system could not operate" (emphasis added). Wad makes it clear that in his view, neonates are authority-acceptors *not* because they are smaller, weaker, and (for the best of biological reasons) dependent on older caretakers, but rather because of an *innate need to obey*! This is literally as well as figuratively incipient authoritarianism.

Waddington explicitly states (EA: 15) that his goal in the book is to justify, in terms of the results, the "scientific revolution" that had brought about the good life that he saw around him. Chapter 14 of the book is an apotheosis to the virtues of science. It is the more ironic that the very technology that he admires so much had made his own informational base badly out of date already when his book was published in 1960, and bizarrely so twenty-five years later. On that same page, Nazis are denoted as evil in their use of scientific technology; no corresponding mention is made of the British role in Ireland (to use an example that must have been as familiar as that of Nazi Germany to Waddington); or in India (which he surely knew better than the Third Reich, because he spent his early childhood on a tea plantation in Southern India, from which his parents did not return to resume residence in England until after his first marriage); or in the Boer War in South Africa (to take an example slightly before his time). On the next page Wad remarks that "it may require quite subtle thought to enable us to decide whether the use of an H-bomb is more or less evil than some other course of action"; and three paragraphs later, he ponders rhetorically: "Is it as evil for a state to order the explosion of a bomb, whose fall-out will ultimately, over several generations, cause the death of, say, a *thousand* people from harmful gene mutations, as it is for another state to order its police to shoot a thousand people personally in the back of the head? I do not think the answer is altogether obvious" (emphasis added). No doubt he might have found the answer easier had he appreciated that his empirical information about thermonuclear blast fallout casualties was off by a factor of 10^3 even at the time that he wrote. Any person reasonably well informed on the subject ought to have appreciated *that* within a decade after Hiroshima; and no doubt now the relevant factor would be closer to 10^5 (depending on demographic, geographical, and climatic parameters of the point of detonation). On page 115 he discusses NVC as "a very primitive kind of social transmission of experience," of a type upon which lower animals than humans may have to rely, but which (in his view) has little importance to the exchange of information in human society; and again, Wad's ideas were long outmoded by considerable scientific research (Birdwhistell, 1952; Hall, 1959; van Hooff, 1962; Andrew, 1963; cf. the bibliography in Hinde, 1972). These examples are provided here primarily to suggest the range and depth of Waddington's proclivity for passing off as scientific knowledge his personal biases and ignorance.

Much more crucial is Waddington's (EA: 204) argument leading "to the conclusion that *biological wisdom* consists in the encouragement of the forward progress (anagenesis) both of the mechanism of the socio-genetic evolutionary system, and of changes in the grade of human organization which that system brings about" (emphasis added). He welcomes (EA: 210) "in the name of biological wisdom the dawning—if it does dawn—of a period of 'co-existence' between the two major cultural forces of the world today . . . Capitalism on the one side, and Communism on the other." If this is the best that biological wisdom has to offer towards the avoidance of nuclear warfare, then heaven help us all. On the other hand, Wad suggests (EA: 216) that too much assistance, *from* the technologically superior and scientific societies, *to* underdeveloped countries, is *against* biological wisdom because it would "result in too great a homegeneity" (i.e., too much economic and social equality among nations). It certainly was a bad fit for the 1960s, at least in the United States; but the book is still in print (in a University of Chicago Press Midway reprint), and it is not impossible that the times may finally have caught up with it. Given the contemporary prominence of "Libertarians," Reaganomics, and the new conservatism that is sweeping America in the mid-1980s, EA may be about to enjoy a rebirth in popularity. No doubt Wad would feel vindicated, notwithstanding the slight annoyance of a pound that almost achieved parity with the dollar at the time this essay was being written.

It is possible to deduce from epigenetic theory a very different prescription for human cultural evolution than what Waddington infers. I (for example) would point to the diversity of alternative developmental futures that any political society (population of human organisms seeking to control each other's behavior) might select, depending upon the explicit characteristics of the particular political environment that such an animal population chooses as its niche. The political genotype for the society, an artifact of cultural ("sociogenic" in Wad's vocabulary) evolution, would be subject to epigenetic development (constitutional "interpretation") as a consequence of environmental stress (cf. Schubert, 1967c; 1975b; ch. 12; 1985a). Waddington (SG: 109) attributes the genetic fitness, of a sufficiently large and freely breeding animal population, to the diversity of its gene pool; but he does not deduce from this analogy—as I would—the indispensability of sociopolitical policies and goals of egalitarianism, in order to protect and preserve for political purposes the precious virtues embodied in the wide-ranging individual differences of any human population.[5] The political—the *constitutional*—fitness of a nation, as one might plausibly argue (and as Jefferson certainly did, and without the benefit of knowledge of any biological theory of epigenesis to analogize from), depends causally upon the maintenance of the diversity and individuality of its people—not upon their increasing conformity and organization into some superorganismic holistic entity. So my first point is that, consistent with the premises and specifications of his theory of epigentic processes, Waddington might have suggested a very different content for a theory of evolutionary ethics (but only if he himself had been a different person than he was, believing different things than he did).

On the other hand, a theory of political change that is based on the relatively slow rate of change of Darwinian phyletic gradualism (Hofstadter, 1944: 7) or even of individual ontogeny (because the consequences of such biological change for an individual organism are likely to be disastrous: immediate death, or, which amounts to the same thing in a natural environment, severe disablement) is set in premises that make either drastic or sudden transformation undesirable. (But for a contrary political scientist view of such matters, see Masters, 1982a.) For a contrasting alternative, based on a different but increasingly popular component of evolutionary theory, one might consider the political implications of punctuated equilibria theory (cf. Peterson and Somit, 1983b; and ch. 12, above): here the biological theory stipulates that change does come drastically, and suddenly, but with biologically desirable results, that is, a phenotypically modified *population* representing a new species that is better adapted (than was its parent population) to the novel requirements of a suddenly altered environment. This is a biological theory for which the metaphor of political change is revolutionary rather than evolutionary; but it has implications for more than one kind of revolutionary change. In its more extreme form (where it might be analogized to speciation) punctuated equilibria theory is a model for such dramatic cultural revolutionary changes as those exemplified by the American Revolutionary War, or by Kuhn's theory of

[5]His espousal (1969: xxxi, in the sentence quoted earlier in relation to his sexism) of a weak-kneed version of Maslowian psychology is done in the name of the creation of a more admirable global "modern civilized personality"—not as the consequence of any biological imperative such as species adaptability, which is of course the posture that he had taken almost a decade earlier in EA.

the displacement of a reigning paradigm in science. It can also, however, serve as a model for more limited but still revolutionary cultural change, such as the American Civil War or the assassination of either of the Kennedy brothers during the 1960s.

The theory of epigenesis, in either its Waddingtonian (genetic assimilation) or Geistian (dispersal phenotype) form, proffers a model of biological change that is much faster than gradualism and perhaps provides one at least partial explanation or operationalization of the mechanisms for (in *geological* time, rapid) change for animals such as birds and mammals. If gradualism provides a biological model for sociopolitical theories of social stasis and if punctuated equilibria is a biological model for sociopolitical revolution, then epigenesis can be understood to suggest a biological model for social change that is transactional, and recursive, with cultural change accelerating genetic (constitutional? regime?) change in bringing human political societies into better adaptation with dynamic environmental change—and vice versa. This seems to be precisely the direction in which contemporary research in cultural (including political) evolutionary theory has been moving (Pulliam and Dunford, 1980; Lumsden and Wilson, 1981; Cavalli-Sforza and Feldman, 1981; Corning, 1983; and see ch. 10, above.

CONCLUSIONS

There is no doubt that Wad fraternized with a broad coterie of leading philosophers and, even more so, of the leading artists of his day (cf. his foreword to Waddington, 1969), as well as all of the major British biologists (and many such in America and on the Continent), plus selected social scientists (e.g., his friend Gregory Bateson's wife Margaret Mead); but of these perhaps his most personal reference group was the poets and painters. If so it is no coincidence that his best statement of organization theory, which was of course central to his early work as an embryologist as well as his ideas about politics, is found in his synthesis of science and art: *Behind Appearance* (1969). But he will not be remembered for his contributions to philosophy; and neither (for the reasons I have discussed above) will he be highly esteemed—save by other conservative ideologues—for his contribution to political and social theory. It has been necessary to discuss his views on both philosophy and politics because he himself entertained such a holistic (Whiteheadian) image of his own intellectual contribution. He claimed (and evidently believed) that his philosophical and political theories alike were *the* logical imperatives of his biological theory and knowledge; logically (given the Whiteheadian premises) he ought also to have claimed "and vice versa"—but he did not do so. It is not unusual to find highly qualified politicians who are lousy biologists; nor indeed, highly qualified political scientists whose knowledge of biological theory or facts would not earn them a pass in a first year biology course at the junior high school level. There are indeed very few Leonardos in either philosophy or biology, and no doubt, not all that many more in politics (to say nothing of political science). So it should come as no suprise that Waddington, although an exceptionally brilliant developmental biologist (and apparently, a modern art critic and connoisseur of importance), was not that outstanding as a creative thinker in some of the other fields (such as philosophy and political theory) in which he maintained a lifelong interest.

My analysis indicates that Waddington was not (as Gould suggested) a "gadfly" in the field of evolutionary theory, but rather that he was a major figure in the construction of one of the several interrelated but alternative critiques of neo-Darwinism/Mendelism. With the exception of punctuated equilibria theorists, most of the proponents of the other alternative critiques acknowledge the important impact that Waddington has had upon their own thinking and approaches. The new paradigm (of which Kuhn spoke) is still in the developmental process of becoming, and the time may not yet be ripe for it to displace neo-Darwinism/Mendelism as the dominant paradigm of contemporary biology. But that reconstruction began with Waddington (1940), barely a decade after the publication of Fisher's *Genetical Theory of Natural Selection*; and it now seems almost certain that when the new paradigm shift does take place, Waddington will then become much more widely recognized (and revered) as one of the half dozen leading geneticists and evolutionary theorists of the twentieth century.

15
Senescent

This chapter reports research designed to test alternative hypotheses about the relationship between the aging and the ideology of political elites. It proposes a three-dimensional biocultural model of political socialization, with a biological abscissa, a cultural ordinate, and a third dimension of time measuring change in the other two dimensions. The model focuses upon human psychophysiological developmental change; it is dynamic by definition, and it requires longitudinal research. A merely cultural model of aging, ideology, and political behavior is limited in its usefulness by its incompleteness (Schubert, 1975a; 1976; Wiegele, 1982). The *bio*cultural approach here postulates, at least as an ultimate goal, transactional analysis of the relationships among both biological and cultural variables (White, 1972) affecting the invocation of values in judicial decision-making behavior (Schubert, 1982a) because the political elite selected for empirical study consists of justices of the United States Supreme Court.

An increasing number of political scientists are becoming engaged in research in political socialization (Schwartz and Schwartz, 1975; White, 1980b, 1981b, 1981d; Peterson and Somit, 1982) compatible with the premises of such cognate disciplines as attachment behavior (Alloway, Pliner, and Krames, 1977), primate biosocial development (Chevalier-Skolnikoff and Poirier, 1977), and neurophysiology (Davies, 1976, 1980; White, 1981a). And more of them are doing so with a focus upon learning throughout the life cycle, including the disintegration of physiology, personality, and learning that is best described as desocialization (Schubert,

---

This chapter is a revision of "Aging, Conservation, and Judicial Behavior," *Micropolitics*, 3: 135–79 (1983). Reprinted with permission of Crane, Russak, & Company, Inc.

1984b; cf. Jaros, 1972; and Jaros, 1973, pp. 68–70). The underlying question is how and why judges (like other humans) learn, modify, remember (and forget), and sooner or later lose, their values (Peterson, 1982b; Schubert, 1981b).

The structure of the discussion below is as follows. First we examine the "bellyache" theory of jurisprudence. Next we shall consider briefly some general psychological theories of conservatism, and then in more detail the contemporary literature on judicial conservatism. That is followed by a discussion of aging and value articulation, with particular reference to the sociology, psychology, and biology of senescence, on the one hand, and gerontological theories of cohort dissonance, on the other hand; the section reviews the use of both approaches in previous research in judicial behavior. The design of an empirical test is then stated in relation to an operationalization of both hypotheses, with time intervals used to cluster the data. The sources of the data are specified, and their statistical properties are summarized. The conclusion is an appraisal of the implications of the research.

THE "BELLYACHE" THEORY OF JURISPRUDENCE

It should no longer be necessary to confront the superannuated bogey of Social Darwinism (Hofstadter, 1944; Halliday, 1971; Gilman, Simon, and Zegura, 1979) as though *that* specter were the only viable alternative to the austere environmentalism that continues to characterize most of contemporary social science (Campbell, 1975; Hines, 1979; White, 1980b). Readers who disagree with this assertion may do so because of their unfamiliarity with the orientation as well as the scope of modern social biology (Brothwell, 1977; Berstein and Smith, 1979; Chagnon and Irons, 1979; van den Berghe, 1979; Lockard, 1980). It is necessary, therefore, to explain certain differences and similarities between bellyache jurisprudence and the interactive biological/cultural hypotheses about judicial aging that we will consider subsequently.

The bellyache theory, as enunciated by its progenitor, seems remarkably in tune with the modern climate of opinion—as the recent translation used here indeed *is*. The young Marquis Cesare Beccaria, who became an instant international celebrity on the heels of the initially anonymous publication of his essay in 1764, asserts that

> each man has his own point of view, and, at each different time, a different one. Thus, the "spirit" of the law would be the product of a judge's good or bad logic, *of his good or bad digestion*; it would depend on the violence of his passions, on the weakness of the accused, on the judge's connections with him, and on all those minute factors that alter the appearances of an object in the fluctuating mind of man. Thus we see the lot of a citizen subjected to frequent changes in passing through different courts, and we see the lives of poor wretches become the victims of the false ratiocinations or of *the momentary seething ill-humors* of a

judge who mistakes for a legitimate interpretation *that vague product of the jumbled series of notions which his mind stirs up.* (1963: 15; emphasis added)

Speaking over a century and a half later, United States Supreme Court Justice Benjamin Cardozo took considerable advantage of the insights made possible by the intervening neo- Darwinian synthesis of evolutionary theory, referring at several points to Jethro Brown's (1920) then recent law review article on "Law and Evolution" and scattering explicitly biological metaphors or discussions throughout the text of his lectures.[1] Indeed, Cardozo anticipates both of the major hypotheses to be tested here, the biological one implicitly and the cultural one explicitly. Implying what today would be called "the limbic system" (MacLean, 1958)—if not directly the effect of aging per se upon cognitive processes and the limbic system—Cardozo remarks (1921: 167–168) that "deep below consciousness are other forces, the likes and dislikes, the predilections and the prejudices, the complex of instincts and emotions and habits are convictions, which make the man, whether he be litigant or judge." Regarding cohort dissonance and the generation gap, he expresses his empathy

> with the doctrine that judges ought to be in sympathy with the spirit of their times. [But] assent to such a generality does not carry us far upon the road to truth. In every court there are likely to be as many estimates of the "Zeitgeist" as there are judges on the bench. . . . *The spirit of the age, as it is revealed to each of us, is too often only the spirit of the group in which the accidents of birth or education or occupation or fellowship have given us a place.* No effort or revolution of the mind will overthrow utterly and at all times the empire of these subconscious loyalties. (1921: 174–75; emphasis added)

Surely Felix Cohen is wrong when he states (1935: 843) that "[l]aw is not . . . a product of judicial bellyaches." The biochemical and neural consequences of indigestion are of recurring importance to middle-aged or older and characteristically sedentary judges—and probably of somewhat greater importance for them than the Freudian id that so titillated legal realists of Cohen's generation (Frank, 1930, 1931).[2] But he is on unimpeachable grounds in his conclusion (Cohen, 1935: 945) that "[l]egal . . . decisions can be understood only as functions of human behavior," behavior that inescapably is the product of *both* biopsychological and sociocultural influences.

[1]Cardozo (1921: 21, 33, 98–99, 103, 110–11, 126, 166–68). With *Human Nature in Politics* (Wallas, 1908) in print barely a dozen years, Cardozo (1921: 13) showed similar precocity in advising Yale law students to "apply to the study of judgemade law that method of quantitative analysis which Mr. Wallas has applied with such fine results to the study of politics."

[2]Today we can distinguish between the substantive paucity of neo-Freudian monotheism as a positive agenda for inquiry into human behavior and its role as a negative foil to mechanical theories of jurisprudence. We ought, also, to distinguish more sharply between the biological basis of Freud's own work (Sulloway, 1979) and the shift to cultural methods and environmental theories of Freud's latter-day followers. It simply is not justifiable today to believe that the bellyache theory, however it may have been restricted to heuristic influence in the past, entails any necessary implication of biological determinism (Geist, 1978; ch. 13, above).

CONSERVATISM

Aging and Political Conservatism

Sociologists have scoffed (e.g., Glenn, 1974) at the proposition that aging, which results in conservatism in behavior for animals generally, in humans leads also to conservatism in beliefs. But the process is well understood and abundantly documented: aging entails increasing physiological incompetence (maladaptability) that begins for humans (typically after the peak of stress resistance is achieved, around age ten) somewhat *before* sexual maturity is reached; decreasing behavioral competence leads (through learning) to even greater conservatism in behavior; and that entails, for humans, additional positive feedback in the form of increased psychological conservatism in attitudes (Bromley, 1974; Weg, 1975; Comfort, 1979a; Riley, 1979; Fries and Crapo, 1981; Riley, Abeles, and Teitelbaum, 1982). Of course there is tremendous individual variation—both within and between individuals—so that with odds of perhaps one in a billion an Einstein in his midseventies, or even a Russell in his midnineties, will be less conservative than more than 99.999+ percent of all humans at any time during the latter's lives. When such geniuses appear, human institutions are modified to adapt to *them*. We need not be concerned to construct retirement systems for such institutions as the Presidency or the Supreme Court in anticipation of the advent of geniuses, whose incumbency in those political roles will be at least as rare as the Second Coming awaited with such enthusiasm by some millennarians (cf. Festinger, Riecken, and Schacter, 1956).

Cutler and Schmidhauser (1975: 379), each of whom has pioneered in the study of the effects of age upon political behavior, have concluded:

> When considering generalizations about age and political attitudes . . . there are several processes which must be taken into consideration. For some issues, age may be quite irrelevant to attitudes, or at least secondary to factors which are most salient. For other issues, the individual's attitude might be a reflection of a more general ideology, and this ideology, in turn, may be related to maturation and to such political factors as political party identification. For still other issues, the connection between age and attitudes may hinge on the fact that a particular age group may receive the benefits to be bestowed by a particular program or policy.

It is true that substantive political beliefs and affiliations do not necessarily change with maturation, but there is a psychological dimension of conservatism, according to Jaros, "that *does* appear to increase with age. Older people are less likely to deviate from the political principles to which they have become socialized, regardless of whether these principles are liberal or conservative. That is to say, an older liberal is likely to be a more inflexible liberal; an older conservative is likely to be a more inflexible conservative" (1973, p. 74; emphasis added). This psychological dimension is Eysenck's (1954, ch. 5; and 1967; Wilson, 1973: 26; and Gow, 1981) tough/tender-mindedness, which has been discussed in previous research in regard to both American and other Supreme Court judges, and also in relation to other types of political actors (Schubert, 1977a, 1977b, 1980b).

A comprehensive analysis of the psychology of conservatism appears in a symposium edited by Glenn Wilson (1973). In his introduction, Wilson distinguishes among four overlapping but contemporary concepts about the nature of conservatism: resistance to change, "playing safe," the "generation gap" (i.e., cohort differences), and the internalization of parental prohibitions. He defines conservatism substantively in terms of religious fundamentalism, proestablishment politics, insistence on strict rules and punishments, militarism, ethnocentrism and intolerance of minority groups, preference for the conventional in art and clothing and other aesthetic questions, antihedonistic outlook and restriction of sexual behavior, opposition to scientific progress, and superstition. These are all familiar topics to those who know the attitudinal research of Adorno et al. (1950), Comrey and Newmeyer (1965), McClosky (1958), Robinson, Rusk, and Head (1968, 1969), and Rokeach (1960). The bulk of the symposium presents evidence collected with survey methods by Wilson and the nine collaborators who are his coauthors; this evidence, analyzed by scaling and factor analysis, is aimed at the development of a scale instrument to measure conservatism and its use with regard to children, religion, racialism, superstition, aesthetics, humor, temperament and personality, within-family differences, mental disorders, and ethnocentrism.

Of particular relevance here is the clear finding (p. 41) "that raw conservatism scores increase markedly in adulthood as a function of age." Wilson's data, based on a survey of several hundred New Zealanders, show (p. 55) that males are least conservative at age 15, only slightly more so at 25, but then increase sharply in conservatism during the next three decennial intervals (i.e., at 35, 45, and 55), after which they continue to become more conservative but at a less accelerated pace. The growth in conservatism is most rapid between 35 and 45, which, we can note in passing, is in phase with general debilitation in physiological structure and function, but also is at least somewhat *before* indicia of cognitive senescence (e.g., forgetfulness) usually are observed or experienced. Males at 65 evince almost half again as much attitudinal conservatism as do fifteen-year-olds; but of course this was not longitudinal research and hence the sexagenarians and the adolescents belong to very different cohorts. At all ages, females were more conservative than males, and this was particularly so in the case of the oldest females (but cf. Schubert, 1987). Wilson deduced from these data (p. 55) that "it cannot be said yet whether the increase in conservatism with age is longitudinal (increasing over time within individuals) or cross-sectional (a constant difference between generations), but it may be hypothesized that both tendencies are probably involved."

In his concluding chapter, Wilson presents a theory of the psychological antecedents of conservatism, depicting genetic factors and environmental ones as joint causes of feelings of insecurity and inferiority, which in turn generate a generalized fear of uncertainty which leads to behaviors of avoidance of both stimulus and response uncertainty and which is expressed verbally in what he terms a "conservative attitude syndrome." The characteristics that he imputes (p. 261) to describe the *genetic* causes of conservatism include low intelligence, unattractiveness, female sexuality, and old age.

Sociologists assert that the "myth" of increasing conservatism with age is attributable to spurious correlation and inappropriate methodologies (to say nothing of paradigms) employed by scholars not properly trained in sociology; or at least that is true of "substantive" conservatism. However, Glamser (1974: 553) does report, as the major findings of his empirical survey research with a cross-sectional sample, that multivariate analysis shows the following: "First, there was a significant positive correlation between age and conservative opinions even when a number of social and demographic variables were controlled. Second, only education [negatively] was more important than age [positively] as a correlate of conservative opinions." Glenn (1974: 177) asserts that "conservatism may be defined in terms of a system of values and beliefs about the nature of reality, *or* it may be defined *simply* [*sic*!] as resistance to change, reluctance to take risks, cognitive rigidity or some similar characteristic" (cf. Cutler, 1977: 1109). Glenn concedes that "obviously, aging may lead to conservatism in one of these [latter] senses."

Judicial Aging and Conservatism

Anyone familiar with American constitutional history in general, and with that of the United States Supreme Court in particular, is aware of the continuing problem that elderly—and that means septuagenarian, octogenarian, and even nonagenarian—justices have posed for their colleagues, to say nothing of their fellow citizens. Typically, the most elderly also were ill for more or less extended periods before they died in office or were requested by their colleagues to resign—as was true even of Holmes at ninety (Fairman, 1938; Goff, 1960). The anonymous student author of a law journal note ("Judicial Disability and the Good Behavior Clause," 1976: 720) remarks that, consequent to the liberalization of judicial retirement benefits, "of the Justices appointed in this century, only William Moody remained on the Court after his disability became evident. Not since 1910 has an incapacitated Justice lingered on the Court." This assertion clearly is wrong, both on the facts and in its premise that the chief motivation for the lingering of justices is pecuniary.

There have been many cases since Moody, including certainly McKenna during his terminal three years on the Court (1922–1925); Holmes during his final three and half months (1931–1932); Stone for the initial three and a half months of the 1936 term; Frankfurter during the four months preceding his resignation in 1962; both Black ("long after an unpublicized stroke") and Harlan (after he had become "nearly blind") throughout the spring of 1971; and Douglas twice: for more than the first six months of the 1949 term after his horse-riding injury, and throughout the ten and a half months following his stroke and preceding his resignation in 1975.[3] More than a few of the justices, ranging throughout the Court's history, have been deemed senile (e.g., Fuller: see King, 1950: ch. 17); and there appears to be no decrease in the tendency for justices to indulge in the puerile game of competing to beat John Marshall's length of tenure. There were two such contestants during the seventies: Black failed by only a few months when he resigned under the conviction that he was in a "desperate condition" and then did succumb to a series of postretirement strokes within the next eight days (Black, Jr., 1975: 248–66)—possibly at least in part in a self-fulfilling prophecy;[4] Douglas then succeeded in besting Marshall by more than two years, although he had to spend almost all of the second year in a wheelchair in order to do it.

Of the thirteen chief justices prior to Burger (who retired in 1986 in his seventy-ninth year), eight terminated their service in their seventies or older. Of these, both Marshall and Hughes were in their eightieth years, and Taney was eighty-seven. Of the eighty former associate justices (with White, Hughes, and Stone included only among the chiefs), more than half were more than seventy years of age—and of these a fourth were more than eighty—when they left the Court. Of the justices in 1983, a majority were over seventy (and that plurality averaged more than seventy-five years in age). Further details, including some descriptive statistics, are readily available in textbooks (e.g., Schubert, 1960: 57–66; Goldman, 1982; and cf. Schmidhauser, 1962a, 1962b).

Among the more noteworthy events in the constitutional history of the twentieth century was FDR's Court-packing plan, a skirmish in which he lost the battle but won the field. But with a little more patience he almost certainly would have won anyhow, without squandering so much political capital on what was a losing cause in the Congress, because all of the *natural* odds were in his favor. In January 1937, when the president launched his proposal, the *average* age of the incumbent justices was seventy-one (as compared to almost sixty-nine in 1983), and six of them (instead of only five in 1983) ranged between seventy and eighty years. In fact, eight justices left within the next four years, and FDR promoted the ninth to the chief justiceship. Roosevelt made two mistakes: he attacked the Court as an entity, thereby confounding his supporters among the justices with his enemies; and he based his assault on the grounds that the justices were too old and *therefore inefficient*.[5] That was both dishonest and a tactical error; his motivation

[3]See the discussion of the condition and behavior of Black, Harlan, and Douglas in Woodward and Armstrong (1979; reported also in *Newsweek*, Dec. 10, 1979: 87–88; and see Schubert, 1984b).

[4]Dunne (1977: 433–34) tells us "a small card on [Black's] desk bearing the years, months and days of the Supreme Court service of John Marshall and Stephen J. Field [which] indicated that within a few short months his own unprecedented term there would add yet another laurel to his record." Regarding this fatuous aide-memoire, Dunne proffers the disingenuous comment that "any suspicion that he had simply been an old man clinging to his office to set a record was dispelled" when Black did eventually resign!

[5]But Schmidhauser (1962b: 121) does state that "actually for most justices who remained on the Supreme Court in their declining years, the most striking manifestation of age was a sharp decline in the number of opinions written in each successive year."

was that he believed them to be too old and *therefore too conservative*. Unfortunately for that thesis, the eldest of all was Brandeis, who remained one of the (relatively) most liberal (Murphy, 1982), although no longer one of the most productive, of the group; but otherwise in that respect the president was correct in his assessment of the situation. The question of present interest is whether the Nine Old Men—or at least the Four Horsemen among them—were conservative because of the psychophysiological effects of aging or because of the conflict between their beliefs and the goals of the New Deal; but that difference was never at issue in 1937. The President's stated charge of inefficiency clearly implied the biological hypothesis, however, as there was then no clear support for the proposition (however true it may be) that elderly liberals are better organized and more energetic than elderly conservatives.

The most extensive and carefully designed empirical study of the effect of age upon judicial voting behavior is Sheldon Goldman's (1975) second survey of the federal courts of appeals. His sample included over two thousand split decisions for seven years, beginning in the midsixties; and he used seven independent variables (partisan affiliation, religion, previous candidacy for elective office, prior judicial experience, experience as a prosecutor, tenure as a federal appeals judge, and age) to predict to ten dependent variables (half a dozen categories; two measures of ideology, political and economic; activism [defined as pro-national in federal relationships]; and dissenting). In relation to sample sizes ranging from 52 to 125 decisions, he observed modest but highly significant simple correlations (ranging from $-.22$ to $-.39$) between age and each of seven of the dependent variables; and he found that "older judges simply tended to be more conservative on the criminal procedures, civil liberties, labor, injured persons, political liberalism, economic liberalism, and activism dimensions than did young judges" (p. 499). He also performed a multiple regression analysis of the matrices of correlations between all of the independent variables and each of the dependent variables in turn. This showed age to be either the first or second best predictor on all ten dependent variables and to be more important than any of the other predictors except partisanship. Indeed, age was more important than partisanship in influencing judicial voting in civil liberties cases and in regard to federal activism.

Lamb (1976) focuses explicitly upon conservatism in the voting of federal appeals court judges, although his study is limited to the District of Columbia circuit (i.e., only one of the eleven courts of appeals with which Goldman worked) and to only one (criminal procedures) of the ten dependent variables of the Goldman study. Much of Lamb's discussion of his data concerns the decisional behavior of Warren Burger prior to his promotion to the Supreme Court. The future Chief Justice is shown to be, for example, a slightly right-of-center moderate in a scale of all split *en banc* decisions during the period 1956-1969 (p. 265). But longitudinal analysis reveals important and systematic changes over that time (p. 270).

> Burger may be described as the court's most moderate member during the 1956 through 1958 terms. His scale score, middle ranking, and voting evenly for both the prosecution and appellants provide the basis for this conclusion. In comparison, Judge Burger behaved only slightly more conservatively from 1959 to 1964. . . . The major shift in his attitudes in criminal issues . . . [came in] the five years prior to [his] appointment . . . as Chief Justice, [when] he appears as conservative as any [other] member of the court. Not only was he tied for last place in scale scores, but Burger also failed to cast a single vote in favor of an appellant in the fourteen *en banc* decisions handed down between 1965 and 1969.

More generally, Lamb concluded (1976: 278) that "in no instance over time did a conservative judge consistently shift to join his moderate or liberal colleagues in voting in criminal cases generally . . . [W]here significant attitudinal changes did occur with age, judges became more conservative."

AGING

Biological Aging

Explicitly in regard to biological aging's implications for political socialization, see chapter 2, above (and cf. Strehler, 1962: 10–19; and Handler, 1970: 709–13) that "'time's arrow' points only and compulsively in the direction of permanent political desocialization." The thrust of that comment about acceleration in the decrements of physiological (including, to the extent that they have evolved for particular species, cognitive) structures is entirely general and applies to all forms of life, particularly once their sexual maturity has been attained. From the perspective of modern natural selection theory (Williams, 1966, 1975), individual cells and/or organisms live to reproduce; and once that task has been accomplished, it is in general inefficient and wasteful of resources for postreproductive animals to survive. In nature they rarely do;[6] and even humans—whose capacity for longevity can have changed very little, in genetic terms, in the past ten thousand years—were limited by the same principle, prior to the agricultural revolution.[7] Postmenopausal females were unknown then; and the skeletal remains of males beyond their thirties are rare, and even "until the past century mankind by and large has lived in a regimen that gave an average life expectancy ranging between 25 and 35 years of age" (Birdsell, 1972: 73, 341–44; Laughlin and Brady, 1978).

[6]According to Comfort (1979a: 31, 39, 46), "wild mice die at a rate that precludes their reaching old age, but mice kept under laboratory conditions have a life-table similar to that of Western European human populations in the year 1900. . . . Aging of the whole organism after a prolonged postreproductive period is a process that is realized only by human interference, at least so far as most species are concerned, and not 'envisaged' by evolutionary teleology. . . . Death from [senility] is itself in many species so rare in the wild state that failure to senesce early, or at all, has little value from the point of view of survival."

[7]"The potential life-span in pal[e]olithic man probably resembled our own: its realization has been possible through the development of a complex social and rational behavior" (Comfort, 1979a: 322; cf. M. Cohen, 1977).

In what remains the leading work on achievement in relation to age, Lehman (1953: 326-29, where he lists sixteen "possible causes for the early maxima in creativity") has concluded that

> a mere increase in man's longevity should not change greatly the modal ages at which man exhibits his greatest creative proficiency since, both for long-lived and for short-lived groups, the modal age occurs in the thirties. . . . Within any given field of creative endeavor: (1) the maximum production rate for output of highest quality usually occurs at an earlier age than the maximum rate for less distinguished works by the same individuals; (2) the rate of good production usually does not change much in the middle years and the decline, when it comes, is gradual at all the older ages—much more gradual than its onset in the late teens or early twenties; (3) production of highest quality tends to fall off not only at an earlier age but also at a more rapid rate than does output of lesser merit.

Bromley, a psychological gerontologist, has described (1974: 173–74; and cf. Raskin and Jarvik, 1979, especially chs. 1, 11) the characteristic manner in which senescent[8] persons attempt to cope with what appears to them to be the increasing complexity of cognitive demands and tasks.

> As age advances, the individual shifts from strategies which are intellectually demanding (but accurate and logically efficient) to strategies which are less demanding (but relatively inaccurate and inefficient). Experiments to determine the effects of age on inductive reasoning (abstraction and generalization) show that older persons are more likely than younger ones to become confused about the properties of the objects or events they are trying to classify. They more often fail to discover the distinguishing characteristics of a criterion class (predetermined by the experimenter), especially if the objects in the criterion class have properties shared by objects *not* in the criterion class.
>
> Older subjects take longer and want more information than the younger ones in order to achieve the same results. They experience greater difficulty in attaching meanings to cues and they lose track of data because of a reduction in short-term memory and mental speed.

Among animals generally (according to the principal text on aging), "the main phenomenon of senescence [is] the decline in resistance to random stresses"; and "senescence, even if it never reaches the ideal state of being expressed as a sequence of chemical reactions and equilibria, must presumably be reducible to a series of definite processes—such-and-such a mechanism leads to the loss of dividing power in such-and-such cells, which then have a life-span limited by the non-renewability of their enzymes to so many chemical operations, after which they deteriorate with [characteristic and specifiable] consequences" (Comfort, 1979a: 35, 180). Consequently, "aging is at root an information loss: in the postadult phases we are studying, first, late developmental processes; then the consequences of deterioration in the system and the subsidiary compensations for these, the partial and often individually irregular failure of homeostasis; and fi-

[8]Senescence means aging or growing old, and it should be distinguished from senility, which is a pathological condition due to excessive and extreme aging (especially cognitive). All mature animals (including humans) are more or less senescent; only a small minority of senescent humans, and virtually no other animals except an even smaller minority of household or barnyard pets, survive to experience senility (Comfort, 1979a: 21; Weg, 1975: 236; Bromley, 1974: 200–201; Schubert, 1984b; Sacher, 1978).

nally, premortal deterioration" (p. 352). As for humans, "senescence in man . . . commences while active growth is in progress"; and "man is by far the most numerous senile animal" (pp. 178, 330).

Cultural Aging

Writing from the perspective that only a former student can bring to bear in memorializing his one time teacher, Grant (1965: 1042; and cf. Hirsch, 1981) contributed to the literature of *Frankfurterweise* a poignant memoir including this perceptive insight: "It has been said of Holmes that he survived into his own generation. It may yet be written of Frankfurter that he was appointed as his was passing into history. He came to the Court beautifully equipped to carry on the Holmes-Brandeis opposition to judicial activism in the economic field. In twenty-three years on the bench he had occasion to write just one such opinion. He came totally ill-equipped, emotionally as well as from his sense of values, to meet the challenge of a new era. Although he came from Vienna rather than from Paris, in a way his history is so French. For France, it will be recalled, on the eve of World War II was so beautifully prepared for World War I." Certainly Frankfurter, rather than Holmes, provides the typical case. Given the average age at time of appointment of Supreme Court justices, which has been fifty-six since the start of the Civil War (except for the two decades 1938–1958, when it dropped to fifty-three, it is *generally* true that they are appointed just as the values that characterize their own generation of age-mates are becoming widely and increasingly perceived, by the bulk of the population (who are younger), as inadequate or irrelevant to contemporary needs and issues (cf. Roelofs, 1979).

Jaros (1973: 64; and cf. Cutler, 1975: 269) has discussed the relationship between political socialization and generational change, using as an example the "'Depression generation' that was particularly susceptible to the adult socialization influences surrounding this event. Party identification, for example, underwent widespread and drastic revision during this period. The age group that just reached adulthood at the time of the Depression is today more Democratic in its party identification than either the previous or subsequent generation. This generation, in its sixties [in the early 1970s], still stands out in surveys of voting." But the principal protagonist of a generational approach to the study of politics is Cutler, who (1975: 276) has proposed that "if we can identify successive generational groups of citizens who have undergone substantially different sets of experiences in their formative years, and if these generational groups can be empirically seen to maintain sets of political orientations grounded in these experiences, then it can be more authoritatively claimed that processes of political socialization are important for understanding adult political orientations and consequently for understanding the operation of a political system."

The concept of "cohort," as developed by sociologists working in the field of demography, and referring to age-mates who have (often presumably) been in a position to "experience the same event within the same time interval," has (in Cutler's view) somewhat more specificity in empirical research than does "the more generic term 'generation'" (Cutler, 1975: 264, quoting Ryder, 1965: 845; and cf. Kirkpatrick and Lyons, 1976). Furthermore, "the notion of critical life-stages . . . complements and enhances the notion of socialization as negotiation between two generational groups, each of its own stage in the life-long developmental process" and thus "the generational context . . . include[s] demographic and sociological as well as political and historical components in which the content and process of political socialization develop" (Cutler, 1975: 269, 278). However, Cutler and Schmidhauser (1975: 378) have added that "individuals of different chronological age do not simply represent individuals at different developmental stages in the life cycle; they also represent individuals born and raised in different historical contexts. Similarly, individuals of the same chronological age are not necessarily homogeneous with respect to the degree that age is salient to them as far as their political outlooks are concerned." Also important is "subjective age" (Cutler and Schmidhauser, 1975: 387–90): the way a person feels about himself or herself, or, alternatively, the reference cohort with which he or she identifies. These same authors report data showing that the chronologically old whose subjective age is "younger" are more *conservative* than the "old" oldsters on government policy; but otherwise the subjectively younger are more liberal.

Aging and Judicial Behavior

The initial suggestion of the cohort hypothesis about the effects of aging, at least in relation to research in judicial behavior during the sixties and seventies, came from Danelski (1966: 17), when he anticipated that the Supreme Court of Japan would become even more conservative than it then was in terms of "the Court's lack of activism in the protection of human rights. My guess is that in this regard things are probably going to get worse before they get better. The reason is that if appointments are made much as they are now, the men coming to the Court in the next ten years will have been born around 1910. Unlike some of their predecessors, they will not have had the experience of the liberal World War I period and the early Taisho Democracy in their backgrounds. In interviews with retired justices and high court judges, the importance of this period of modern Japanese history often came up, especially in regard to the genesis of their liberal ideas which, in some cases, they wrote in Supreme Court opinions." Subsequently, Danelski (1969: 149) restated the proposition: "What the hypothesis suggests is not that age makes for liberalism, but that liberalism and conservatism are related to the climate of political opinion prevailing during the critical periods of the justices' political socialization."

Danelski's field survey research in Japan stemmed directly from a seminar in which he and I jointly participated, together with a group of Japanese and other Asian lawyers and political scientists, on cross-cultural analysis of judicial behavior, for several months at the East-West Center in Honolulu. I was then conducting research on the High Court of Australia, so it is not surprising that the cultural hypothesis was articulated also in my contemporaneous observations on that work. "Age signifies what pattern of cultural norms was dominant when the judicial subjects were growing up and were being socialized into a particular phase in the development of Australian political society," and "age is potentially relevant because it is an index to the pattern of dominant cultural norms of a person's youth and socialization, and to his direct experience (and therefore, internalized understanding) of major upheavals for his society (such as those accompanying a major war or depression)" (Schubert, 1969c: 340; 1969a: 6; and 1968b: 35).

A second cultural variable also was used in that research, in an attempt to distinguish possible variation in socialization due not to time but to place (i.e., the effects of the principal subcultures [Schubert, 1980b] of Australian political society): "Domicile signifies the variation in subcultural differences that have implications for political ideology, such as the cultural parameters of the milieu in which adult learning and reinforcement took place for the judges in my sample. An American example of this variable in operation would be to compare the boyhood of [Tom Sawyer: the original reads "Mark Twain," an apparent Freudian slip] with that of James Russell Lowell, or to compare the boyhood of Robert E. Lee with that of Ulysses S. Grant" (Schubert, 1969c: 340–41). But these mutually and severally iterated suggestions of the cultural hypothesis were greeted with benign neglect by other judicial behavioralists and public lawyers alike a decade or so ago.

That posture stands in sharp contrast to the indignant cries of alarm evoked by my contemporaneous mention of the biological hypothesis—not as *the* explanation for Hugo Black's reaction to the revolt of the sixties, but rather as a possible alternative or interactive explanation together with the cultural hypothesis (Schubert, 1984b; Dunne, 1977: 418). I had stated the biological hypothesis much more explicitly earlier (1969b: p. 489): "Has Hugo Black 'changed his mind' in the quite literal sense that synaptic deterioration has changed it for him? Of course, no one knows, and least of all perhaps does Mr. Justice Black."

It is understandable that some social scientists have reacted indignantly to the suggestion that they ought to take other people's—and therefore, by implication, their own—biology more seriously. Possibly their consciousness raising about aging should start with a popular work which takes them directly into the confrontation between aspirations for immortality and the evidence against it, in the evolution—both cultural *and genetic*—of human cognition. Such a work might be Lionel Tiger's *Optimism: The Biology of Hope* (1979). Or they might begin with Sagan (1977) or Shepard (1978) and then read a more advanced work such as Crook (1980) or Reynolds (1981; see also ch. 8, above). One can only marvel at a morality that appraises the luck-of-the-draw of acculturation as somehow less denigrating than the luck-of-the-draw in genetic inheritance—the inheritance that has such a major impact upon the quality and quantity of neurophysiological deterioration, which, together with cultural deprivation, is responsible for pathology even in normal aging.

SUPREME COURT DECISIONS

Clearly both the cultural and biological hypotheses have been at large in the public (and public law) domains for more than a dozen years; and we turn now to research designed to adduce empirical evidence that would permit a choice between the two hypotheses.

Research Design

The first step in such a test is to construct an operationalization of our biocultural model. We shall take as our population the justices of the United States Supreme Court, beginning with the end of the 1970s, working backwards to draw as large a sample as can be supported by the available published research. For each included justice, we shall observe the dates of his birth, of his appointment to the Court, and of his termination of tenure (either by resignation/retirement or by death). This chronological information is readily available; and from it we can construct certain time-based variables described below. The dependent variable is a dichotomized classification of judicial ideology (Schubert, 1967a, 1977a) derived from voting; this is one of three secondary measures of time. The overall time period spanned by the study—beginning with the earliest birth of any judge included in the sample—is partitioned into subperiods, each of which is classified according to its predominant character as an ecological source of national cultural influence, using the same metric of ideology employed for the dependent variable measuring each judge's voting behavior. The other two secondary time measures are independent variables, of which one observes the modal age of each incumbent judge, during the portion that he serves of each decisional time period, while the second denotes the range of time during which youthful socialization of the judge is imputed to have taken place.

For each judge in relation to each decisional period of his tenure, we can measure the relationship between the ideology into which he was socialized as a youth and the ideology that he espouses as a judge. The observations for individual justices can be aggregated and the resulting data used to compute a correlation between socialization and decision making: this is an empirical measure of the degree of support for the cultural hypothesis (H_c).

The modal age of each judge can be classified nominally, as either older or younger in relation to the mean for his set of colleagues, for each of his decisional periods; and that observation then can be related to his decisional score for each such decisional period. Aggregation of the results of those tabulations constitutes data that can be used to calculate a correlation between age and decision making:

this is an empirical measure of the degree of support for the biological hypothesis (H_b).

The correlations, or sets of correlations, can be tested for statistical significance in relation to the sample sizes, and they can provide the basis for confidence in the extent to which either, or both, of the two hypotheses should be deemed unsupported by the empirical data of the study.

Data

The list of the fifty-nine Supreme Court justices who comprise the sample for analysis, beginning with Miller and, concluding with Stevens, is contained in an appendix,[9] together with their scores on the relevant independent and dependent variables. The sample begins with Mr. Justice Miller because, although he died in 1890 (at the age of seventy-four), he served on the Court for enough of that year to be included in the decisional period defined to commence on 1 January 1890. All justices whose tenure terminated during the Court's initial century are omitted from the sample. The decisional period beginning in 1890 is the earliest for which reported research supports ideological classification of voting in decisions, on the basis of equivalent (cumulative scaling) analyses of the Court's decisions; and this proved to be the limiting criterion. My archived data (Inter-University Consortium, 1979; Schubert, 1965b, 1974; and cf. Ryan and Tate, 1981) cover the period from 1946 through the retirement of Earl Warren in 1969. Pritchett (1948), as reinterpreted in Schubert (1962b), provides data for the period from 1921 through 1946.[10] Handberg (1946) covers the later period of the White Court, 1916–1921; and a paper by Leavitt (1974)—which extends in fact through 1945—is relied on for the period 1893–1916. It was necessary to extrapolate, with the aid of the Congressional Quarterly *Guide* (1979) and such standard references as Blaustein and Mersky (1978), Swindler (1969), and King (1950), for the three years from 1890 to 1892. For the Burger Court, extrapolation was based on research reported by Wasby (1976) and Schubert (1972).

The decisional periods are a subset, the most recent eight periods, of the set of fourteen eras that were defined, spanning the more than two centuries from 1765 through 1979;[11] these are listed in table 3. The table shows the alternation

[9]The appendix (together with the detailed report and analysis of the data, including other tables), most of the findings, and most of the critique of the design, data, and findings are not reproduced here; see the article as originally published (cited in the source note, p. 223).

[10]Inter-University Consortium (1979, p. 210) describes #7289, a class 1 study. These data, in combination with Pritchett's as he reported them and as I reanalyzed them (Schubert, 1962b), make possible similar analyses of all the other justices incumbent during the past half century or so. Many of those justices did *not* become more conservative as they aged; many remained conservative throughout their tenure, and a few remained liberal.

[11]In a fairly recent study of the effect of age upon political alienation, Cutler and Bengtson (1974, pp. 165, 174) examined both the generational and maturational hypotheses, and also "a third hypothesis, that of historical or period trends." They found that their "analysis appears to strengthen the period effect interpretation. . .and indicates that each of the generational groups responded to the events of the 1952 to 1968 period in an identical fashion—that is, the same pattern as already noted for the population as a

Table 3: Eras of Decisional Periods

Sequence Number	*Dates*	*Characterization*	*Classification*
1	1972–1979	Post-Watergate stagflation	C
2	1960–1971	New Frontier	NC
3	1940–1959	Militarism	C
4	1930–1939	New Deal	NC
5	1917–1929	World War	C
6	1901–1916	Progressivism	NC
7	1897–1900	Imperialism	C
8	1890–1896	Working-class revolt II	NC
9	1879–1889	Industrial expansion	C
10	1873–1878	Working-class revolt I	NC
11	1849–1872	Civil War	C
12	1801–1848	Frontier democracy	NC
13	1789–1800	Federalism	C
14	1765–1788	Revolutionary constitutionalism	NC

C = conservative; NC = not conservative.

between conservative and nonconservative eras—definitionally, each era was continued until a change (i.e., reversal in trend) occurred. The eras are defined strictly in calendar years; for present purposes the conventional periodicities, of the terms of the Court or the tenure of chief justices, are irrelevant. The delineation of what constitutes an era, how long it lasts, and its predominant ideological character as a cultural environmental influence is based on standard historical treatises, such as Paxson (1929) for the epoch from the end of the Civil War to the beginning of the Depression (1865–1929), after which the author's personal recollections and interpretations become more useful. More specialized studies such as Nye (1951) also were consulted.

The definition and classification of eras was certainly subjective; but it was not arbitrary. Only the third through the twelfth of the eras are used in this study to define socialization periods: the socialization of Samuel Miller is presumed to have occurred during the twelfth, while that of William Rehnquist picked up the end of the fifth, all of the fourth, and half of the third. The time of socialization for each justice was defined to take place over a period of twenty-one years, beginning with his own fifth year of age; this makes fifteen years of age—about midadolescence for boys—the median of each socialization era.

Evidently the socialization era (SOCERA) of each individual can extend over one or more of the general eras, each of which is classified as either "conservative" (C) or "not conservative" (NC) in its socialization-effects character—"not conservative" rather than "liberal," because the category includes moderates as well as

whole is repeated." This finding supports the use of eras here to define both (1) the ideological influence of the times of their (respective) socialization upon decision makers, and (2) the times of their judicial behavior.

liberals. The SOCERA scores of individual justices are straightforward derivatives of the classification of general eras and are weighted to reflect primarily their modal and secondarily their more recent experience.

Eras are classified quite generally as SOCERAS and entirely personally as DECERAS. A DECERA score reflects a justice's own voting behavior. Though individual tenures can span one or more eras, every DECERA for each individual is scored (as C or NC) according to the classification of his voting behavior for whatever portion of the era is covered by his tenure. (This voting classification is arrived at independently, at least in a methodological sense, even for my own data.) Hence each justice receives a DECERA score for each decisional period included in his tenure; but each such classification of a DECERA as C or NC is, like each justice's SOCERA score, personal to the justice (as distinguished from the general classification of eras). Each general era also is characterized as either C or NC, according to its predominant ideological thrust; and C eras are defined to alternate with NC ones.

Findings

The older justices tended to vote conservatively and the younger ones nonconservatively; while there was a very slight tendency for conservatively socialized justices to vote nonconservatively, and vice versa for the others, thereby producing a low negative correlation for the cultural hypothesis.

For the combined conservative eras there is little difference in support for the two hypotheses ($H_c = +.08$; $H_b = +.13$). But there is a much sharper difference for the combined NC eras: there is no support here for the cultural hypothesis, for which the composite correlation is slightly negative (-.05); but the correlation between age and decisional ideology is a modest but statistically significant phi of +.26. This indicates the possibility that NC eras may have had an important "period" influence (Cutler and Bengtson, 1974) in overcoming the persisting effects of youthful conservative socialization.

The effect of combining all eight eras is of course to cancel out the difference, just observed, in the "period" effects of eras as C or NC. However, it does provide an alternative basis for a decision concerning the overall support for the two hypotheses. The composite correlation is only +.01 for H_c, but H_b has a significantly larger phi of +.20 (and a chance probability of only about .02). This requires us to conclude that, in their most general portent, these data indicate no support for the cultural hypothesis of communality in early socialization; but there is a modest tendency for older justices to vote more conservatively and for younger ones to vote less conservatively. The latter tendency is strongest when the justices are being exposed to a less conservative ideological environment; although older justices tend in both C and NC eras to be conservative (and younger ones not), it is the *older* justices whose behavior is most accentuated by NC era stimulation—particularly so when we take into consideration the relatively smaller size of the older subsample then. The most stimulating era was that of the New Deal during the 1930s; but political lag (Schubert, 1965a: ch. 6; Roelofs, 1979) postponed the peaking of NC for the Supreme Court as an institutional group until the middle sixties (NC ERA 4) when the dominant conservative bloc was disintegrating, and a majority liberal bloc was established for the first (and so far, only) time in the history of the Court. At that, it lasted barely the three years of Goldberg's tenure.[12]

There are twenty-eight justices who are classified as consistently conservative in their behavior, twenty-four who are consistently nonconservative, and six others. Of these, the three appointed prior to 1937 (Stone, Roberts, and Hughes) were all conservatively socialized justices who voted nonconservatively while they were younger, but conservatively when older. The era in which all three changed to more conservative behavior is number 3, a conservative period beginning in 1940. The behavior of all three fails to support the cultural hypothesis during the earlier portion of their tenure, but does support it during the latter portion. Yet the behavior of all three is more strongly supportive of the biological hypothesis, because all three changed from nonconservative to conservative behavior as they became older. The other three justices were FDR's initial three appointments. All of them are classified as having experienced nonconservative socialization. All three of them voted nonconservatively while they were classified as younger, during era 4; and all of them changed, when older, to a conservative position. The cases of these three justices are of particular importance because they clearly fail to provide support for the cultural hypothesis, while they do provide strong support for H_b. For these reasons these data are better understood as strongly supportive of the biological hypothesis.

More generally, it is apparent that, of the half-dozen individuals who explicitly did change, none went from conservative to nonconservative; all changed, in confirmation of Lamb's (1976) observation, in the opposite direction, from less to more conservative. In view of the circumstance that H_c can be supported by evidence in *either* direction (although, for the reasons explained above, these data in fact fail to support it)— whereas the biological hypothesis, like time's arrow, is only unidirectional, and these half-dozen cases all *do* support it—it seems fair to conclude that, at this more micro level of analysis of individual change, as well as at the more macro level of correlational data, the evidence of this study all points in the direction of support for H_b rather than for H_c.

[12]There has been not infrequent discussion of the concept of "token" seats on the Court: token Catholics, token Jews, token blacks—and now, since the appointment of Sandra O'Connor, we can add token *female* justices. We should add also to this litany the concept of the token leftist justice, because two at the same time appears to be more than the conservative modality of American political society can tolerate. There have been only three justices who might, with some empirical warrant, be classified as leftists: Clarke, Fortas, and Douglas (cf. Schubert, 1970: xv–xvi). Of these, the first was harassed into retirement by McReynolds; the second resigned under threat of impeachment; and the third was the repeated target of impeachment attacks throughout his last ten years of tenure. (Both notoriously and unsuccessfully, McReynolds also waged moralistic aggression against Brandeis for more than two decades.)

Critique

In comparing the general patterns of correlations during the conservative and nonconservative decision-making eras, H_b is better supported than H_c in every one of the NC eras; the opposite is true of the C eras (except in the third, when both H_b and H_c are $+.33$). The composite correlations for the two sets of eras confirm that H_c is better supported in C than in NC eras, whereas the opposite obtains for H_b. This must reflect the importance of the "period" effect discussed by Cutler and Bengtson (1974), and it can be interpreted to signify that the acceleration of change during NC eras—which are more activist than C eras—intensifies perceived general differences, thereby reinforcing *more* conservative behavior in the *older* justices, while stimulating *less* conservative behavior from the *younger* ones. The C eras, to the contrary, can be assumed to be less threatening to older judges but also less stimulating to younger ones, thereby permitting greater influence to be exerted by their respective prior leaning (i.e., SOCERA orientations)—with the effect, of course, of making H_c seem more important. But this analysis, in turn, suggests at least a partial insight into what must be the dynamics of interaction between cultural and biological influences upon both conservatism and activism. This is therefore a direction that might fruitfully be explored in future inquiry.

The experiment described here did take into account period effects by using them as a control in testing the other two principal hypotheses. However, we can now see, with the benefit of hindsight, that it would have been better to formulate a third interaction hypothesis and modify the model accordingly. This would require a distinction between two cultural hypotheses, one relating to youthful socialization and the other to ongoing or contemporary adult—we could even call it "senior"— socialization. Such a revised model, necessarily a more complex one, would feature a design that would facilitate not *competition* between youthful socialization and adult biological aging, as alternative explanations of ideological change and/or constancy in the aging, but instead a transactional analysis of the interrelationships among the persisting effects of youthful socialization, the ongoing effects of biological aging (see Schubert, 1987, fig. 2), and the ongoing effects of the ideological context interposed by the thrust of contemporary society and its dominant culture.

PART V
Political Thinking

There is a long tradition, among political scientists, to take political thoughts very seriously, but to pay only ephemeral and superficial attention to the psychological processes by means of which individuals think (including their political thoughts); and they tend to ignore completely (e.g., Lane, 1962) the neurophysiological structures, functions, and processes in which any individual's thinking and thoughts necessarily are imbedded. Whatever the explanation for this professional intellectual myopia prior to a dozen years or so ago, the explosive growth of the brain sciences during the past generation and the emergence of interdisciplinary life-science journals—such as *Behavioral and Brain Sciences*, *Journal of Social and Biological Structures*, and *Political Psychology*—accessible to political scientists and psychologists alike, have combined to delegitimize continuing ignorance, of students of political thought, about how the human brain works.

Part 5 focuses explicitly on how humans think, both politically and in relation to other animals (who think, but do not talk; and whose thinking *must* be—for reasons relating to the structure of human vis-à-vis all other animals' brains—much more emotional and autonomic [which is to say, lacking self-consciousness] than our own).

Chapter 16 examines several key questions of cognitive ethology, in the context of a symposium published in the first volume of *The Behavioral and Brain Sciences* on "Cognition and Consciousness in Nonhuman Species," for which the protagonist contributors of target articles came from such leading proponents of teaching language to apes as the Rumbaughs, David Premack, and Donald Griffin (see also Terrace, 1979; Sebeok and Umiker-Sebeok, 1980; and de Luce and Wilder, 1983; and ch. 6, above). The chapter is empathetic to continuing and expanded study of language use and functions in other animals than humans (and not only in other primates) as a basis for better understanding of human language *and thinking* as well as that of other animals, but it criticizes the exaggerated claims of animal trainers whose use of what data they report indicates primary concern for guiding their apes rather than comprehending them.

The next chapter is on human thinking, and it examines many of the strikingly innovative developments in brain science during the past two decades, reflecting advances in a dozen different biological disciplines (such as biochemistry, biophysics, endocrinology, neuropsychology, genetics, and human development), which have created a new psychobiology that nevertheless has had no impact on mainstream political science theory and research. Chapter 17 discusses the psychobiology of the brain in terms of human consciousness and memory and then examines the epigenetic and recursive relationships between brain structure and political perception; between brain lateralization and dynamics, and political thinking and decision making; and between brain development and political equality, with particular regard to health, age, sex, intelligence, and race.

Chapter 18 is a critique of an "evolutionary theory of discovery and innovation," proposed by Findlay and Lumsden (1987) for a symposium in the *Journal of Social and Biological Structures* organized and edited by Edward O. Wilson. Their discussion of "The Creative Mind" is an attempt to explain creative activity and its effects—together with those of unusually high "productivity"—on behavior and society. To accomplish this, they propose what they consider to be a new evolutionary approach to human creativity, relating it to the functional architecture of the brain by their own construction of mathematical models of macroevolutionary social change and studying symbol systems within a single multi-dimensional space of "basic properties." In their own words (in their draft paper), they seek "to begin construction of improved formal frameworks, the so-called superspace theories, which can better accomodate [*sic*] unforseen [*double sic*: and likewise unspelled?] discoveries."

This chapter discusses the authors' nineteenth-century epistemological stance toward "creativity," their abuse of epigenetic theory in their concept and use of "epirules," and their continuing efforts to define cultural evolution as an epiphenomenon of mathematically defined genetic influences. I appraise the authors' naive notion of interest-group politics and political decentralization, from the perspective of mainstream political science during the past thirty-five years (cf. Corning, 1983), concluding that their advocacy of entropy as the spur to human creativity puts their own thinking right in the heart of pre-Renaissance Europe—somewhere near Lower Slobovia.

The concluding chapter in part 5 presents a comparative analysis of major paradigmatic change in physics, biology, and political science. Modern physics (relativity, quantum, and particle theories) is described as a science that embraces a paradigm invented in the twentieth century, whereas biology remains attached to nineteenth-century Darwinism plus Mendelism; but mainstream political science arrays itself theoretically across the two millennia extending from Plato to Newton (i.e., the Aristotelian biology of the fourth century B.C., to the classical physics of the seventeenth century A.D. that displaced it). The emergence within post-World War II American political science of first behavioralism and then biobehavioralism is reviewed in relationship to the implications for political psychobiology (the study of political thinking) of relativity theory's contra-evolutionary concept of *time*, in combination with quantum mechanics' uncompromising stance on the perceptual question of observer bias. This is in the context of discussions of the biophysics of the (human) brain and of such entailments of the quantizing of political theory as the equiprobability of past and future; the substitution of pervasive *in*determinism for the fascination of the professions of both political science (through econometrics: see Simon, 1983) and biology (through sociobiology) with rational models of decision-making behavior. Similarly, quantum theory requires the rejection of metaphysical dualism in favor of a neuromaterialistic understanding of how even political scientists make their political and other choices—even when they prefer to do so unwittingly. There *are* more things in heaven and earth than are dreamt of in the political philosophy accepted by most political scientists.

16
Animal

To a social scientist interested in the biology of human behavior, it is astounding that primatologists formulate their research inquiries in terms of such questions as whether "the chimpanzee's concept and use of tool names [provides] insight into the factors that determine and promote evolving, complex forms of *word usage*"—presumably among chimpanzees (Savage-Rumbaugh, Rumbaugh, and Boysen [henceforth SR&B], 1978: 553; emphasis added); or "whether or not the chimpanzee imputes mental states to others," which evidently is deemed by its authors to be an operationalization of their speculation "that the chimpanzee may have a 'theory of mind' . . . not markedly different from our own" (Premack and Woodruff [henceforth: P&W], 1978: 515, 6). What is wrong with both questions is that they appraise the relative excellence of nonhuman cognitive abilities by measuring the extent to which these conform to those characteristics of our own species—which impresses me as a very unbiological approach. And that raises questions about what we should understand to be the relevant theoretical significance of what these human experimenters, as well as their chimpanzee subjects, are doing.

I shall undertake to comment upon each paper in turn, in the same sequence in which they appear in the symposium. But before doing so, two prefatory caveats are needed. Whatever may be true of chimpanzees, it certainly is the case that humans can and do make many imputations, in complete innocence of any kind of cognitive theory, to say nothing of a theory of mind. On the other hand,

---

This chapter is a revision of "Cooperation, Cognition, and Communication," *Behavioral and Brain Sciences*, 1: 597–600 (1978). © 1978, reprinted with permission of Cambridge University Press.

theory of the human mind has yet to achieve the degrees of cohesion, integration, and consensus meet to its functioning as the criterion for modeling the cognitive processes of other species, at least at the grandiose level of "theory of mind" and "states of mind."

TRAINING PRIMATES TO COOPERATE

"Words" are elements of either human speech or else human written language; and their postulation as a variable in chimpanzee cognition is neither necessary nor sufficient, nor indeed even helpful, to the elucidation of how two such animals were trained to cooperate in the sharing of tools and food. The latter kinds of social behaviors surely were critical in the evolution of hominids and eventually the human species, but there is no evidence that either tool or food sharing depended or depends upon the prior acquisition of verbal language; to the contrary, language probably developed to meet the social needs of bands of protohumans whose ancestors already had been sharing tools and food for millions of years (Lancaster, 1975: 72–73; Pfeiffer, 1977: 50). Moreover, both tool and food sharing sometimes occur within feral bands of chimpanzees (Lancaster: 1975, 77; McGrew, 1977; Wilson, 1975: 128, 207; Pfeiffer, 1977: 48–49), so the training here enhances what are already evolved abilities (if not traits) of *Pan troglodytes*.

SR&B state (1978: 543) that Austin and Sherman "distinguished these *words*, as well as additional tool *words*" (emphasis added); and I take exception to this kind of linguistic anthropomorphizing, particularly in a paper that purports to discuss cross-species language transfer. The paper is replete with statements that exemplify my point, such as several in its conclusion (SR&B: 552–53): "The chimpanzee's comprehension of single words" is discussed there, together with "word acquisition" by chimpanzees, and then the remark about "complex forms of word usage" quoted in the opening sentence of my introduction. This kind of attribution of "word" as a concept of chimpanzee cognition may not be characteristic of the authors' own thinking but it certainly permeates their discussion here.[1] Yet it is abundantly clear from the data presented that what Austin and Sherman responded to, and used to initiate their other than nonverbal (including oral nonspeech) communications, were the lexigrams of the customized computer terminal that was their constant robot companion. These lexigrams are symbols, color-coded combinations of from one to four among nine design elements, all linguistically arbitrary from the perspective of human language; but they are not words in either English or any other natural language. As native speakers of the English language, SR&B apparently found it easiest to transliterate the lexigrams, which

[1]The Rumbaughs are much more careful to distinguish between words and lexigrams in their concluding chapter to the book that he edited (Savage and Rumbaugh, 1977); so it may be that what we confront in the present paper is a writing problem—but that makes it no less a problem. The difference between Yerkish lexigrams and English words is explicitly discussed (Savage and Rumbaugh, 1977: 96–97), and the design of lexigrams is explained in detail in von Glasersfeld (1977: 92–95).

identify the keys that symbolize the computer language that the keys activated, to the English in which it is most probable that they thought about the behaviors with which the lexigrams had been associated by or for them. The lexigrams were words to the humans who directed and participated in the research and who had a sophisticated understanding of the artificial Yerkes language and the computer programming that some of them had spent much time and effort to develop, as the medium through which chimpanzee-human communication might better take place.[2]

But irrespective of whether chimpanzees have a "theory of mind," no evidence is presented to suggest that chimpanzees have a theory of *computer* minds. Pushing a lexigram key and monitoring one's own (or another's) performance by observing a lighted display of the same lexigram is a function qualitatively indistinguishable from that of pigeons pecking or rats pressing levers to obtain food (or surcease, or whatever). The observation and interpretation of more than one lighted lexigram, in sequence and in context, certainly does have implications of notable import for any theory of chimpanzee cognition; but the chimpanzee's role in the interaction process neither requires nor entails understanding on its part about how computers work. The Cs (chimpanzees) know lexigram symbols as identification signs not atypical of the strange, mechanically human environment in which they find and adapt themselves. It is not helpful, in discussing such behavior, to attribute to them the cognitive concepts of natural language.

An explicit demonstration of the kind of unnecessary difficulties that SR&B's perspective towards words invites is found in their having attached, for their own purposes (and conceivably, psychic benefits) the word "straw" to the piece of plastic tubing that is photographed in their figure 1. This word usage, I submit, is not merely semantically incorrect; what is worse, it is patently an affectation. It is semantically incorrect because flexible plastic tubing is not referred to as "straw," at least not by native speakers of the English language. It is true that a tertiary meaning of "straw" is a hollow tube, used in drinking certain beverages such as an ice cream soda; but the primary meaning is a single stalk or stem, especially of certain pieces of grain. It is an affectation because the latent image implied by the use of the word, in the context of the research, is that of a feral chimpanzee, indulging in the frequently reported tool behavior of using a plant stem, twig, or long stick, to dip for termites or ants. This is a natural tool use by the animal (McGrew, 1977). So here we have an example of what Kenneth Burke (1937) once called "secular prayer," as SR&B—whether intentionally or not—cash in on the overtones of the word, at least for other primatologists (although probably not for either Austin or Sherman). Calling their plastic tubing a straw permits SR&B to take a free ride in the minds of their human readers, on the associated idea of

[2]All three of the authors, plus several other researchers involved in the design or administration of the LANA project, spoke at a symposium on "Language Formation Studies with Apes and Children," on September 7, 1978, at the second annual meeting of the American Society of Primatologists convening on the campus of Emory University. I was present and heard the presentations.

how natural it is for the research animals to be using the tubing. Of course, Austin and Sherman could not care less whether or not SR&B call a spade a spade; they would have responded no differently if SR&B had decided to call the tubing "snake" instead of "straw"—unless and until, of course, both the analogy and the research were extended to include live representatives of the order *Ophidia*. It would probably have been much simpler, as well as more accurate, to have called the plastic tubing "plastic tubing."

There is other evidence of a proclivity for hyperbole, as in SR&B's claim (1978: 544) that if Austin and Sherman "could comprehend the function and *intentionality* of their communications and, through joint symbolic communication, *share* their access to tools and the food obtained through tool use—then, by all definitions of human culture, they would surely have taken a large step" (emphasis in original). This claim of "all" definitions of human culture is both literally and figuratively a gross exaggeration. Even a rudimentary acquaintanceship with cultural anthropology shows that not only many, but indeed most, definitions of human culture demand more than the sharing of tools and food through nonverbal communication—which is what Austin and Sherman did, no matter what "words" were attached to lexigrams by the experimenters. What cultural anthropologists typically mean by culture is an elaborate complex of cognitive associations together with both linguistic and other symbolic representations thereof built upon the use of natural human language. A critical element in anthropological definitions of culture, which is both conspicuously and oddly missing from the one proffered by SR&B, is evidence of transferability from one generation to the next. It may be that Austin and Sherman will succeed by some means of social communication in teaching their progeny how to cooperate in the sharing of tools and food obtained through their use. But they haven't done so yet, and the requirement that they succeed in doing so is a minimal element in most definitions of human culture. We surely can speak of the culture of preverbal hominids, or of other primates, or indeed of other animals (Mainardi, 1980); but to identify with *all* definitions of culture the exchange of tools and food through the use of nonverbal communication is unacceptable to most anthropologists (Hall and Sharp, 1978; Weiss, 1973; Chapple, 1970), to say nothing of other social scientists or humanists (who think that they also know something about the meaning of culture).

A more serious problem concerns the administration of the major test. In part the exciting questions raised (p. 15) are rhetorical: "Could [Austin and Sherman] perceive the necessity of requesting tools from one another?" This question is not answered by the tasks in fact performed because the experimenters evidently chose not to be daring enough. What would have happened if, after the animals had been trained in the requisite skills and placed in the experimental situation (as described in SR&B, 1978: 530), one crucial variable had been omitted: the presence of the experimenter in the baited room? The chimpanzees' performance would have been extraordinarily more impressive if they themselves were able to figure out, interactively with the computer and each other, how to solve their problem. Such a performance would indeed have answered the question raised by the experimenters about the test that they in fact administered; but then one more thing would have been essential. It is remarked (bottom of p. 16) that before the test situation, Austin and Sherman "had never observed one another use tools or employ tool symbols to request tools, and thus had no reason to presume that the other animal knew and used such symbols or would cooperate with requests for tools." No explanation for such a research policy is given; one infers that an assumption was made that such foreknowledge would "contaminate" the experiment—but it is by no means apparent why this would have been so. All social skills are learned and shared through social interaction by feral bands of chimpanzees; and it would be unnatural in the extreme for one such animal to be able to use tools (or do anything else) without some other animal in the band having observed his behavior. I think that a mistake was made in the research design, at precisely this point. It would have been much better to have gone the other way and socialized this pair of animals in their reciprocal knowledge about skills in the requesting and use of tools in interaction with the experiments; and *then* the crucial experiment should have been set up to ascertain whether they could voluntarily transfer that knowledge and induce equivalent interaction with each other. If that stage of the experiment failed, even after a fair test with role reversals and an adequate number of trials, it would still have been possible to undertake the explicitly directed training in cooperation that in fact was done. Indeed, it is entirely possible that the principal accomplishment of that training was to show the two chimpanzees each other's abilities to use tools and to use the keyboard symbols to request them.

One's puzzlement that these animals were never given a chance to show what they might have done is enhanced by noting Menzel's remark "that there is already good evidence that group-living chimpanzees are capable of communicating a good bit of information to each other about the environment even without the benefit of extensive human training" (1978: 891; 1973). The spontaneity of such "cultural" innovations as sweet-potato washing and grain cleansing by *Macaca fuscata* (Itani and Nishimura, 1973) is of course an example familiar to thousands of undergraduates through courses in comparative psychology and introductory ethology; and because *Pan troglodytes* is considered to be more intelligent than the macaques (Chevalier-Skolnikoff, 1977), grounds for the missing experiment in serendipity surely were present in the professional literature of primatology (cf. Menzel, 1972). However, as Menzel (1978: 890) has remarked, "the problem of 'animal genius' has . . . received almost no scientific attention . . . [although] I have been repeatedly impressed . . . by how often one can find a single oddball monkey that does something quite out of the ordinary for other members of [the same] population. . . . [But the] spread of a behaviour requires good receivers as well as good senders,and for this reason the group-as-a-whole went nowhere."

This leads us into a closely related point, not discussed by SR&B but manifest in their data: the significance of individual differences in the intelligence (Gibson, 1977) of these juvenile research animals. SR&B do recognize (1978: 542–43) that "Austin. . .did attend to the tools closely from the beginning" and "Austin attended closely to E's statements and did well from the beginning of the task"; and their table 2 certainly supports these appraisals. But Austin is one year *younger* than Sherman; and it is therefore unlikely that the difference in learning speech between them can be explained on developmental grounds, or at least those of age. It is true that other differentials in the training or experience of the two animals might be sufficient to account for Austin's superiority (Chevalier-Skolnikoff and Poirier, 1977); but no hint of such information appears in this paper. On the evidence presented, Austin is a more intelligent chimpanzee than Sherman; and if this is true, that finding could have been—but apparently was not—used in the design of the "tool transfer" tests, irrespective of whether these were to be used as I have argued they should have been for an exploration of the ingenuity and intelligence of these animals. Intelligence could have been a useful variable even if the Cs were only going to be taught the standard operating procedure, for which purpose it would seem reasonable to have anticipated a higher standard of performance from Austin—both in speed and accuracy—than Sherman could have been expected to achieve.

SR&B claim that their tests are "blind"; but to what extent does this appear to be true? Implicit in the role of the unidentified experimenters (henceforth: Es) during various of the tests are two problems. One concerns possible differences in the emotional attachment of each of the two Cs towards each of the Es. The other is a Clever Hans problem and involves the possible effects of unwitting nonverbal cues, communicated differently but perhaps consistently by different Es. SR&B acknowledge (1978: 542) that "changes in personnel inevitably resulted in performance decrement at all stages of training"; this implies not only that there was some information loss (mostly, no doubt, nonverbal) when one E replaced another; it implies also that different cues were communicated by different Es. P&W point out (1978: 521) that "Sarah's choice was affected by the actor's identity"; and their paper reports details of the results of her fondness for her regular trainer and her lesser affection for Keith's substitute. We are not informed about the dispositions of either Austin or Sherman toward either E1 or E2; and neither do we know the identity of either experimenter for purposes of the roles discussed in the reports of the tests (P&W, 1978: 518–23). Randomizing the assignment of Es, as well as reversing Cs, in the test roles during their repetition would at least have tended to control for whatever effect differential attachments may have had. SR&B state (1978: 543) that "both animals did very well on these tasks (Table 3), thus demonstrating that their abilities were not dependent upon cues from E." But aren't affective cues "cues"? Moreover, "in the naming task, E again stood outside the room and held up a tool so that it was visible to C through a lexan wall, although neither C nor E could see one another." But if C could see the tool well enough to distinguish it, he could also see at least part of E's hand, and perhaps part of his arm(s) too. How much more information does a chimpanzee need to identify which human (among the small sample of available alternatives) he was dealing with? There is also an entailed problem of equity in the invocation of standards for appraising the reliability of these particular research results. For reasons that are not divulged, each of the blind tests "was administered only once" (SR&B 1978: 543). Of course, this magnified the opportunity for either of the two types of extraneous communication discussed above to have influenced the results; but there is the additional consideration that the Gardners are criticized (p. 541) for their acceptance of "a weaker criterion of initial acquisition . . . [with only] *one* correct spontaneous usage per day," the apparent point being that little confidence should be placed in data of such putatively low reliability. Evidently, what is sauce for the goose is not sauce for the Gardners.

SR&B recognize that there is an obvious alternative explanation, and one that ought to be preferred on philosophy-of- science and methodological grounds: that Austin and Sherman cooperated as they had been trained to do, and after they were trained to behave so by the experimenters. Perhaps, say SR&B (1978: 544), Austin and Sherman's "use of the keyboard merely reflected the continuance of behaviors that they had been conditioned to emit by E"; but this straw man is immediately knocked down, or so the authors presume, by the next sentence and those that follow it. They remark that "observations of trials on which Cs were in error suggested that this was not the case"—but the adjective "anecdotal" is omitted from the pride of place that it evidently deserves in the sentence quoted, because all of the evidence presented is precisely that. If these authors are to discount the research findings of other primatological linguists on the grounds that the evidence supporting the latters' work is merely anecdotal (SR&B, 1978: 550–52), there seems no reason why we should accept a lesser standard of proof for their own claims.

DO PREMACK AND WOODRUFF HAVE A THEORY OF MIND?

Theories of mind get pretty esoteric even for human subjects; and they are more so in relation to a species whose natural language is at the more difficult level (Menzel, 1978: 891) of nonverbal communication. Having endeavored to employ a theory of mind to study certain social behaviors of elite human political decision makers (Schubert, 1965b, 1974), I am convinced that there must be a more parsimonious way to tune in on the cognitive signification of the photographic preferences of a particular chimpanzee, without vaulting to the more transcendental levels of cognitive theory.

There is at least a logical problem with the explication of the theory proffered (P&W, 1978: 515: "can be understood"), as a possible "explanation" of Sarah's photo choices. The authors' preferred "theory of mind" is defined as Sarah's imputation of "at least two states of mind to the human actor, namely, intention or purpose on the one hand, and knowledge or belief on the other." Yet in the concluding remarks, in the context of further speculations about the possible findings from as yet unanalyzed data comparing chimpanzees with both normal and retarded children, it is conceded that chimpanzees may well be *in*capable of making imputations about knowledge (P&W, 1978: 526). On the basis of the data reported, therefore, it is more parsimonious to reject these authors' "explanation" of Sarah's behavior, in favor of one more consistent with the authors' own supportable imputation (P&W, 1978: 521) about chimpanzee mind: that Sarah was guided in her choices primarily by her affective stance towards the human actors.

Alternatively, and these authors to the contrary notwithstanding, it remains possible that Sarah's discriminations were based on what is described as the simple matching of physical elements. It is conceded that one of her three correct choices might be explained by physical matching (the more upright posture of the actor); but it is argued that the same explanation cannot account for her other two correct choices, *because the actor's posture was not upright in them.* Surely this misses the point of what is implied by "the same explanation." Because the authors' own discussion (P&W, 1978: 516–78) is anecdotal, it seems fair to point out that there are an infinite number of other possible physical matches between the content of the photographs and the videotape, depending upon what is perceived to be relevant and important; and it is Sarah's perceptions, not the authors', that count. But even if no such unwitting cues were present (perceived by Sarah) in either of her other two correct choices, her discrimination (which the authors call her "comprehension of problem solving") is down to two right and two wrong for this series, on the authors' own concession; and success ratios that match chance are not impressive evidence of comprehension. Nor does the remark that "physical matching is ruled out . . . even more [for other series to be reported] later" save *this* series. It may well be that "chimpanzees . . . can solve problems with strategies more sophisticated than simply matching physically identical or similar items" (P&W, 1978: 516); it may also well be that, not unlike humans, chimpanzees do things the simple, easy way when they can. In any event, we ought to assume that if Sarah could have solved these problems by physical matching she did so, leaving her keepers to worry about the complexity of the primate mind.

THE EVOLUTIONARY CONTINUITY OF COMMUNICATION

I associate myself wholeheartedly with the general tenor of Griffin's thesis. In particular I agree with his admonition that "defining mental experiences as uniquely human discourages inquiry into the possibility of their occurrence in other species" and with his suggestion of the more useful hypothesis "that thinking and experiencing are related in comparable ways to the functioning of central nervous systems in various species" (Griffin, 1978: 529, 530). Such cross-species analyses of psychoneural relationships would of course need to be done in the context of all that we know about the brain systems as well as the behavior systems of each species concerned, pursuant to Menzel's advice (1978: 892) that in "conducting a field study of communication" among chimpanzees, one should "start at the level of general cognitive, societal, or even ecological considerations and then work backwards toward the data that might be of more concern to" cognitive ethologists. Furthermore, "communication is part of general information-processing activities of an organism . . . [and] beneath the 'deep structure' of human language and human thought there are indeed 'deep-deep' structures that we share with other species, and . . . it is on these structures that our linguistic abilities are predicated" (Menzel and Johnson, 1976: 140; and see especially Lenneberg, 1967, and Denenberg, 1981: 8).

The qualitative evolutionary continuity discussed by Griffin refers to development that is species-specific consequent to divergence, even for such closely related species as humans and chimpanzees. To take as an example our own experience, for which the evidence of dynamic change during the past ten million years is more impressive than for chimpanzees, we can distinguish among three levels of communicative ability: nonverbal communication, verbal speech, and written language. We can appraise these three modes of human communication in terms of two dimensions: (1) recency of evolution and (2) relative efficiency (and/or use) as a carrier of affective, as distinguished from effective, messages. The ordinality of these modes is indubitable, and even the interval estimates for the first dimension are sufficiently disparate to preclude extended discussion for present purposes. Nonverbal communication developed prior to verbal communication among humans, and at least some aspects of it are very much older than the six million years to which our knowledge about some characteristics of protohominids can arguably be stretched. Oral speech evolved much more recently, perhaps a million to half a million years ago and in phase with the doubling in size of the brain. At least on our evolutionary time scale, written language is an exceptionally recent acquisition, probably less than ten thousand years old; the earliest known writing dates from 5500 B.P., but that threshhold may be pushed back by future discovery and research. Conversely, and although their number and variety are diminishing rapidly, many bands of preliterate but feral human societies have been studied throughout the present century: all of these lacked *written* language, but were articulate in both verbal speech and nonverbal communication. Our interest in, and possible admiration for, the adaptive or aesthetic virtues of such other modes of animal communication as whale or bird song, or bee dancing, cannot sway our appraisal that all of these—indeed, all known modes of non-human communication are nonverbal, at least in the sense of written language. On the other hand, many mammalian oral communications, and certainly those of many canids (Harrington and Mech, 1978) as well as non-human primates, may very well exemplify some of the cross-species psychoneural relationships adumbrated by Griffin (cf. ch. 6, above). Certainly we should anticipate at least a broad spectrum of such relationships for nonverbal human communications, because of the potentially larger number of homologous species with which they might be shared.

MacLean's model of the brain (1958; and cf. Lancaster, 1975: 62) implies that affective messages must have been communicated for a very long time before any animals had developed brains of sufficient complexity to receive or transmit messages with effective content. Perhaps this is what Premack and Woodruff mean when they say that "motivational states seem more primitive than cognitive ones" and that "inferences about motivation will precede those about knowledge, both across species and across developmental stages" (P&W, 1978: 521, 526). Assuming that to be true, we might hypothesize that the reaction of affective to effective message content will in general be maximal for nonverbal human communication and minimal for messages in written language, with verbal speech in the middle;[3] it can also be expected that components of *both* affective *and* effective content usually will be present in all three modes for human communication. Certainly the hypothesis is not inconsistent with observations that "the nonverbal communication of nonhuman primates has traditionally been characterized as species-specific, emotionally-based, and nonintentional" (SR&B, 1978: 539) or that "there is little evidence that the ability to devise non-verbal codes derives from the prior possession of verbal ones" (Menzel, 1978: 890).

Griffin certainly is correct that "ethology has made a contribution of fundamental importance by discovering a rich variety of nonverbal communication in many kinds of animals," and "many social animals communicate by systematic codes which convey information and often lead to predictable changes in the behavior of the animal receiving the message" (1978: 531, 533). We now have available the work of more than a generation from the students of Tinbergen alone, as exemplified by the sophisticated models of social communication constructed by the first of these students, based upon a lifetime of field studies of herring gulls and related species (Baerends, 1975, 1976a, 1976b) and also by the flourishing of studies of human nonverbal communication (Hinde, 1972; Argyle, 1975; Key, 1979; Morris, 1977; von Cranach, 1979). There is even a beginning of studies that undertake to analyze all three modes of human communications (Schubert, 1982a).

To the extent that we can study the communicative abilities of chimpanzees and other animals, with our own minds open to the possibility that in the process of doing so we may learn as much about our own modes of communication as we do about theirs, we can begin to explore the evolutionary continuity of mental experience that is the subtitle of Griffin's book (1981) and that he has postulated to be the proper task of cognitive ethology.

[3]The modal qualities of speech are illustrated by voice stress analysis (e.g., Wiegele, 1978a; Wiegele et al., 1985), which provides information about emotional arousal that is manifest in speech even though undetectable in written transcripts of the same communications.

17
Human

It is now a full eighty years since Graham Wallas called for a political science based on the study of human nature; and more than sixty since Charles Merriam (1925: 171) placed human biology front and center—not at the wings—of his prescient scenario for a biobehavioral science of politics:

> Is it not possible that the real relationship of students of politics is with biology or neurology rather than with psychology? Do we yet know what changes may be wrought in the individual through biological modifications or through biochemistry? How far may attitudes and behavior be influenced or determined by biochemical processes which we do not yet thoroughly understand, as for example through modification of glands, physiological functions, or neural mechanisms? To what extent is it possible to condition and determine these attitudes by conscious biological or biochemical processes?

Of course, the state of knowledge available then precluded answers based on the empirical, scientific evidence that Merriam's political behavioralism demanded as a criterion. But in his view the time would come when psychobiology might play a significant role in political science just as (as he assumed) it always had done in the practice of politics. He believed it entirely possible that, when that time arrived, neurobiology might be even more important than psychology to the understanding of political consciousness and thinking.

That time is now at hand. The explosive growth of modern biology in general, and the brain sciences in particular, has begun to evoke a nontrivial political-science response, as we shall see below; but it is the fruit of collective professional

---

This chapter is a revision of "Psychobiological Politics," *Canadian Journal of Political Science*, 16: 535–76 (1983a). Reprinted with the permission of The Canadian Political Science Association.

action that began two decades ago (Somit, 1968; J. Davies, 1969, 1976; White, 1972; Laponce, 1975, 1978).

The objective here is both to describe a model depicting how and why the human brain behaves as it does and to relate that modern psychobiological paradigm of brain structure and dynamics interactively to political inquiry and the professional political science understanding of political behavior. The latter is done by discussion based on contemporary biopolitical research, both theoretical and empirical, focusing upon several of the most basic components of human brain function, which include any and all types of political behavior and action.

There is a striking isomorphism between the physical structure of the human brain and the processes of political perception: humans create a political world in the image of the structure of their own minds. This involves the dynamics of political thinking, relating current neurobiological models of consciousness, memory, and cognition to such central concerns of political theory as rationality, positivism, and determinism, vis-à-vis projection, subjectivity, and emotionality as explanations for political attitudes and decision making. Also involved is development—the development of the human brain in relation to political development—but not in the sense of bringing to the Third World the glories of industrial civilization; instead we shall consider development in the sense of potentialities for political contributions of all humans, including those who populate North America, Western Europe, and Eastern Asia (South Korea, Japan, and Taiwan), as well as the rest of the world; and we shall discuss the development of the brain as a function of lateralization and sexual differences in political behavior. But the transactions between brain and behavior typically are highly recursive rather than unidirectionally linear, and that kind of positive feedback is exemplified by the impact of language-culture ethnicity upon brain development and dynamics. Similarly, the nutritional deficiencies that starve and stunt the developing brains of fetuses, neonates, infants, and young children are not acts of God but rather are the characteristic consequences of political decisions and behavior. Such decrements in brain development exert powerful negative feedback on political socialization, participation, culture, public policy, and behavior, completing a vicious circle that constrains the possibilities for effective political life of a majority of humankind today. This universal biosocial phenomenon is exemplified by reexamining a frequently discussed question in educational and social—and it also should have been political—psychology: the causes and cures for the IQ differential observed among American blacks during the past three generations.

BRAIN STRUCTURE AND POLITICAL PERCEPTION

On the basis of two decades of work studying human intelligence as a factor in political learning and socialization, Elliot White asserts (1982a: 279; and 1980a: 11) that "the process of political learning—and the continuity and the change it implies—will be at least in part explicated by an emerging human neurobiology"; and he adds that "the most sensible course for social scientists to take is to adopt the basic theoretical model of the brain now propounded by leading neuroscientists." If political scientists are to do that, they are going to have to take more than psychobiology seriously, because the basic neurobiology of the brain can be credible and comprehensible only for persons with at least an elementary understanding of such allied life science fields as evolutionary theory, animal behavior, primatology, physical anthropology, and physiological psychology. This is no small undertaking, but it is one on which several score political scientists have been engaged for the past generation.

A reinforcing reason in support of such substantial investment of professional time and energies is that the *practice* of politics is inherently *bio*cultural, *bio*social, and *bio*psychological. Indeed, this is true of virtually all human activity and, as primatologists and ethologists assert, of many other social mammals in addition to humans (e.g., Alcock, 1983; Griffin, 1981; ch. 5, above). The program of contemporary biologically oriented political scientists never has been and is not now that political behavior is genetically or otherwise biologically determined: the latter is prima facie an idiotic posture unsupported by any and contradicted by all the evidence about *human* behavior. Rather it is that politics and political behavior cannot be adequately understood, explained, or predicted on cultural grounds *alone*. What humans become and do—politically or otherwise—is the consequence of complex, continuing, and dynamic interrelationships between the (individually unique) genetic inheritance of each person, and her or his (equally individually unique) experience of life between the moments of conception and death. Such transactions among genotype, phenotype, and environment are quintessentially the basis for *psychobio*logical politics.

The Structure of the Human Brain

To understand the structure of political perception, we first must know how the human brain is constructed. Neurobiologists distinguish three primary dimensions of brain topology. The first of these is rostral/caudal, denoting the commonly observed linear separation between the nose and the tip of the tail in vertebrates. Humans are quite exceptional because for them the rostral dimension tends to be co-aligned with the *vertical* (instead of the horizontal), at least much of the time for awakened people. This is a position that characteristically puts the conscious human head in relatively maximal opposition to the force of the earth's gravity. Other primates also typically assume upright posture to a greater extent than do mammals generally, but less so than do humans—or at least all except, perhaps, twentieth-century humans. The second dimension is ventral/dorsal, or front-to-back. The third is the dimension of laterality, side-to-side or left-and-right. Full three-axis symmetry is fixed in humans by the end of the first week, and visibly so by the end of the second week, of embryonic development (Boklage, 1980: 129). Together the three dimensions define an Euclidean space, and they are equivalent to a set of Cartesian coordinate parameters (or can be used in that sense). It is notable that the rostral dimension is extremely asymmetrical; the ventral is generally asymmetrical; but the lateral is extremely symmetrical (Gardner, 1964; Corballis and Beale, 1976).

In humans the functions of integration and control, thinking, and consciousness are concentrated in the uppermost and most recently evolved sector of the brain, the neocortex (J. Davies, 1980: 33). There are extremely intimate and complex interfaces between the cortex and the limbic system, and consequently with emotional and actional (or motor) systems (Panksepp: 1982), which necessarily project downward throughout the spinal cord. The "reptilian" (phylogenetically the oldest) portion of the brain, predominantly in the spinal cord and therefore lowest in relation to the limbic system and cortex, is concerned with the control of such primarily autonomic activities as breathing and digestion. Even in the latter respect, however, as much relatively recent work in biofeedback has demonstrated (Schwartz and Shapiro, 1976, 1978), the human brain is hierarchically structured and the neural bases for the most complex volitional controls and cognition are concentrated at or near the (front of the) upper extremity of the rostral dimension (and also typically in the left hemisphere).

The cortex itself is partitioned between the cerebrum (or ventral portion) and the cerebellum. In the front are located sensory perception, memory, anticipation, and concepts of distance and external space; the back cortex is preoccupied with the control of internal processes (Livingston, 1980). Uttal (1978: 317) has credited brain surgeon Karl Pribram with the suggestion that the frontal lobes act to inhibit mutual interference that otherwise would occur among a series of nearly simultaneous brain events and that the linkage of a sense of mental state into a smoothly flowing stream of consciousness and behavior is due to frontal lobe effects. The forebrain areas exert inhibitory control over the hindbrain mechanism, while forebrain development appears to modulate both primitive reflexes and behavioral arousal (Dawson, 1977: 433).

The brain is most conspicuously bifurcated, laterally, into left and right hemispheres, although the two halves of the cerebral cortex are linked together by great bands of axons, the corpus callosum and the anterior commissures. One hemisphere tends to dominate the other in humans; and for a preponderant majority of individuals—some 70 to 90 percent (Uttal, 1978: 319)—it is the left that does so. For that majority, speech, language, and analytical and serial processing (Pribram, 1980a) depend primarily upon the structure of the left hemisphere, while the estimation of spatial relationships and abstract patternings and configurations is structured more in the right, together with parallel processing of information. However, as Uttal (1978: 318; and cf. 142) remarks, "language is not a single process or act but is a complex of a number of different sensory, response, and integrative functions"; and so too, of course, is the recognition of a human face or any other complex function of pattern integration.

The Structure of Political Perception

"We do not have to teach that power comes from above," says Jean Laponce (1978: 393), "but we [do] need to teach that power comes from below." This is because "the up-down dimension is so basic to our perception of the physical environment that it inevitably spills over from the physical to the social and the cultural. . . . If power, and consequently the symbols and the institutions which incarnate it, is positively valued, as will normally be the case except of course in countercultures, then, power, and more specifically legitimate power, will be located Up rather than Down."

Laponce is the political scientist who has done the most work and provided the most sophisticated discussion of both spatial models and symbolization of political relationships, in regard to political perception and the relation of brain lateralization to behavioral and ideological laterality. His graphic discussion (1978: 386, following Mirabile et al., 1976) of spatial dimensions provides striking confirmation of the neurological model of brain topology just outlined above (cf. Livingston, 1978), but Laponce emphasizes even more strongly the basis in *physics* (cf. ch. 19, below) for the hierarchy of precedence that characterizes Euclidean spatial dimensionality, once asymmetry begins in the form of deviation from the protean perfection of spherical shape. As he has remarked earlier (Laponce, 1975: 11), "our perception of space in terms of up and down, left and right, close and far, front and behind has universal or near-universal characteristics related to biological and physical constants such as the pull of gravity, the concentration of the major sensory organs in the front part of the body, and the slight imbalance between the two hemispheres of the brain as well as between the two sides of the body. We can then propose universal explanations of the applications of spatial symbols to social, and more specifically to political phenomena."

Thus a biologically determined cognitive bias toward verticality in perception does *not* imply biological determinism in culturally mediated conceptions—although it may influence them in an important way.

Verticality appears first, for reasons closely linked with the force of gravity (as noted above in regard to the rostral dimension of the brain), hence implying what we might denominate as a *vertical* gradient in maturation. Laponce's theory follows the neurological model in asserting that the ventral dimension is of secondary importance and that right-left differentiation (laterality) is tertiary. Because (Laponce, 1978: 387) "up-down has none of the ambiguity that front and back share with left-right and close-far—a lack of ambiguity due to the fact that gravity orders both self and landscape in terms of up and down, while there is no such congruent ordering of both self and environment on any of the horizontal axes—up-down became man's privileged spatial measure, *the* spatial dimension par excellence. . . . Control of the notion of verticality comes first to the infant, then control over the notion of front and back and much later control over left and right—so much later that many individuals never master the latter completely. The more asymmetric the spatial dimension, the sooner it is comprehended."

Laponce's remarks are true of humans in most "natural" environments; but the clarity of the vertical is quickly dissipated for the airplane pilot and the SCUBA diver—to say nothing of the inebriate or the day tripper. As many physical and cultural anthropologists have informed us, the drunk and the drugged define alternative states of consciousness that humans must have besought long before the transition to domesticity (of plants, animals, and ourselves) (Pelletier, 1978; ch. 8, above). But the systematic exploitation of the heavens above and the oceans below the earth's surface is a phenomenon of the twentieth century (and to a much lesser extent, the nineteenth). These artificial environments, in which the natural human orientation with the vertical becomes easily disorganized, have increasingly preempted the thought and activity of our species.

The "vertical reordering [in recollection]," says Laponce (1978: 388, 390–92), "tends to subject the landscape to the logic of gravity. . . . [Thus] the vertical is truly an antirevolutionary dimension, it is a dimension that constrains, it is indeed the most constraining of spatial dimensions; its strength is its invariance. . . . The relating, the binding that occurs in up and down" [that is, to rank and link the elements of a common structure, which Laponce describes as "an essentially political task"], occurs in a dimension that is already loaded with meaning for its being marked positively at the top, negatively at the bottom." Laponce assumes "that the spatial dimensions we are compelled to use in order to relate to the physical environment are unavoidably used to classify and clarify our social and cultural landscapes . . . [and] man cannot but perceive his social environment in terms of an up-down hierarchy."

If Laponce is correct that the vertical is the natural dimension for human social (as well as individual) orientation on earth, then it ought to follow that this no longer will be true in environments where Roederer's (1979: 100–1) "classical physical representations" no longer are impressed as our engrams, and even our macroscopic sensory world comes much more closely to "resemble the underlying reality [of astronomy and cytology] with which we must cope" (Pribram, 1980a; Zukav, 1979; P. Davies, 1980; Jahn, 1981; Wolf, 1981; Pagels, 1982; Wilber, 1982). Hence our escalating trend—on earth (skyscrapers, urban welfare slums, London Bridges in the desert), as well as in liquid or gaseous habitats—to move into and dwell in artificial environments that we create, underscores the need for revolutionary (that is, antivertical) changes in our politics, to bring our political paradigms into closer correspondence with our contemporary understanding of how humans think in natural environments (J. Davies, 1963: 104–40; Schumacher, 1973; Rodman, 1980); how humans think and behave differently in artificial environments; and how their thinking and behavior in both natural and especially artificial environments can take better and more creative advantage of artificial intelligence and thinking.

BRAIN DYNAMICS AND POLITICAL THINKING

At least since Harold Lasswell's (1930) return from his postdoctoral grand tour of European universities in the midtwenties, political science has vacillated between the irrational environmental determinism of neo-Freudianism (Etheredge, 1979; Horowitz, 1979; Sulloway, 1979) and the rational environmental determinism of logical positivism reinforced by econometrics (Axelrod, 1981; Axelrod and Hamilton, 1981; Gianos, 1982: ch. 5). The positivist paradigm has been and remains overwhelmingly dominant in a variety of guises in the political science of the past half century, whether as philosophy, legalism, historicism, institutionalism, behavioralism, or postbehavioralism (for example, Marxism, antienvironmentalism, Reagonomics). In his provocative discussion of cultural (as distinguished from disciplinary or intradisciplinary) paradigm change in political science, Rodman (1980: 55) has pointed out that "classical philosophy reflects . . . a politics of virtue that divides the human personality and ecosphere into good and bad elements, identifies bad with wild and good with tame, and struggles to bring wildness under the governance of a monoculture of rational control"; and in Rodman's view, the most fundamental aspect of the shift to the modern paradigm consists in substituting the acquisition of property for the pursuit of virtue. Now it is nature, rather than humans themselves, that must be brought under rational control; but that can only be accomplished through the even more disciplined exercise of the human mind and will. Clearly some humans can be trained and indeed motivated to think rationally, at least part of the time; but the question remains to what extent human rationality, and hence rational political thinking, typifies the way in which the human brain is designed (by the evolution of the species) to operate.

Consciousness

"The intrapersonal self-awareness that each of us has of his own existence," says William Uttal (1978: 216), "is the most primitive, most compelling, and most powerful force in human society. Everything else in human culture flows from this 'superfact.'" That assertion comes not from a proponent of mind transcendent such as Sir John Eccles or Roger Sperry; Uttal (1978: 685–86) is instead a self-professed materialist and reductionist, and his statement therefore can serve as a convenient baseline for, as well as prelude to, the consideration of models of dynamic brain function, with particular regard to consciousness and intentionality (Scher, 1962; Schwartz and Shapiro, 1976; Globus et al., 1976; Pelletier, 1978; Edelman and Mountcastle, 1978; Davidson and Davidson, 1980).

The alternative Eccles model deems self-consciousness and intentionality to be initiated and integrated, and to achieve conscious awareness, in the dominant hemisphere, which for most persons will be the left one. Eccles considers the cerebral cortex to be composed of vertical neural columns (or "modules") that interact as a complex mosaic, with projections to other modules constructed of qualitatively different types of neurons. Modules thus act as basic circuits, which in interaction can be analogized to integrated circuits. Eccles (1977) asserts that left-hemispheric dominance derives from its verbal and ideational abilities and its liaison to consciousness, which he illustrates with a schematic drawing and tabulation of functional differences between the "dominant" and the "minor" hemispheres. Thus the "pyramidal tracts" of the dominant hemisphere determine voluntary actions and consciousness of self, whereas minor hemispheric events enter consciousness only after transmission through the corpus callosum to the dominant hemisphere. Specialized modules in either hemisphere concerned with muscular-skeletal movement are subject to the control of the linguistic and ideational centers of the dominant hemisphere. Hence the minor (which typically is the right) hemisphere remains unconscious (see Jaynes, 1976); consciousness and intentionality depend upon and characterize the dominant hemisphere.

An alternative model of synergistic "multipotentiality" has been proposed by E. Roy John (1980: 131, 138–39), whose approach derives from statistical analyses and derivatives of extensive observations and measurements of electrophysiological phenomena, which John describes as "statistical configuration theory." According to that theory, information in the brain is distributed diffusely, in a process that is sensitive to the average firing of neural ensembles. Specific information about diverse sensory stimuli is dispersed throughout the brain regions sampled by his data. The memory of corresponding events ("endogenous processes") similarly is displayed in all regions. Hence any region may contribute to the mediation of a diversity of functions although this does not mean that different regions are functionally equivalent; the signal-to-noise ratio of averaged evoked potential, for a given input (or the memory that it invokes), can vary among regions on a magnitude scale of twenty to one. But any region participates in any given sensory neural input, whether visual or auditory, at the same relative degree of activity, with the neural firing level of the memory of the event proportional to that of the corresponding present sensory input.

A third model is provided by Karl Pribram's hierarchical model of attentional systems and hologic theory of brain functioning. His composite theory was developed over a period of about two decades (Pribram, 1971, 1978, 1979, 1980a, 1980b), utilizing both cybernetic and optical analogues. Pribram asserts that substantial neurochemical data support his hypothesized mechanisms of motivational readiness (or "activation") and of emotional arousal, which control the ranges of effort expended and comfort experienced by an active organism. The cortical events of activation and arousal (Pribram and McGuinness, 1975: 133) are controlled by independent subcortical systems denoted as "effort." Activation may occur with or without accompanying arousal, and vice versa; while in the absence of both, behavior is autonomic and is based directly upon stimulus-response contingencies that do not require or command attention.

Pope and Singer (1978: 106–7) define attention as the process by which the material available is screened and selected for introduction into the stream of consciousness. They note a distinct bias toward attending to sensory material, which they consider to be a characteristic of obvious survival and adaptational value. But "when the environment becomes predictable, dull, or barren, the tendency is for consciousness to move toward the more private end of the continuum, for memories, associations, and imaginary materials to flow into the stream of consciousness."

Pribram asserts that *self*-consciousness is a higher level involving intentions and is characterized by a structural organization that is feed*forward* (rather than feedback) in operation. He hypothesizes (1978: 2271, 2273) several states of consciousness, each defined by a unique and relatively stable configuration of chemically active core-brain systems. As in John's theory, such neurochemical configurations act as *templates*, that is, as attentional controls on sensory input. Hence it is the *pattern* of the electronic processing of signals in the cerebral cortex that defines the matrix determining the content of consciousness. But Pribram hypothesizes that such patterns encode neural signals by the *frequency* (rather than the amplitude) of the waves of electronic transmissions. Therefore, image construction or reconstruction is based on holographic-like representation of the encoded signals, which Pribram distinguishes from information processing based on feature extraction through the operation of the association cortex. Hence this model posits a bimodal sensory processing mechanism, a parallel processing linear frequency analyzer, and a hierarchical nonlinear extraction process, which results in discriminable alternatives in conscious content (information), while the frequency analysis produces images. Points within a source being imaged emit signals whose wave fronts transect, and the encoding of the frequency of the interference among such wave fronts generates a hologram. In Pribram's view the holographic hypothesis of image processing is particularly compatible with the *projective* nature of human conscious experience (cf. Pelletier, 1978: ch. 4).

Speaking both intro- and retrospectively, Pribram (1979: 69–70) has explained that initially (back in 1966) he had deemed holography to be only a powerful metaphor, useful to explain the distributed nature of memory traces in the brain. Before that insight, the problem had been how to explain learned behavior's resilience to brain damage. Once the analogy was noticed, the optical theory of holograms pointed to a solution: a hologram is a blurred record of an image or object; it encodes ripples made by a disturbance such as a sensory input; ripples are vibrations or waves; hence individual cells in the brain cortex encode the frequency of waves within a certain bandwidth. Pribram (1971: 153–55) now takes hologic theory much more literally, and he has discussed in detail how the brain could work physically as a three-dimensional hologram, with the ability of neurons to propagate received signals permanently enhanced by frequent use (reinforcement); indeed, he suggests (1980a: 27–34; 1980b, 56–59) that the equivalent of a two-dimensional hologram can perform a fast search for the requisite engram, which when located becomes readily accessible though stored three-dimensionally.

Memory

McGuigan (1978: 85, 90) has proposed that there are critical nonoral muscular components of cognitive activity that are related to language behavior. This implies that cortical control over, and the use of, language may involve many other aspects of the muscular skeletal systems; and conversely such nonverbal behaviors may be an integral component of oral language behavior. Gazzaniga and LeDoux (1978: 135, 137) make the further point that, although there has been much emphasis placed upon the relationship between memory and verbal processing, there may be many other systems of memory involving gestures, movements, and other nonverbal modes of both sensory input and response. Thus they remark that "when a past experience is to be remembered, it is prima facie a multidimensional experience involving time, space, colors, sounds, smell, temperature, and a variety of other stimuli. Many of these are not activated when one is called upon to relate an old memory *verbally*, and as a result, the verbal memory is limited in extent. When a person reenters the physical circumstance of the memory, however, the ability to recall verbal aspects of the event is usually increased . . . [and] being in the physical surroundings has a tonic effect on the verbal recall" (emphasis added).

Long-term memory apparently involves independent systems for the storage of cognitive engrams, however widely these may be dispersed throughout the cortex, and affective engrams, which entail much greater involvement of the limbic system and midbrain. It is probable that the cognitive or informational content of memory is much more directly dependent upon cortical language processing and control; affective or emotional memory, on the other hand, probably is more closely associated with the sensory and nonverbal components of multidimensional experience (Roitblat, 1982). Izard (1980: 195, following Tomkins, 1962, 1963) asserts that it is only as a function of the emergence of emotions that consciousness develops and realizes its highly complex organization; and he notes (1980: 212) that the available evidence indicates that neonate consciousness consists *primarily* of emotional responses to endogenous stimuli—affective experiences devoid of images and symbols. Izard (1980: 195) points out that Piaget and Inhelder (1969: 158) also took the position that the cognitive and intellectual development of children depends upon the motivation of emotions; and he quotes their statement that "there is *no* behavior pattern, however intellectual, which does *not* involve affective factors as motives" (emphasis added). Pope and Singer (1978: 107, also following Tomkins) agree, adding that "the basic set of affects operates with telling effectiveness to bring to consciousness material from the body, from the environment, and from entirely private sources (memories, dreams, fantasies, etc.)."

Emotion comes *first*, both phylogenetically *and* ontogenetically (Panksepp, 1982); and metaphorically speaking, it was and is subsequently overlaid with consciousness, just as physiologically the neocortex overlays the hippocampus and the rest of the limbic system (O'Keefe and Nadel, 1979; Olton et al., 1979; Gray, 1982b). Emotion thus screens what enters consciousness from sensory perception and memory, as well as from other physiological processes of the body.

Cognition

Human brains, although critically different from even those of other primates in regard to language as well as other aspects of culture and plasticity of behavior, nevertheless remain constrained by many of the characteristic operational modes of mammalian brains generally. This is a subject that has been appraised by Roederer (1979a: 100–101), and the remainder of this paragraph and also the two following it attempt to summarize his excellent discussion. As he points out, the quantity of information sensorily perceived is utterly beyond the capacity of any animal brain to process or store; therefore only information deemed relevant (ultimately, to the animal's survival) is screened for long-term memory storage, or (as the present writer would add) for conscious thought. A cognition of "subjective relevance," by which Roederer means that which is associated with higher emotional arousal (and therefore will be stored simultaneously with a stronger affective trace-component) is more likely to get information into memory than "objective truth," by which he means a cognition that the information corresponds to expectations about empirical reality and has no particular apparent implications for the needs and interests of the animal and therefore is not very arousing.

No animal brain can afford to produce accurate statistics about causality in observed empirical relationships, because in evolution a much higher value is placed on maximizing the survival of the organism, which necessitates expedition in detecting and responding to the environment. Hence animal brains do not process information in the sense of statistical analysis. Instead they substitute a biased predisposition for evaluating causal relationships optimistically and frequently mistake coincidence for causality. But there is also a "fundamental motivational drive"—especially among primates, although quite generally among mammals—to seek out environmental information exploratively, even in the absence of any manifest physiological need of the organism; such "curiosity" and "playfulness" are notably characteristic, because critical to the learning and development of, young animals (Bruner et al., 1976; Baldwin and Baldwin, 1977a; Symons, 1978; A. Smith, 1979; P. Smith, 1982). Such environmental explorations are guided by *random* operations of the animal's brain; environmental data are not typically acquired as the consequence of planned explorations.

Access to long-term memory is hierarchically ordered, so that information is maximally accessible to the extent that it relates to an important (emotionally significant) event and/or an abstract one (low in information content). Vast quantities of information are storable in any given spatial region of the brain, but only one program (of the many stored in the same region) can be displayed (resonated) at the same time; and this is what accounts for the coherent, serial mode of conscious experience. Because the animal brain evolved long before the extremely recent (in either biological or geological time) human invention of telescopes and microscopes, it is the "macroscopic"—really, it would be better to say the *mesoscopic*—world of natural sensory perception with which the brain interacts; and that is the world of classical, pre-Galilean as well as pre-Newtonian physics (as well as pre-Darwinian and pre-Mendelian biology), *not* the world of Einstein or Bohr or Heisenberg or Planck, nor that of Rosalind Franklin and Francis Crick and James Watson. Paradoxically, therefore, information is stored in animal memory, including the human brain, in the sense of Euclidian geometric representations that conform to the *Weltanschauung* of pre-Renaissance classical physics, with putatively irreversible causal relationships among macro[meso]scopic objects, according to the rules of traditional logic.

In the lecture notes for his course on the "Physics of the Brain" (as discussed in this and the following paragraph), Roederer (1976: 30, 35–37, 52; and cf. his 1978, 1979a) has provided a useful overview of how brain science views *human* learning, memory, thinking, and intelligence. In operational terms, learning depends upon *templates*, and a template is a particular grid (or engram) of four-dimensional paths of interconnected neural columnar structures, the stability of which has been determined both qualitatively (by the strength of its associated affective trace) and quantitatively (by the frequency of its repetition). An engram is activated—that is, the electrical strength of brain waves through the particular set of paths is increased—when the configuration of perceived sensory signals produces in short-term memory a structure that sufficiently resembles the subject engram. An extremely rapid search of the set of possible engrams leads to a *satisficing*—not necessarily optimal—match to a particular engram, and thereby the perceived object then is recognized. This process is the physical counterpart of what learning psychologists call "reinforcement." But this oversimplified explanation must be qualified by the recognition that few engrams will consist of a well-defined, spatially limited circuit of neurons; instead, typically they will involve extraordinarily complex networks including millions of neurons, which conduct neural pulses in many directions throughout many parts of the association as well as the interconnecting tissues of the cortex. It follows that long-term memory retrieval requires the replaying, at least in part, of the neural activity that occurred when the information was stored. In terms of our three-dimensional model of the brain's structure, sensory input from "below" stimulates the continuing display of engrams "on the left" which have been selected (that is, have associated neural connections) on the basis of what the person's experience indicates is likely to come next (that is, what produced the associated neural connections).

"Thinking" is defined as such acts of internal information processing as are described in the preceding paragraph. Roederer presumes that, irrespective of whether or not any (by definition, maternal) language has yet been learned by a person, the serial activation of neural activity that a continuing display of engrams constitutes occurs *in the language networks*, which will be in the left hemisphere for most persons. Other animals think in the sense discussed so far, although they do so with much simpler cortexes and with vastly simpler language networks than humans are preadapted for; but humans may be unique in the degree of their development of the ability to think *in the absence of direct sensory stimulation.* Thus the human brain can recall and alter images or representations, and then put the revised engrams back into long-term memory, by placing the recalled image into short-term memory and then matching to that internally generated image instead of to a construction from presently perceived sensory input. Knowledge acquired neither through the senses nor by genetic transfer is a synergistic product of the human thinking process. Obviously, either mode of thinking—the processing of information from sensory input, or from remembrances of things past—results in the storage of (what will generally be partially) new information. Both human and other animal brains "can acquire, analyze, organize, and selectively store information and make predictions about the outcome of alternatives posed by the environment," but "only the human brain can consistently create new information by making predictions about the self-generated alternatives." Roederer believes that "the capacity to respond in non-random fashion to unexpected, unforeseen circumstances" is one that "lies at the very root of the concept of 'intelligence,'" although of course many animals besides humans are intelligent in that sense (see Denenberg, 1981; and Griffin, 1978, 1981, 1982, 1984).

Political Thinking

The positivist paradigm in political science presumes that political elites generally *ought* to make decisions rationally; to the extent that political masses fail to achieve this degree of political perfection, better civic education ought to improve their performance. Rational political actors, in turn, are presumed to think analytically; or in the conceptualism of psychobiology, rational political actors are thought to be strongly lateralized for left-hemispheric brain function—which also implies (as we shall see presently below), not inappropriately for times past, that they would be males. Any sizable infusion of female actors, at present and in the future, would of course contradict that presumption; but so do hologic theories of consciousness (for males as well). If thinking indeed is a process of matching and reshuffling images, on the basis of frequency-modulated resonances of microcolumnar neural structures dispersed throughout the cortex, and if such images (and their constituent components) entail explicit affective traces, so that remembered emotion is associated with the construction or reconstruction of the image, then it is understandable why our rational, analytic models fit so poorly the empirical data of what political actors actually do.

Political scientists have tended to be interested in facets of consciousness related to linguistic phenomenology and analyses of language demonstrating how and why communication is biased by the use of language (Shapiro, 1981; Tanenhaus and Foley, 1982; cf. Cohen, 1981). There is, however, more recent work, much of which has been done or discussed by political scientist Steven Peterson, that relates political cognitive psychology to its base in psychobiology. In one set of papers, Peterson and his psychologist colleague Robert Lawson (Peterson and Lawson, 1982a, 1982b) have discussed cognitive psychology and political science in both substantive and methodological terms. But Peterson was investigating also the psychobiological structures and processes that underlie the experimental and survey research upon which cognitive psychologists base their findings; and in subsequent papers (Peterson, 1981a, 1981b, 1982a, 1982b) he integrates these two domains, the psychobiological and the cognitive, of psychological research.

Peterson points out that preliminary evidence from several studies of reaction time in response to political stimuli tends to show that the speed with which verbal political information is processed is related to the clarity and habituation (familiarity) of the cognitive *schemata* (neural templates) that previous experience had made available to the respondents through their respective long-term memories (Lodge and Wahlke, 1982: 145 for reaction time, 149 for schemata). He also discusses cognitive *heuristics*, or short-cuts to decision making under conditions of uncertainty of information. These include: (1) confirmation bias, (2) availability, (3) vividness, (4) representativeness, and (5) attribution. "Confirmation bias" involves the evasion of facts if these contradict preexisting beliefs. "Availability" is reliance upon the most easily accessible information (stimuli) *or* schemata (recollections that can guide expectations about what stimuli are likely to be encountered). "Vividness" involves overgeneralization from small but affectively important remembrances. "Representativeness" involves overgeneralization from a small number of experiences that do not happen to have been typical of the population category to which they are projected. "Attribution" error ascribes the motivation of other persons to presumed attitudinal or other predispositional causes, while ignoring situational or contextual causal variables that might provide an independent or a better explanation for the observed behavior.

A widespread example of attribution error among American politicians, political scientists, and the general populace alike was the long-standing and popular explanation for the fact that females participated considerably less than males in elite political office holding (in terms of both the numbers of participants and the level of participation), namely, on the grounds of "feminine mystique" (that females lack interest in politics, they do not have the natural ability of males, it is unfeminine of them to be involved in politics, etc.). Both situational and structural constraints (that is, female biological and social role burdens of infant production and child care and employment discrimination against females, especially in such politically relevant professions as law) were ignored (cf. Schubert, 1985c).

An increasing body of empirical political research demonstrates the political importance of schemata in political decision making; and these data fit like hand in glove for the psychobiological theory of political cognition that has been proposed by Peterson and that is presented in the present article (see also Schubert, 1981b, 1983c: 117–18; and ch. 19, below). Research by Doris Graber (1982b: 3, 11, 17; and 1982a) demonstrates the extent to which voters ignore political information to which they are exposed but which does not agree with what they already "know" and are prepared to accept. "Much available information is ignored from the start. Information scanning is done carelessly and unsystematically [so] that only a fraction of the information supply is incorporated into the average individual's knowledge base. . . . Most person[s'] judgment schemas apparently did not include the idea that inconsistencies might involve rational readjustments to changing conditions. [T]he bulk of political news is never processed at all. When processing takes place, news is shaped to satisfy the individual's needs which, in turn, are shaped by past and present experiences and by personality factors." Confirmation bias is exemplified also by the research in cognitive mapping in international politics of psychologist Gerald Hopple, who (as quoted by Peterson, with Hopple himself quoting Shapiro and Bonham) claims (1980: 101) that "in the decision-making process, beliefs act like templates for channeling information and for relating possible policy options to perceptions about the intentions and behavior of other nations, and also to the policy objectives of the decision-maker." Peterson (1982b: 1) reports that "humans consistently overestimate their rational abilities."

Let us consider briefly some of the leading characteristics of the model of human consciousness and thinking indicated by contemporary psychobiology, so that we can contrast them with the assumptions of rational political decision making, and also so that we can better consider what might be done about the hiatus between the nature of our brains and the demands made upon them by our political systems and theory alike. First, we as political scientists must accept the engram—a particularly stable configuration of chemically active core-brain systems that define a given state of consciousness and serve as an attentional control upon sensory input. Such states of consciousness account for the highly *projective* nature of our thinking, including our political thinking; Harold Lasswell was right on target when he defined public policy as consisting of private interests displaced (projected) onto public objects that are then rationalized in the public interest, except that with another half century of neurobiological research we now have a much more detailed and empirically based comprehension of projection than the Freudianism that was Lasswell's premise.

Subjective relevance is much more important to us than objective truth, and for the very good reason that for at least hundreds of thousands of years, *Homos* who relied upon subjective relevance survived to reproduce offspring carrying genes biased toward subjective relevance in thinking; they did better than hunter-gatherers whose passion was for objective truth, so that the latters' genes tended to disappear from the genus and species genome. Even today, when we no longer are (at least, for the time being) primarily hunter-gatherers, we retain the species genomic preference for subjective relevance; and because "statistical analysis . . . is alien to brain processing" (Roederer, 1979a: 101), we characteristically employ the statistics more to justify than to "guide us" in the making of our political decisions (see Livingston, 1978; Shepard, 1978; Crook, 1980). "Lying with statistics" thus comes naturally to us; stochastic processes and game theory have to be learned, usually through formal education. Subjective relevance has been the subject of considerable research inquiry during the past half century, especially on the part of social and political psychologists who have studied human values and attitudes (Eysenck, 1954; Eysenck and Wilson, 1978; Manheim, 1982; Rokeach, 1968; Schubert, 1974, 1975b, 1967a, 1967b, 1977a, 1977b).

Nevertheless, because our brains interact with a macroscopic political world, our engrams necessarily are based upon the "classical physical representations, Euclidean geometry, irreversible cause-and-effect relationships, and classical logic" (Roederer, 1979a: 101) that we have learned through our cultures; this creates a continuing need for us to rely upon additional *culturally* processed and produced information and stimuli that can correct our misperceptions of the underlying reality that must be coped with—if we are to remain adapted under anything resembling present circumstances (Pribram, 1980a: 32–33; Filskov and Boll, 1981, section on "emotions"). There is a tension between the processes by which we think and the macroscopic data of sensory input that provide the substance of what is thought. But the natural (to the brain) mode of environmental exploration is random, as exemplified by the curiosity and playfulness of young mammals generally and juvenile primates in particular: planning for a scheme of exploration is an artificial mode of thought that must be culturally imposed upon political (and other) decision makers. Even optimism, which must have been essential to the differential survival of the human species during the millions of years of our hominid evolutionary past, may under present and probable future environments prove to be a lethal political predisposition, especially among political leaders at the highest levels of governance, unless constrained by culturally imposed inhibitors (Tiger, 1979b, 1981; Peterson, 1983a, 1985). Thus fail-safe devices and a host of other servo-mechanisms are familiar artifacts of military and industrial technology; but we have conspicuously failed to be correspondingly ingenious in our political technology. Our need for complementary intelligence escalates as we move into increasingly artificial environments; and that need is no less great for political than for other decision makers, as two biopolitical scientists recently have asserted. "One of the strongest arguments favoring application of the schema approach to political research found in the theoretical literature on semantic memory is its link to contemporary models of human and artificial intelligence" (Lodge and Wahlke, 1982: 148). But to make effective use of artificial intelligence, we must first accept the limitations of what is natural to ourselves in thinking.

We cannot yet say on the basis of present *empirical* knowledge what kind of political and other learning would be most likely to produce individuals with the experiential and personality traits that ought to be deemed prerequisite for particular political positions; but the work that several biopolitical scientists have undertaken, in recent papers dealing with the neurophysiology of political learning and socialization (White, 1980b, 1981a, 1981b, 1981d, 1982a; Peterson, 1981a; Peterson and Somit, 1982), surely is launched in the necessary direction.

BRAIN DEVELOPMENT AND POLITICAL EQUALITY

The orthodox social science standpoint of the past generation has been that there are few important differences among humans, but to the extent that there are, they can be completely explained and, if necessary, reformed in terms of cultural and environmental changes. The contrary assumption here is that neither heredity nor environment is destiny for humans and that both biology and culture are relevant to any realistic discussion of political equality. Moreover, this is no one-way street: not only is it the case that genotypes influence the phenotypes that produce culture, and hence genetic differences affect culture through those phenotypes; but it is also true that given enough time and enough control over breeding—which is difficult to sustain very long in regard to humans—culture produces genetic differences among humans: controversy in sociobiology today concerns how much time and how much control (Baldwin and Baldwin, 1981a; Gilman et al., 1979; chs. 9–12, above).

The brain is every human's most unique and individualistic organ; and each brain is the product of the interactions of a unique genetic inheritance that has been uniquely modified by a unique experience of environmental effects. (Phenotypically there are slight epigenetic differences even between monozygotic twins because of differences in the intrauterine environment, such as position and proximity to the maternal heart, birth order, and of course postnatal differences in experience even for those reared together in the same household.) Among the many variables that directly and decisively influence brain development, two have been selected for discussion here: lateralization, or differences in the degrees and types of asymmetry in hemisphericity, and nutrition. Both, as we shall see, have crucial and pervasive effects upon political equality as well as brain development.

The strong consequent biological impetus is toward individual variation, which among humans is exacerbated because of the capacity—unprecedented among other animals—for plasticity in the development of the human brain, in particular due to cultural and social influences. One such sociocultural influence upon brain development and behavior has been the powerful—and probably also unprecedented—thrust of culture, through ideologically based public policy to maximize human equality during the past two centuries, especially in the West but more recently of global scope. One major component of that equality has been political; and the drive for political equality has involved a continuing attempt to minimize (or at least to reduce) basic human differences, in relation to several social (biological) variables of political equality including health (survival), age (development), sex (reproduction), race (subspeciation), and intelligence (adaptation). In the discussion below, we shall consider these social biological variables of political equality, relating them to *biosocial* variables of lateralization and nutrition.

Lateralization

Lateralization refers to differential development in the structure of the two hemispheres of an animal brain. From an evolutionary perspective, hemispheric specialization evolved in humans in synchrony with speech; but hominid communication was gestural and in other respects nonverbal for a very long time before elements of speech were added (Corballis, 1978; Kinsbourne, 1978; Calvin, 1981, 1982; Masters, 1976, 1982b; Schubert, 1981a, 1982a; Bruce, 1983; Mateer, 1982). Indeed, nonverbal communication *still* carries much of the message (especially the *affective* component) in human speech, including political speech and transactions involving it (Henley, 1977; Izard, 1980; Schubert, 1982a). But political scientists almost never study nonverbal communications, notwithstanding our penchant for anecdotal references to the importance of baby kissing, pressing the flesh, and assorted fraternal hugs and pats on the back. Hominids were nonverbal communicators for millions of years before the recent appearance—no more than a hundred thousand years ago—of our present species (which has evolved very little physiologically since then), and of our immediate ancestral hominid species, in whom hemispheric lateralization and speech alike must already have been well advanced (Washburn, 1978; Stanley, 1981: ch. 7; Parker and Gibson, 1979).

Written language, as both a cultural phenomenon and behavioral trait confined to a tiny minority of humans during only the latest five to six thousand years, and the establishment of governmentally supported public education even much more recently together vastly increased the proportion of literate humans—although even today they include only a minority of the species. Biolinguists hypothesize, on the basis of such empirical evidence as that there are only about forty distinctive sounds associated with all known languages, that the human genome includes an evolved and genetically determined "deep structure" or preadaptation in the brain—in addition, of course, to the several genetically determined physiological structures relating to such organs as the soft palate, tongue, and larnyx, which also are prerequisite to the production of human speech—that facilitates the learning of language by human infants and juveniles (Chomsky, 1975; Von Eckardt, 1978; Lenneberg, 1967). The indispensability of both Wernicke's area (comprehension) and Broca's area (motor control of speech organs) of the left hemispheric neocortex to the production of normal human speech was of course a discovery of nineteenth-century psychobiology; but the discovery of a related structure of speech preadaptation in simians is much more recent: evidence of specialized development of auditory structure in monkeys, in the left hemispheric locus that is precisely where Wernicke's area is found in humans. Moreover, generalizing from research on chicks, songbirds, and rodents as well as *non*human primates, Denenberg (1981: 8, 20) has hypothesized that for efficient communication to develop, hemispheric asymmetry "is an initial condition of brains. The left hemisphere is preferentially biased to receive and to transmit communications; the right is selectively set to deal with spatial and affective matters; and both often interact via activation-inhibition processes in the affective-emotional domain."

The basic dichotomy of hemispheric lateralization is between language on the left and manipulospatiality on the right. The prevailing consensus among many brain researchers, working independently, is that the specialization of the left hemisphere is *analytic*, defined typically as serial, focal, difference-detecting, time-dependent, spatial, global, and synthetic or gestaltic (Gazzaniga and LeDoux, 1978; Bradshaw, 1980; Bradshaw and Nettleton, 1981; Sackheim and Gur, 1978; Pelletier, 1978; Eccles, 1977). But the right hemisphere is not irrelevant for speech; and neither is the left unimportant to nonlanguage functions, although this obviously depends upon extensive interhemispheric communication via the corpus callosum and the commissures.

Once speech had become primarily centered (phyletically speaking) in the left hemisphere for the human species genome, so likewise did the conscious cognitive control over behavior generally—not just speech behavior because (Gazzaniga and LeDoux, 1978: 79, 81–82, 92) "the unilateral representation of the mechanism by which speech is programmed and executed provides a final cognitive path through which behavior can be organized and controlled. Though both sides of the brain may comprehend and store linguistic information, the unilateral control over speech provides a common point through which the various cognitive activities related to language can be channeled and through which their relative importance is ranked and motor commands are programmed and executed. Therefore, comprehension generally follows where speech goes."

The neural structure for language emerges bilaterally during early development, to become consolidated (typically) in the left hemisphere only after several years of postnatal experience, when speech already has begun. However, there is evidence (Kocel, 1980: 299–300) that lateralization does not generally increase but rather becomes weaker with maturation as interhemispheric communication becomes, through continuing reinforcement, more efficient as a consequence of aging; this leads to greater isomorphism in the two hemispheres, due to a relatively sharper *decline* in right-hemispheric abilities (thus making the right less different from the left). The consequence is increasing reliance with age on *left*-hemisphere strategies (causing, exacerbating, or in response to right-hemisphere deterioration).

Behavioral Laterality and the Lateralization of Ideology According to Stuart Dimond (1977: 491), "people showing different patterns of handedness possess different properties of intellect, and . . . they can be regarded as basic biological variants." Laterality is positively, although not consistently, associated with lateralization. Sinistrals are not as strongly left-lateralized for speech, although the majority appear to have speech represented primarily in the left hemisphere (Morgan and Corballis, 1978: 270). Most individuals with either right-hemispheric or bihemispheric language representation are sinistrals, and some studies indicate that the most strongly left-handed individuals have right-hemispheric speech representation (Witelson, 1980: 82). These findings imply a linear correlation between two continuous variables: the strength of sinistrality in handedness and the degree of right-lateralization in language hemisphericity, with lateralization causing laterality—although not necessarily in the evolutionary sense of causation (see Kinsbourne, 1978). But this conclusion leaves open the question as to what extent hemisphericity is epigenetic and what interactive relationships between genetic and "environmental" (including cytoplasmic) variables account for degrees of both lateralization and laterality in individual cases (Fincher, 1977; Barsley, 1970).

Levy and Gur (1980: 200) point to evidence that manifest handedness is affected by both prenatal and perinatal factors that increase the frequency of sinistrality, while most postnatal (cultural) influence is in the opposite direction; but they conclude that none of these factors is sufficient to account for the greater part of the variance in laterality and for its association with other traits (e.g., about 70 percent of dextrals, but also about 50 percent of sinistrals, are right-eyed: Levy and Gur, 1980: 203). It is also notable that environmental pressures against sinistrality have apparently decreased in recent decades (Corballis, 1978: 335).

Morphological asymmetries in both the frontal and occipital regions lead LeMay (1977, as summarized by Witelson, 1980: 94) to describe dextrals as humans with a counterclockwise torque in their brain structure whereas sinistrals are characterized by tendencies (of varying strength) toward a clockwise torque. The available evidence suggests that sinistrals, although much fewer in number, constitute a much more heterogeneous group than do dextrals. But the completely right-lateralized sinistral, and exclusive right-lateralization for speech are both very rare (Boklage, 1980: 116).

The inference of dextral homogeneity may be at least partially, however, an artifact of the research bias heretofore in testing dextrals with stimuli that are more specific to left-rather than right-hemisphere activity. Witelson (1980: 105–6) suggests, for example, that perhaps the difference between musicians and others lies not in cognitive strategy but rather is due to a difference in cerebral dominance (e.g., symmetry in bihemispheric representation), so that musicians are not necessarily left-lateralized, but rather are *less* lateralized (Roederer, 1979b, 1979c). For example, recent research on the tactile sensing of braille shows that the overwhelming majority of right-handed blind children that were born blind read braille much more efficiently with their left-hand middle or index fingers than with their corresponding right-hand fingers (Harris, 1980: 309, 322). Presumably this is true because the right hemisphere analyzes the spatial information supplied by touching braille alphanumeric characters more efficiently than the left hemisphere; and it does so most efficiently with direct (left-handed) stimulus input, which is then transferred commissurally to the left hemisphere for language analysis (Harris, 1980: 310).

Laponce (1976: 51; and cf. Laponce, 1970, 1972a, 1972b) deals explicitly with the relationship between left-handed people and the left orientation in politics; and he explains why we ought to expect to find a positive correlation between minority behavior (at least in the United States and in Canada, where Laponce resides) and left-wing political attitudes and behavior.

> Left-handed people live in a more or less hostile or at least in a more or less unfriendly surrounding. Of course many will overcome the difficulty and the awkwardness resulting from their being wrongly "bent," and this very overcoming may well strengthen their ego, may even give them a certain sense of superiority, but we should expect nevertheless that, on the average, the left-handed would be at a constant disadvantage, a disadvantage leading him to be relatively dissatisfied with the present and hopeful of a more equalitarian future. . . . [Since] left-handedness has . . . biological determinants . . . the likelihood of there being specific political correlates to left-handedness is even greater. . . . The left-handed . . . is a different individual with different reactions to self, others and environment.

Laponce's hypothesis awaits empirical investigation with adequate data.

If Laponce (1978: 393) is correct that the order explicit in verticality has priority due to the weight of authority (in more senses than one), then it is understandable that there is a need for political education about the importance of more widespread awareness of the necessity for the perspective of the left and consequent speeding up of political cultural change (Schubert, 1982c). For at least two centuries, it has been the case that in both the study and the practice of politics, the metaphor of the left is understood to lead in change, new growth, and development; the right, on the other hand, stands for conservatism, stasis, and an attachment to roots rather than to futures (Laponce, 1981). Laponce has remarked (1975: 16, 17) that "the Left-Right dimension, which is now so widely used to express political contrasts was not applied to politics until the end of the 18th century. . . . [F]rom the French revolutionary Assemblies, where they first took their political meaning, Left and Right were then transmitted to other countries. . . . The Left-Right terminology is likely to have succeeded for two major reasons (a) in all language-cultures, the Chinese being the only major exception, the notion of left was already associated in religious and social customs with the notions of opposition, challenge, and secular ideas; (b) Left-Right provided a useful 'equalitarian' substitute to the more strongly hierarchical notion of up-down." Hence the relationship is not merely metaphorical; the French Chamber of Deputies (with its slogan "toujours à gauche") naturally comes first to mind as exemplary of a long series of political structures whose laterality is isomorphic with the biological understanding of permanence and change (cf. Monod, 1971).

The contemporary trend toward adaptations to artificial environments (discussed above in this chapter) demands reexamination of our traditions and practices in hierarchical politics, which evolved on earth but (as we must assume, from a biological point of view) constitute potential maladaptations and therefore may provide very inappropriate models for the political behavior of humans, who are as literally out-of-touch with the earth as those who fly (or soar) above it or who swim (or glide) in its fluid depths. Conceivably, the planting of the United States flag (symbolizing territoriality and at least inchoate hierarchical control) by American moon-walkers was an anachronistic attempt to cling to a rapidly dying past and not at all the symbolization of future politics that it was intended to be. And no less bizarre are our misguided (if not pathetic) endeavors to project via outer-space vehicles cornerstone relics of our emphemeral artifacts so that creatures "out there" can admire such guided obelisks of the hauteur and grandeur of our accomplishments. The probability that extraterrestrial life exists in some form somewhere else in space is matched only by the reciprocal improbability that our space needles in the dark will encounter it.

The Politics of the Sexual Brain Kocel (1980: 298) states that "females and left-handers have similar cerebral organization." The general finding with regard to brain organization (as distinguished from function) is that females are more symmetrical than males: that is, females have more RH (right-hemispheric) language facility and more LH (left-hemispheric) spatial conceptualization than do males, for whom language is more specialized in the LH, while nonverbal cognition is in the RH, especially after puberty has been attained, due to the influence of testosterone upon subsequent hemispheric organization and development. Yet it is also true that females show considerably greater motor asymmetry than males (Annett, 1980: 227; Waber, 1980: 250), but are better at fine motor activity while males do better at gross motor actions (McGuinness, 1981: 103). Boys are more left-handed—which is consistent with their being more highly lateralized—than girls (Herron, 1980); but girls are more right-handed than boys, which as Bradshaw (1980: 230) remarks, is inconsistent with the assertion of greater female symmetry. Males evince greater verbal deficits than do females, after LH lesions; and this finding *is* consistent with greater female symmetry. Females do markedly better in cross-modal sensory integration, which makes little sense in terms of classical hominid evolutionary theory: until ten thousand years ago, it would have been the presumed "hunting bands" of *males* (under that hypothesis) who needed quick and accurate read-out of sensory input (about prey, predators, each other, etc.) from all modalities available to them.

It is possible that what is perceived to be the less-lateralized female brain is due, in whole or in substantial part, to better commissural (interhemispheric) communication. It is also possible that female verbal superiority is due not only to female precocity in LH development, but that it also is enhanced by optimal sharing of the verbal function in *both* hemispheres, in lieu of the more extreme lateralization attributed to males (cf. Bradshaw, 1980: 230). In any case, McGlone's (1980: 251) remark that "lateralization of one function (e.g., speech), may in turn determine the degree of lateralization for other functions such as spatial orientation" seems to put the cart before the horse. It makes much better evolutionary sense that the RH developed *first* asymmetrically, to facilitate the development of the better spatial skills upon which hominids perforce had to rely for adaptation a million years ago; and when verbal skills subsequently evolved, they *had* to become overspecialized in LH because no uncommitted space remained in RH (see Whitaker and Ojemann, 1977; and Whitaker, 1978: 324, citing Jerison, 1977). Similarly, the overcommitment of male RH to nonverbal functions (according to the suggested hypothesis) would explain how females could have exploited the more equipotential development of language, in RH as well as LH, thereby getting the head start in constructing the language superiority that they enjoy. Indeed, McGuinness (1981: 104) thinks that "it is quite possible that females invented language."

The evidence from psychobiology (Katchadourian, 1979; McGlone, 1980; Parsons, 1980; Wittig and Petersen, 1979; Levy and Levy, 1978; Kocel, 1980; Newsweek, 1979) is that females think differently, physiologically, from males. "Sufficient information has . . . now been accumulated for the male and female brain to be regarded as basic biological variants of brain type" (Dimond, 1977: 491). Very much to the contrary, many social scientists, reflecting their strong policy commitment to sexual egalitarianism and equal opportunity, reject the very idea of male as distinguished from female brains as subversive even to discuss (Grady, 1977). Psychobiological research indicates, however, that human males are more strongly lateralized than females; that females develop earlier, stronger, and better language ability, in part due to the bihemispheric representation of language in yearling females; that males develop earlier, stronger, and better manipulo-spatial ability, which is primarily right hemispheric and related to the more extensive and intensive exploration of the environment by yearling males; that males at puberty equal and then exceed females in left hemispheric language lateralization, explicitly because of the hormonal effect of greatly increased testosterone secretion in males at that time; that females are better at fine motor coordination, probably due to their better commissural (inter-hemispheric) communication; and that all of these effects are independent of cultural influences (Harris, 1978; Whitaker and Ojemann, 1977; Taylor, 1978; Schubert, 1983f). The Y chromosome slows down the rate of male development by leading to a more detailed read-out of information from the genotype, with the consequence that males are more neotenous than females—they spend more time than females do in any given immature and vulnerable state (Gualtieri and Hicks, 1985); and in comparison to the relatively more advanced left hemispheric maturation of two-year-old girls, boys are retarded in the development of that aspect of their brains.

There is much more variation in lateralization *within* than between the sexes—but this is especially true of males. From an evolutionary perspective, the existing differences (in whichever direction) evolved because at some time they proved adaptive for the species (or its precursors); and even if presently they are not subject to strong selection pressure, these characteristic differences in the species genome, which evidently contribute to its heterogeneity, may become important to the future adaptability of the species under changed environments unpredictable now.

Other sex-linked differences among humans, relating to personality, emotionality, social skills, and sociality, as well as in modal size, strength, *and* reproductive behavior per se, also are important, in addition to the cognitive differences. These questions I have discussed in a separate paper, which asserts that the modal female brain is structured to support better balanced and better integrated thinking than that of the typical male, just as the female hormonal system supports less competitive and more cooperative and nurturant behavior than does the male hormonal system (Williams, 1975; Levy, 1980; Springer and Deutsch, 1981: ch. 6; McGuinness and Pribram, 1978; McGuinness, 1979; Schubert, 1983f). Females are more likely to use language skillfully to seek verbal accommodations to perceived conflict situations, while males are more likely to attempt direct solutions through action that demonstrates their physical superiority over competitors. The lateralized male brain recreates in its own image conceptions of politics that quickly polarize under stress; the female brain has greater natural ability to synthesize across multiple sensory modalities, which, in combination with the capacity for more balanced use of both cerebral hemispheres, should be highly facilitative of holological political thinking (Pettman, 1981; Watts, 1981: ch. 5). This implies a model of politics that would entail more complex and synergistic consideration of the working out of the interrelationships of consequences in the searches for solutions to social conflict (Corning, 1983; Watts, 1983).

Our present theory and practice of politics is the evolutionary product almost entirely of male brains epigenetically programmed for highly competitive behaviors that require in-group authoritarianism in support of aggression against outgroups; and neither Golda Meier, Indira Gandhi, Margaret Thatcher, Sandra O'Connor, nor Jeane Kirkpatrick typifies the female politician who would become prominent in a more androgynously constructed polity. During the first half of the 1980s Petra Kelly typified the new type of female political leader, of the future that had not yet socioculturally evolved; but by 1986 Cory Acquino already had arrived, as the future woman of the present. (On Kelly, see Hill, 1983. On the male political model, see Chagnon and Irons, 1979; Tiger and Fox, 1971: 32; Crook,

1980: ch. 5. For contrasting female perspectives, see Gilman, 1911; Firestone, 1970: ch. 8; Lerner, 1979: vii, 1986; E. Fisher, 1979; Gilligan, 1982; Boulding, 1977: 230–32; Hershey, 1977: 277–78; and Schubert, 1987, 1988b.) To implement polities designed—or at least more influenced—by female political thinking will require massive, not merely token, changes in the present ratios of female-to-male political participation, for leaders as well as followers. But this can and should be justified not on the negative grounds of eliminating discrimination against females, but rather on the positive psychobiological grounds that there are many political roles that women can perform *better* than men, not just with equal ability.

There is increasing evidence that the effects of such a reversal of the presently universal and putatively "natural" masculine political hierarchy will be qualitatively as well as quantitatively significant. Women have made a major political contribution in many policy areas besides education at the local level of government, as exemplified by their leadership in activist roles in groups protesting both the establishment and continuation of nuclear power plants and widespread governmental lethargy in cleaning up existing toxic-waste environmental pollution. The public opinion polls of the past decade consistently show women less willing than men to sacrifice quality-of-life issues to economic growth; to support larger expenditures for national defense; or to back Reaganomic cuts in social programs (Katz and List, 1981; Goodman, 1982; Frankovic, 1982: 448; Schubert, 1985c). Indeed, "for perhaps the first time, women are bringing their values into politics and sticking with them. For the first time, men are the followers"; and "studies of politically active individuals have found that once women cross the threshold into political activity, women tend to be *more* active than men" (Baer, 1980: 25). Not only are American women (and especially unmarried women) now more liberal than men in voting in national elections, a majority of the national electorate are females. These findings suggest that the correct political feminist strategy undoubtedly is that by making more women aware and by reinforcing their comprehension of existing sexual differences in political participation, more women may become catalyzed into the role of political activist.

To the extent that a more fully sexually egalitarian politics can be achieved on a global basis, there are sound psychobiological reasons why the result should be a less hostile, more cooperative, and better integrated performance of both political thinking and political behavior. An entailed byproduct would be the probability of postponement, for yet a while longer, of what evolution teaches is the inevitable extinction of our species (cf. ch. 12, above).

Does Brain Lateralization Vary across Cultures? A leading Welsh psychologist researcher into laterality (Dimond, 1977: 490) asserts, in his concluding remarks to a booklength report of an international conference on the subject, that "there are three main areas where the search for basic biological variants or graded differences of brain yielding different mental qualities could be expected to exist: these are race, sex, and handedness." Just a few years later came the claim that in the "Japanese brain, the difference between East and West may be the difference between left and right" hemispheres; the commentator made his remark in the context of his discussion of Japanese audio-technician Tsunoda's work, which attracted considerable interest and attention in Japan due to its assertion that because of environmental influences in learning to *talk*, native speakers of the Japanese language become much more extremely lateralized than Westerners do generally (Sibatani, 1980a, 1980b; Maruyama, 1980a, 1980b; Sasanuma, 1980). That finding is premised completely on cultural rather than genetic causes for the claimed differences in brain development: unlike Americans raised in the mainland United States, it is claimed that those brought up in Japan who as children become fluent in the Japanese language (rather than in English) as their first language also develop typically Japanese (that is, over-lateralized) brains. Tsunoda's research was based on dichotic listening tests measuring the speed of differential hemispheric response; but because of the (neurologically) low-level crossover of sensory input to *both* hemispheres for auditory as well as visual input from *either* side, plus the unmeasurable speed of commissural transmission between the hemispheres of such information, several expert critics have impeached on methodological grounds claims such as Tsunoda's (Waber, 1980: 250; Corballis, 1978: 335; Uttal, 1978: 327). However, subsequent research with commissurotomized patient subjects demonstrated that Japanese language maternals do *read* phonetic writing (*kana*) with left-hemispheric processing, but the Chinese ideographs (*kanji*) with the right hemisphere; and this type of brain organization was peculiar to Japanese and was not found among either Chinese or Koreans (Mecacci, 1981).

In American psycholinguistics, the Whorf-Sapir hypothesis of linguistic relativity has received striking confirmation in recent multidisciplinary research building directly upon Whorf's own pioneering ethnolinguistic study of the Hopi. The new research was premised upon the finding of fundamental differences between the structure of the Hopi language and English: Hopi emphasizes what is concrete within the sensory field of the speaker and auditor and establishes the immediate involvement of both with the perceptual field, so that direct attention is paid to nature and speech is placed within that explicit context. English is much more abstract and establishes a linear (as distinguished from a holistic) organization of time in relationship to speech, so that attention is directed to culture (as distinguished from nature) and to a context-free universe of discourse. What English conceptualizes as verbs and nouns are treated alike in Hopi as eventuations and are spoken of as differences in duration within the perceptual field. So it was hypothesized that bilingual Hopi children (who learned English as their second language in reservation schools) would process the English language left-hemispherically and the Hopi language on the right; this hypothesis was supported by field investigation (on a reservation) using EEG instrumentation (Rogers et al., 1977). In related work it was found that urban blacks were (as hypothesized) more right hemispheric than urban whites and that farmers were more right hemispheric than urbanites; and partial confirmation was adduced for the hypothesis that persons in positions of social subdominance (black, low and middle SES, female) would be more right-lateralized than social dominants (white, high SES, male), who would be left-lateralized (Ten Houten et al., 1976; Ten Houten, 1980). It should be emphasized, however, that in these studies (like those of the Japanese language) the subject of inquiry is *not* how the human mind is racially determined by genetic differences; to the explicit contrary, this research purports to show how language and other *cultural* differences directly affect how human brains develop epigenetically.

Psychobiology helps to explain how and why language differences play such an important role in political thinking and behavior, because communication problems are rooted in the structure of not merely individual, but also of *populational*, differences among human brains. The idea that *culture* itself is thus imbedded in human *biology*—but only ontogenetically, not genomically, of course—may come as a surprise to some social scientists. However, according to the editor of *Behavioral and Brain Sciences*, "neuroplasticity, or how environment modifies the brain, is virtually the name of the game in neuroscience" (Harnad and Steklis, 1976: 322). This suggests that to have any important effect, political change must be addressed to more complex neuropsychological-social levels of communication than those that have attracted the attention of social scientists in the past (such as, for example, the simultaneous translation of speeches at the United Nations). If culture can change the brain, why not *political* culture?

Nutrition

There is a direct, temporal, and causal relationship between the nutrition of the brain and its development, both quantitatively and qualitatively. This is true in the gross sense that early embryonic nutritional deprivation is more injurious than late foetal deprivation; prenatal more than postnatal; infantile more than juvenile; and childhood more than adult. There are especially critical periods within several of these developmental stages; and there are qualitative nutritional differences that have differential neural growth effects that vary among such stages. Nevertheless, the generalization holds that the earlier the incidence of nutritional insult, the more pernicious its effects upon the human brain are likely to be (Shneour, 1974; Pryor, 1975; J. Schubert, 1983a).

Data based on experimental research on rodents show that if an increase in either calories or protein is begun during a critical period of rapid proliferation of brain cells, the result is an increase in the rate of brain cell division and a permanent increase in the number of brain cells. Given growth retardation as a consequence of malnutrition, rehabilitation through reversal of that trend has to begin soon enough during the critical period of brain cell growth to allow time for change in growth rate and cell production to occur. Post-mortem examinations of the brains of first-year human infants killed by severe malnutrition revealed deficiencies in the number of cells, and the size of that deficiency was negatively correlated with birth weight. According to the medical authors relied on here, kwashiorkor is a disease due to protein insufficiency which can be expected to be highly positively correlated with weaning in the African locales and societies where it occurs; but its yearling victims suffer from deficiencies in the *quality* rather than the quantity of brain cells because after the first year of human life, malnutrition affects the *size* rather than the number of brain cells. Such cellular growth topologically of course varies considerably between human and rodent brains: severe postnatal malnutrition in humans curtails cell division in all areas studied; the effect is equally severe in both cerebrum and cerebellum; the human cerebrum is affected earlier and much more severely than the rat cerebrum; and there is definite retardation of cell division in the human brain *stem*—but not in the brain stem of rats (Winick and Rosso, 1975: 44, 47–48).

The human brain is the organ that grows at the greatest rate during the first months of foetal life; typically it weighs 350 grams at birth, which is almost a third of its maximal adult size; and the brain of a fourteen-month-old infant already has grown to 80 percent of its adult size. The central nervous system is the *last*—not the first—organ to lose weight due to malnutrition; nevertheless, in nursing rodents (among whom malnutrition was experimentally induced by artificially increasing litter size), within three weeks from birth the deficit in brainweight is 30 percent (in relation to an overall weight deficit of 70 percent). From a qualitative point of view, important changes in the composition of different brain areas were apparent after only two weeks of malnutrition, involving the diminution of DNA, RNA, proteins, and total lipids; and these deficiencies were even more evident by weaning time. In Chile (where the research was carried out with both rodent and human subjects), malnutrition among humans was observed to begin at an early age, due to a remarkable decrease in breast feeding during then recent years (Monckeberg, 1975: 23; Davies, 1963: 14n.18, 1980: 45–49). The chronically poor quality of socioeconomic, cultural, and sanitary conditions affecting the lives of the women—and therefore their babies—of the clientele of the urban Santiago clinic supplying the subject patients precluded success in the artificial feeding of human infants of nursing age; hence infants aged six to seven months were admitted for hospital care with virtually no increase beyond their birth weight and with increases of height of barely one to one and one half inches.

The same researcher undertook to investigate the reversibility of malnutrition in humans by making a follow-up study of children whom he had treated as infants for severe marasmus, a disease due to *caloric* deficiency among infants during the first postnatal year. The patients, between four and seven years old, appeared clinically normal and all showed biochemical indices of contemporary nutrition within normal limits. But their average IQ tested to sixty-two, and the highest was seventy-six—values significantly below average for lower SES preschool children in Chile, from which Monckeberg (1975: 29) inferred that his data were consistent with other animal and human experiments indicating "that the effects of early malnutrition are permanent up to the 7th year of life despite improving nutritional conditions."

A commentator on the above and other studies of the same symposium suggested the general finding that permanent mental retardation can result from perinatal malnutrition, in underprivileged countries; he added that there appears to be, during the first six months of postnatal life, a critical period during which advanced malnutrition inflicts permanent deficits upon both physical and psychological development. From a therapeutic perspective, treatment imposed during that early period "when many parameters are developing rapidly will have the greatest effect on brain function and possibly behavior" (Pryor, 1975: 108–9). Treatment begun thereafter necessarily would be less effective.

Recent work along similar lines indicates that the mylenization of neural cells in the brain continues throughout the period from birth through the fourth year; moreover, cerebellar growth seems maximal during the last few weeks before, and the first few weeks subsequent to, human birth. Therefore the maternal diet during the perinatal period is crucial to brain development of the infant, with malnutrition resulting in decrements in the infant's motor coordination. Animal studies of PCM (protein-caloric malnutrition) demonstrate a range of negative effects upon vision, emotionality, sociality, fear, and aggression. Among humans, PCM acts as an environmental factor affecting intellectual development, leading typically to IQs of less than seventy, and individuals who are non-empathetic, self-centered, sedentary, and irritable. Even mildly malnourished children lag in development as measured by Piagetian criteria. Because diet is correlated with SES, so also are a host of learning disabilities among schoolchildren—and these data are for American, not Chilean, victims of malnutrition. Among females, a particularly vicious cycle is generated, with malnourished female infants maturing into adult females with deficient placental development, which in turn converts their pregnancies into spontaneous abortions or else fetuses that survive but do so at high risk. There are even genetic consequences entailed, so that a malnourished mother's seemingly normal female offspring can herself give birth to a female infant with a defective placenta, in a cycle that skips phenotypically—but not genetically—alternative generations (Stern, 1981).

Fully one-third of the pregnant women in San Diego are undernourished, according to a distinguished neuroscientist who lives there (Livingston, 1981). The low caloric intake of these women precludes the storage of proteins; all proteins available to them must be and are used for immediate energy needs. From the point of view of rehabilitation, intervention should begin with pregnant mothers (as early in pregnancy as possible) and should continue with their infants for at least two, and up to four years; most children, even after rehabilitation, remain in a deprived environment. The IQ deficits of infant victims of severe malnourishment typically are fifteen to twenty points, *even after controlling for SES*.

A review of clinical and experimental studies of undernutrition and brain development (Chase and Metcalf, 1975: 279) concludes that "undernutrition is currently estimated to be present in two thirds of the children of the world. Although the United States is considered to be an affluent country, there are population groups in which undernutrition is found with great frequency. These groups include the children in the ghettos of large cities, in rural poverty populations, the American Indian population, and the children of migrant farm laborers." Thus domestic malnutrition, certainly of undernutrition but including also that due to overconsumption of junk food, is highly correlated with SES. There ought, therefore, to be direct implications for class differences in intelligence—not because the upper classes are born brighter, but because they eat better, especially when young, thereby making the best (or at least, the better) of whatever might be their respective individual genetic potential for the development of intelligence (Geber and Dean, 1957), whereas the opposite is apt to be true of the poor, the uneducated, and the sociopsychologically dispossessed.

Consider, for example, Arthur Jensen's (1980, 1985) reiterated claim that the fifteen-point modal IQ decrement, for American blacks as compared to whites, cannot be explained in terms of test bias. Alternative cultural explanations have been proposed, but these have been much less controversial than biological explanation has been. To my knowledge, however, the only biological hypothesis disputed in previous discussion has been a genetic one, which is without empirical support and is highly implausible from the point of view of human evolutionary theory (Macphail, 1985; and Rushton, 1985. See also Gustaffson, 1985; Johnson and Nagoshi, 1985; Poortinga, 1985; Rabbitt, 1985; Wilson, 1985). An alternative biological explanation, which seems prima facie to lend itself to both empirical testing and preventative social therapy (at least as concerns future generations) stems from research in brain function and malnutrition, especially prenatally and among infants during the first two years of postnatal life, as exemplified by the pernicious effect (noted above) of substituting bottle for breast-feeding of infants upon the long-range mental development of the urban poor of the slums of Santiago. Also available, of course, is direct evidence regarding the positive benefits of breast-feeding for African infants (Geber, 1958). But it was American corporations that pioneered, together with West German and Swiss international capitalists, in *that* Third-World-wide experiment in food-aid-in-reverse; and by the time the rest of the world finally was ready to take formal action to constrain such macabre profiteering, the United States had ensconced the Reagan administration, with the consequence that its representative to the World Health Organization cast the *only* dissenting vote, in the face of 118 ayes, to propose "a voluntary commercial code designed to encourage breast-feeding by banning advertising and restricting marketing practices of the $2 billion-a-year international baby-formula industry" on the utterly specious grounds that the American vote was compelled by the Constitution of the United States ("The Breast vs. the Bottle," *Newsweek*, 1981).

If malnutrition can have such consequences for the developing brains of poverty-stricken children in the slums of Santiago, it seems plausible to hypothesize that equivalent conditions of deprivation, which in the United States involve high correlations between race and class in the slums of Washington, Detroit, Chicago, Los Angeles, and all other major American cities, have had similar effects. If so, such a modest modal deficit—at least in relation to the findings for both Chile and San Diego—as fifteen points can easily be accounted for by the explicit differences in the then modal perinatal and infantile nutrition of the IQ-tested blacks and whites in the United States. At a time when Reaganomics had produced drastic curtailment of food and medical aid to millions of American babies whose mothers already were well below the poverty line, with all the reinforcing effects of the world-wide depression of the early 1980s, it is not inconceivable that a greater appreciation of the extent to which malnutrition also begins at home could have a salutary effect upon the redefinition of the relevant social welfare, and more generally, food production and distribution policies in the United States, as well as abroad.

Beyond the specific hypothesis about racial differences in brain nutrition in the United States, and beyond the medical and moral questions of the effect of malnutrition upon brain development, there are even much broader questions of global political consequences about the impact of malnutrition upon political participation, political change, and the very possibility of a political future for our species. Attention was redirected to this subject a decade and a half ago by Robert Stauffer (1969); it also became the focus of work by psychobiologically oriented political scientist James C. Davies (1963: 45–49, 53–56), who pioneered in the establishment of the psychobiological approach within political science, and James N. Schubert, who has reported extensive empirical research in a series of papers and articles, on the biopolitical roots of world malnutrition (1979); the impact of food aid (1981b); the effect of malnutrition on political violence (1981a); the politics of famine (1983a); and the postulation of explicit dynamic models of psychobiological processes, nutritional deprivation, and political violence (1982). But political science generally remains committed to a culturally deterministic understanding and ignores the critically relevant psychobiological and biosocial developmental questions of the politics of food.

18
Creative

> The difference between the mathematical mind (*esprit de géométrie*) and the perceptive mind (*esprit de finesse*): the reason that mathematicians are not perceptive is that they do not see what is before them, and that, accustomed to the exact and plain principles of mathematics, and not reasoning till they have well inspected and arranged their principles, they are lost in matters of perception where the principles do not allow for such arrangement. . . . We must see the matter at once, at one glance, and not by a process of reasoning. . . . Mathematicians wish to treat matters of perception mathematically, and make themselves ridiculous. . . . the mind . . . does it tacitly, naturally, and without technical rules.
>
> —Pascal, *Pensées*

It is one thing for a Jansenist mathematician who died 325 years ago to speak (in his posthumous "An Apology for the Christian Religion") in terms of the mind. It is quite a different matter for authors Findlay and Lumsden (1988; henceforth F&L) of this symposium's article to espouse what purports to be a biologically based evolutionary theory of human creativity; but they indulge instead in mentalistic dualism—apparently not self-consciously (Libet, 1985; Wood, 1985; and cf. the implications ch. 17, above) in the choice of words for their article's title. Hence I have chosen for these remarks a title that explicitly distinguishes between brain and mind; invention and discovery; and innervation and innovation. "Discovery" implies the characteristic natural-science presumption that there really *is* some preexistent phenomenal secret to be found (ch. 19, below), whereas "in-

---

This chapter is a revision of "The Creative Brain: Toward a Behavioral and Brain-Science Theory of Invention and Innervation," *Journal of Social and Biological Structures*, 11(1): (1988a), as part of a symposium organized by Edward O. Wilson, "The Creative Mind: Toward an Evolutionary Theory of Innovation." Reprinted with permission of the publisher.

vention" implies a novel consequence of animal cognition (Crook, 1980; Davidson and Davidson, 1980); and "innervation" locates "invention" in the functioning of an animal brain (Libet, 1985; Wood, 1985), whereas F&L use "innovation" (p. 7) as an unoperationalized and epiphenomenal byproduct of "culture."

CREATIVITY

F&L are themselves not very creative in their tautological definitions of *creative process* and *creativity*, which they postulate to be the functions of *creative individuals* whose ratiocinations produce both *novel* and *successful* solutions to *social* problems. Neither is such thinking very biological: Why isn't just any nonorthodox response creative? What's wrong with an inappropriate solution (from the point of view of "the problem"), if it is fitness-enhancing for the acting animal? From an evolutionary perspective, shouldn't it be the *range* of diversity in alternative responses that may be most important in populational terms (ch. 9, above)? Aren't "problems" better defined (in terms of contemporary behavioral ecology) as "obstacles to needs gratification," instead of psychologically (nineteenth-century style) as "motivat[ions] to respond"?

Other than Lumsden himself, D. K. Simonton is F&L's most frequent reference (e.g., their table 2); but F&L evidently reject Simonton's (1984) main finding: that social recognition of meritorious behavior (in the form of ascription of "genius," "creativity," "leadership" status, and similar denotations of excellence) is best understood as a statistical consequence of the *quantity* of individual production. Humans engaged in the activities surveyed by Simonton (musicians, mathematicians, physicists, politicians, etc.) do various things of which some are deemed by appraising persons—in judgments poorly correlated with the actor's view of merit in her or his own work—to be good, and others bad. *Persons who produce quantitatively the most over the longest period of time are most likely to produce something that others deem meritorious.* It is ironic that creativity *thus* defined is much more consistent with F&L's sociobiological premises than is the softcore humanism that apparently represents their image of what social science *really* is about—no matter what social scientists say or do empirically.

F&L's table 1 (fig. 5) constitutes a reasonably comprehensive description and analysis of the half dozen social-science and/or biological disciplines defined by them as major approaches to the study of creativity. On the whole the listed references provide a representative sample of the implicated research; but the organization of these approaches in relation to each other in the table appears to be random. In contradistinction the figure presented here suggests a hypothesis about the structural configuration of these six fields, in terms of a primary (substantive) dimension distinguishing biology from culture and a secondary (methodological) dimension of quantification, with artificial intelligence denoting the positive direction of relatively maximal emphasis upon quantification (while relatively neu-

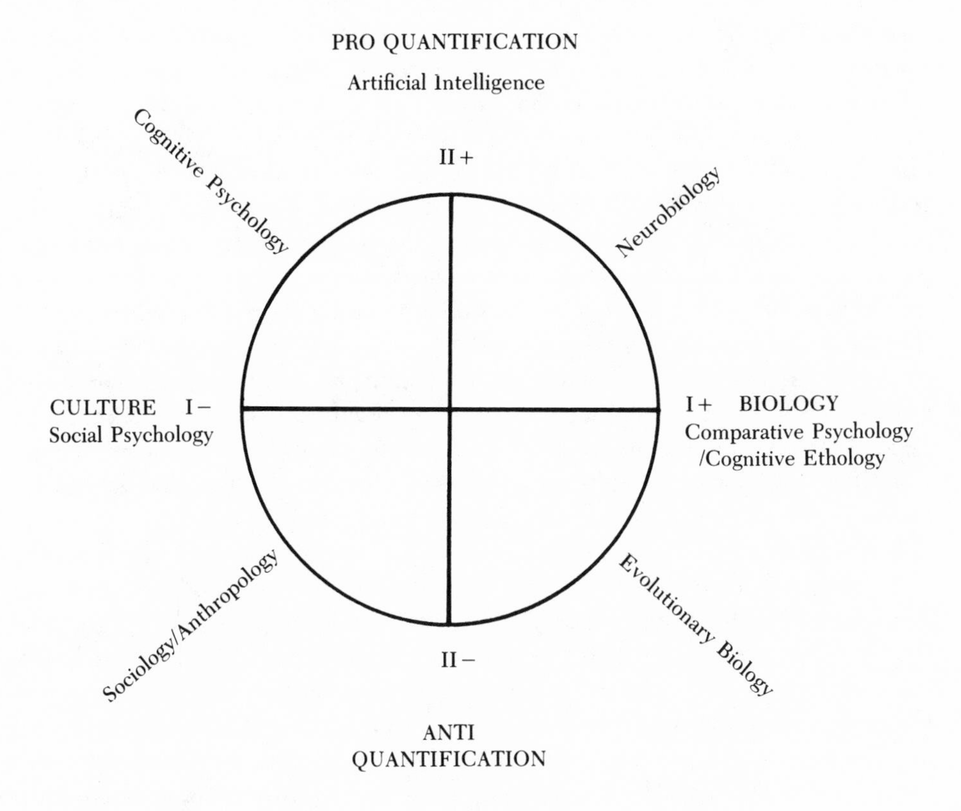

Figure 5 A hypothesis about the structural configuration of Findlay and Lumsden's table 1.

tral as between biology and culture; i.e., uncorrelated with either of them). Neurobiology and evolutionary biology of course define the positive direction of the first dimension, with neurobiology manifesting more, and evolutionary biology less, reliance upon quantification in methodology. In the cultural direction of the first dimension, social psychology is depicted as maximal, while cognitive psychology is more, but sociology/anthropology are less, quantitative. While accepting the table on which it is based for its data, the figure implies a more rigorous (social) scientific basis for the launching of future research on creativity.

EPIRULES

F&L assert that "all discoveries, from relativity to the hulahoop, reflect the operation of a common set of high-level cognitive procedures. . . . The end result (the creative work) . . . admits only properties subsumed by the constituent elements and their interactions"—i.e., the "epigenetic rules." This cannot be; *every* creative work is the fruit of the entire previous life experience of its creator, by epigenetic necessity. F&L claim that such intrinsically computational procedures as the "epigenetic rules" they postulate signify "a rule-governed process involving operations on syntactically encoded semantic representations." Evidently this represents an extremely *left*-brained way of thinking, whereas creativity in vivo undoubtedly involves considerably more reliance upon *right*-hemispheric than left-hemispheric operations (cf. Springer and Deutsch, 1981).

F&L highlight their eclecticism by ostentatiously comparing Darwin with Beethoven (cf. Roederer, 1979c); but then they extend the comparison to include the pneumatic tire, which provokes the gravest doubts: Harvey Firestone's airy wheel demands a different level of analysis from the theory of natural selection or the general theory of relativity or even Piaget's theory of cognitive development. Building a better mousetrap is not at the same level of creative diversity as the building of the universe, or the development of the human brain.

F&L repeatedly box themselves into corners with mathematical deductions (e.g., p. 32). "Given a particular set of conditions, those individuals whose semantic networks support a geometry characterized by high [psi] are likely to be the most creative . . . [but] an extensive background knowledge is not sufficient for discovery. It is . . . the way information is organized and accessed—that influences the potential for creative thought." This is an argument about innovation, not about creativity; and it is trite. The history of mathematics—as well as that of genetics and dozens of other disciplines—is replete with examples such as the simultaneously independent rediscovery in 1900 of Mendel's laws of the segregation and independent assortment of genes by a Dutchman (de Vries), an Austrian (Tschermak), and a German (Correns).

F&L endeavor to build upon Edelman's important hypothesis of neuronal group selection; but again they become entrapped by preoccupation with their own idiosyncratic concept of "epigenetic rules": "genetically encoded algorithms that modulate . . . developmental responses." F&L substitute those rules for the process of epigenesis as defined by Waddington and as understood by Waddingtonians, stating that such higher brain functions as learning, memory, and directed [*sic*] cognition are the "products of a program involving both *pre*-experiential epigenetic differentiation and experiential selection" (emphasis added), for which they wrongly give credit to Edelman. What is wrong is F&L's restriction of epigenetic differentiation to preexperiential animal life, evidently including foetal life; but there they are wrong again because humans experience continuing sensory inputs prior to birth (Hofer, 1981) notwithstanding the constrictions of life in utero. F&L are most wrong in failing to acknowledge experiential *epigenetic* selection, explicating that epigenesis describes transactional—not merely interactional—relationships between experience and a developing genome *after* birth (Ra, 1986). So they err on both scores: (1) by not recognizing the importance of environmental influences (i.e., "experience") on *pre*natal epigenesis and (2) by ignoring the importance of epigenesis to human development *post*natally.

This bias permeates the prior work of Lumsden and Wilson (henceforth "L&W") as well as of F&L here; bizarre rationalization of epigenesis (as "epigenetic rules") vitiates their theory. F&L mention Waddington only once, and then not for substance but rather for his graphics and metaphor of "canalisation." There is no evidence in F&L or in anything that L&W previously have published to indicate that any one of these three authors has even read Waddington beyond a most superficial skimming of his best known book (cf. ch. 14, above). Moreover, Lumsden continues to ignore the vital relevance of the work and theories of Valerius Geist (see especially Geist, 1978; ch. 13, above; and ch. 10, above). F&L refer, for example, to "environments where rapid, novel responses provide an *occasional* adaptive edge, as appears to have been the case in human evolution" (emphasis added). Geist proclaims such environments as the norm—not as exceptions to it—for human evolution from 100,000 BP to 10,000 BP.

CULTURAL EVOLUTION

Boyd and Richerson (1985; henceforth "B&R") are merely cited by F&L, whereas B&R devote considerable attention to L&W's books and articles. B&R present a much more comprehensive and dispassionate synthesis of work generally in the field of cultural evolution, than do L&W or E. O. Wilson; and B&R give much greater scope to nongenetic sociocultural processes of change; F&L and Wilson try to interpret culture as an epiphenomenon of mathematically defined genetic influences. F&L state, for example, that "at the heart of gene-culture transmission are the *epigenetic* rules," so their "cultural evolution" reduces to little more than an indirect expression of genetic determinism (cf. the epilogue to Dreyfus and Dreyfus, 1986: 182–86; and Grings and Dawson, 1978).

"A typical evolutionary model includes structural elements such as a specific mating system, a pattern of linkage between loci, and rules for translating genotypes into phenotypes. It also includes environmental elements that determine which phenotypes reproduce, how population size is regulated, and so forth. . . . The synthetic theory of evolution is justifiably regarded as the centerpiece of biology" (B&R: 3). No doubt; but in social science there is nothing remotely approaching a synthetic theory of evolution. *Naturally*, the neo-Darwinian synthesis understandably leads, for many persons trained in biological evolutionary theory, to sociobiological thinking (but cf. Lewontin, Rose, and Kamin, 1984). By comparison, *social* evolutionary models (B&R: 4–5) do *not* lead to a clear and simple analogue of "cultural fitness" (cf. Simonton, 1984), and *most* certainly not for modal human animals (as distinguished from creative geniuses and elite leaders). Consequently, social-science studies of cultural in relation to biological evolution lead to an emphasis upon social biological, *not* sociobiologial, criteria of cultural fitness (cf. the subtitle of the *Journal of Social and Biological Structures*: "Studies in Human Social Biology"; see Lewontin, Rose, and Kamin, 1984; and B&R: 12–14, and chs. 6–8).

F&L claim (p. 18) "that a wide variety of cultural traits. . .are known to influence reproductive success (see [L&W], 1981, for a review). A distinguishing feature [of what?] is the directive link between genetic fitness and innovation adoption." It certainly distinguishes F&L from B&R, who are misleadingly cited in the preceding sentence as *supportive* of F&L on this point. Two pages later F&L speak repeatedly of "innovator alleles," reinforcing L&W's insistent habit of purporting that specific genes have a one-on-one relationship to complex human social behaviors (such as "innovation.") Only lay persons utterly naïve about such hardcore behavioral genetics phenomena as pleiotropy and polygenes can take seriously such scholastic fantasies (cf. B&R: 81, on Francis Bacon's experience as an undergraduate at Cambridge University, four hundred years ago) as the postulation of "innovator alleles" (cf. Ehrman and Parsons, 1976; ch. 11, above).

F&L argue that the effect of interest-group organization in a society is to limit rather than to expand the prospects for creativity. This conclusion is contrary to the overwhelming thrust of empirical research in political science during at least the past thirty-five years (Cigler and Loomis, 1983; Manheim, 1982); F&L deduce it from their postulation of a (completely nonempirical) "completely open information market" as their only mentioned alternative to "the cultural repertoires of individual groups [which are] more accessible to members than non-members." So "as a consequence, individual creativity and innovation rates are both reduced by [closure due to] the synergistic relation between them." F&L seem utterly unaware of any possibly relevant research work in political science bearing upon their discussion of social science and cultural evolution; but they might well have tempered their unrealistic discussion of interest-group effects by awareness of such work as Corning's (1983) study of sociocultural synergism, from an explicitly political and evolutionary point of view (cf. chs. 6, 7, 10, above).

F&L infer, from equations that they state on the preceding page, the whimsical cultural conclusion that the environmental enhancement of genetic creative potential is "most likely to be realized under a decentralized form of political organization wherein legislative power is dispersed throughout a collection of quasi-autonomous local bodies," a discovery for which F&L hasten to give more than due credit to Simonton for his "historimetric investigation of the influence of political fragmentation on creativity." Thus, "over a period extending from 700 B.C. to A.D. 1839, political fragmentation was found to be the single best political predictor of creativity." That, however, is a trivial inference and highly atypical of the thrust of Simonton's thinking and findings alike; see Simonton (1984: 121–33 and ch. 9) for Simonton's *own* conclusions about the political implications of his studies. Nevertheless, F&L suggest that "for a specified encounter rate the structural complexity of the associated semantic network at any time during development is maximized if all cultural traits along each dimension are present in equal frequencies, and if these frequencies are constant over time." I understand this particular epigenetic rule to signify that the more static the environment, the more creative the animal. Again, they have things backwards, as the Middle Ages testify so poignantly (see Hannon, 1985). For both the stimulation of creative behavior in individual animals and for human survival generally, the major ecological problem confronting our species is that of environmental entropy.

19
Quantum

PARADIGMATIC CHANGE IN NATURAL AND POLITICAL SCIENCE

Politics is an aspect of both cultural and biological modes of behavior among humans, and by definition, political evolution is in part a component of cultural evolution (Lumsden and Wilson, 1981; Corning, 1973; Masters, 1970). From a biological perspective our concern is with the phylogeny of political science as an academic discipline, which involves the academic ideology of the profession (Schubert, 1967b) and hence the political behavior of political scientists.

The Prescience of Physics and Politics

Many political scientists would date the inception of the systematic study of politics no later than Aristotle, and this is a standpoint with which I am not entirely unsympathetic (Schubert, 1973). However, for present purposes I shall compare the state of development of contemporary political science with that of medieval physical science, which also, of course, can and did claim a direct affinity with Aristotle. Political science has been an organized academic discipline for little more than a century, irrespective of whether we attribute its nineteenth-century origins to Francis Lieber or John Burgess (Haddow, 1939; Somit and Tanehaus,

---

This chapter is a revision of "The Evolution of Political Science: Paradigms of Physics, Biology, and Politics," *Politics and the Life Sciences*, 1: 97–110 (1983), which was presented as the inaugural lecture of the "Dialogues Panel" series sponsored by the Association for Politics and the Life Sciences, at the joint Annual Meeting with the American Political Science Association, in Denver, CO, Sept. 1982. Reprinted with permission of the publisher.

1967). We can therefore skip over the intervening twenty-two centuries since Aristotle, during which no noteworthy changes took place in either political theory or political methodology as applied in the systematic, empirical, scientific study of politics.

According to Crombie (1961: 13), a biologically trained historian of science,

> a quantified science . . . comprises not only quantified procedures but also quantified explanatory concepts, each applicable to the other within a theoretical system. The development of a science then takes place through a dialogue between its theories and its procedures, the former offering an exploration of the expected world through predictions and explanations made by means of technical procedures, and the latter confronting these theoretical expectations with the test of quantified data.
>
> A dialogue of this kind requires that both sides should speak the same language. We are so familiar with the close and precise adaptations of conceptual and procedural language to each other in modern physics that it may come as a surprise to find authentic scientific systems in which this is not the case. Yet we do not have to look very far to find examples. In the contemporary social sciences and in psychology, they are notorious. . . . The main interest of medieval physics in this context seems to me to be that it provides the earliest example in the development of modern science in which we can study the state of affairs when the dialogue between concepts and procedures was incomplete or absent. Then we can study the difference it made when clear and exact communication was opened, as it was in the seventeenth century.

For physics of course, Crombie's point is well taken; for political science, any such event can only fall in a future rather than a past century. However, as we shall see, for many political purposes the future ought to be viewed as accessible to cognition, at least as usefully as is the past.

Crombie's analysis (1961: 29–30) of medieval physics suggests some striking parallels with many of the most conspicuous characteristics of contemporary political science:

> The primary internal, intellectual need felt by medieval natural philosophers [and for political science today, substitute "political philosophers"] was for rational, theoretical clarification and understanding rather than for knowledge acquired through observation. This need arose in a desire to understand rationally and clarify (a) the features of cosmogony and cosmology accepted as having been revealed [then, as Crombie implies, "by the grace of God," a presupposition that obtained notoriously not only for Galileo and Newton but subsequently for Darwin and Mendel as well; although for political science, the non-Creationists among us would substitute "by modern natural science" as the source of the revelation. As I remarked earlier (Schubert, 1976: 165) in regard to antiquarian Greek science, political scientists now "would not wish to rely upon either Hippocrates' or Aristotle's knowledge of human anatomy or medicine"], and (b) the system of natural philosophy presented in the Latin translations [then; now, English or some other modern language] of Aristotle and the other Greek and the Arabic philosophers and mathematicians. . . .
>
> This theoretical emphasis in intellectual interests was supported and maintained by the aims, content, and methods of the education provided by the medieval [for physics; modern for political science] universities, where the basis of both the arts course and of the higher courses in theology [then; sociology now], law [then and now], and medicine [then; political science now] was the making of a critical study and commentary on theoretical problems

> raised by standard texts [viz., contemporary hermeneutics in political science]. . . . As a result, medieval physics [like political science to date] never escaped its Aristotelian framework.

Hence, what contemporary political science has in common with sixteenth-century physics is that it has not yet developed a disciplinary paradigm. Like sixteenth-century physics, it is prescientific in its theory and methodology.

Crombie spoke at a conference that convened in substantial measure on the initiative of a political scientist, Pendleton Herring, who acted in an institutional role that was launched six decades ago due substantially to the academic-political efforts of Charles Merriam, a political scientist whose exploits in political theory and praxis alike need no retelling for political scientists (but see Karl, 1974; and Schubert, 1983a, "Introduction"). Merriam's design (1921, 1925) for a new science of politics has been on our profession's agenda throughout the greater part of the twentieth century. His students have provided the leading role models for political scientists for more than two generations (Somit and Tanenhaus, 1967: 113), including the profession's only Nobel laureate (H. Simon, who earned it for work in organizational theory after having made a quantum leap of sorts to industrial management, barely half a dozen years after completing his political-science doctorate at Chicago). It is the more ironic, therefore, that it remains not merely possible but justifiable to assert that as a *science*, political science is best characterized as a preclassical or antiquarian discipline (cf. White, 1980b, and 1982a). What follows is an attempt to explicate the grounds and the process for further developing the study of politics as a more modern science.

Natural Science Macroparadigms

During the past generation no work in the philosophy of science has influenced political science more than Thomas Kuhn's (1962) theory of paradigm change in science (for a precis of the theory, see Landau, 1972: 56–57). Apart from its other merits, no doubt the denotation in the book's title of "revolution" as the subject of study was an attractive feature. In any case, during the sixties (as Landau, 1972: 57–58 has recounted), a panoply of our profession's leaders leaped agilely on Kuhn's bandwagon and strove mightily to cling to his coattails—albeit with indifferent success. Their problem is our problem, as we shall see below. Moreover, whatever controversy Kuhn's thesis may have engendered during the sixties both within and beyond the overlapping fields of the history and philosophy of science (see the postscript to Kuhn's second edition, 1970; and Landau, 1972, ch. 2), there can be no cavil about the pertinence of his theory to an explanation of paradigm change in the first of the epochal shifts to which we now turn.

For both physics and biology, three principal eras can be distinguished from antiquity to the present. The fourth century B.C. is a convenient marker for what we conventionally interpret (given the differential survival of writings) as the climax of Greek biology, physics, philosophy, and politics, as represented in works attributed to Aristotle and the Academy in which he was trained (although mathematicians such as Euclid and Archimedes as well as the apogee of Greek physics came slightly later; see Sambursky, 1975: 37–45). Antiquarian physical science persisted for two millennia. It was finally displaced by classical physics, which began in the sixteenth century with the work of Copernicus and Galileo, but which is now identified principally with Newton's work in the last third of the seventeenth century, and which terminated *as a paradigm* with Einstein's major work, in the first two decades of the present century. Elementary particle, subatomic physics began in 1900 with Max Planck's initial postulation of the quantum concept and competed for several decades with neoclassical (i.e., relativity theory) physics (Gamow, 1966). Einstein's theory, like Newton's, is deterministic and objective; quantum theory is indeterminate, subjective, and probabilistic. Nevertheless, the paradigm that has dominated both theoretical and applied research in the physical sciences during the past half century has been a synthesis of quantum and relativity physics.

What we have just considered is change in the *major* emphases at the cutting edge of scientific inquiry into the nature of physical phenomena. It is nevertheless true that Newtonian physics remains extremely useful and sufficiently correct as a predictor of macroscopic physical events on or near the surface of the earth—although no round trips to the moon could have been accomplished in exclusive reliance on classical physics. Moreover, Einstein built (almost immediately) upon Planck's discovery, just as special relativity theory contributes importantly to subatomic as well as to astronomical research today.

With better justification than in the case of physics, antiquarian biology also is identified with Aristotle, "whose original research was in the field of biology; his whole conception of nature was biological and influenced by a way of thought which today does not seem applicable to inanimate objects" (Sambursky, 1975: 40–41). Aristotle's biology survived even longer than his physics—until the advent of Darwinism about 125 years ago marked the inception of the classical paradigm in biology. However, like physics, biology underwent a major change in paradigm with the 1900 discovery (independently by Karl Correns, Hugo de Vries, and Erich von Tschermak) of both Mendel's original article and his principles of genetic inheritance. Together, these constitute in effect a quantum theory of biotic change (Margenau, 1950: 453–54). In twentieth-century biology the Darwinian and Mendelian theories merged four decades ago in a "modern synthesis"—almost a decade after the synthesis in modern physics (Handler, 1970; Futuyama, 1979, ch. 1). This synthesis now provides the dominant macroparadigm (but see chs. 12–14, above) for such diverse subfields as population, developmental, and molecular biology as well as for others such as behavioral genetics and animal behavior. Gradualist natural selection theory continues to dominate these and other fields of biological inquiry such as evolutionary theory (Williams, 1966) and sociobiology (Ruse, 1979), although not without significant and increasing dissent, especially from paleontologists (Stanley, 1979, 1981; "Enigmas of Evolution," *Newsweek*, 1982: 46).

In both physics and biology, the shift from the classical to the modern paradigm has been much more evolutionary than revolutionary. The revolutionary change came when classical physics and classical biology became differentiated from antiquarian natural science. Both the theory and the substantive knowledge of the macroparadigms that prevailed before Newton and Darwin were displaced by *their* classical works. Then the resulting classical macroparadigms became synthesized into their respective neoclassical forms. Notwithstanding their classical heritage and components, in macroparadigmatic terms, both modern physics and modern biology are primarily twentieth-century sciences. The persistence of Aristotelian politics and philosophy (and a host of other such antiquarian and preclassical sources) in contemporary political science, not as an aspect of the history of the discipline but rather as the foundation of "theory" in the field, undoubtedly would be an anomaly *if* political science were an integrated science, but of course that has never been true. A major segment of political science remains, for purposes of teaching and research alike, humanities-oriented. Thus the antiquarian paradigm persists because of political science's prescientific structure.

Political Science Pseudoparadigms

About a hundred years ago the profession of political science did begin with a paradigm. It involved humanities-oriented, "traditional" (Schubert, 1965a: 159–61), institutional studies focused in public law, history, and philosophy, and it lasted for about forty years. There were, of course, early challenges such as those of Bentley (1908) and Wallas (1908), but the disintegration of the traditional approach as a paradigm directing and dominating the activities of the profession began in the twenties (Somit and Tanenhaus, 1967: 87–89; and with particular reference to the public-law component of traditionalism, see Schubert, 1966, part 1). The result was a substantial reorientation of the activities of many political scientists toward the social sciences (particularly sociology and economics). The reorientation was accompanied by a strong emphasis on empiricism in research design and statistics in methodology. This marks the beginning of the importance of the "conventional" approach in political science (Schubert, 1965a: 161–63; 1967b: 6–7, 11). The consequence was *both* post- and preparadigmatic: postparadigmatic because the traditional approach ceased to dominate political science as a paradigm about sixty years ago; preparadigmatic because no single paradigm has dominated the profession since traditionalism was displaced (see Hines, 1979). Instead, throughout the twenties, thirties, and forties (described as "The Middle Years" by Somit and Tanenhaus, 1967, part 3), the traditional and the conventional approaches competed with each other (and with the much smaller and still nascent behavioral- and policy-science minorities) without ever combining in the kind of synthesis that has characterized modern physics throughout the past half century.

As an example of this synthesis, the electromagnetic transmission of light is described completely independently by Heisenberg's quantum matrix-mechanics and by de Broglie-Schrodinger's relativity wave-mechanics (of which one direct byproduct was the invention of the electron microscope)—both of which were translated by Dirac and von Neumann into the same mathematical "transformational theory" that became the basic for quantum mechanics (Margenau, 1950: 329–30; Rao, 1962: 27). Yet in political science, the traditional and conventional approaches were (and still are!) practiced by different persons. During the period 1920–1950, the dichotomy was such that they often were practiced by different *departments* of political science.

The behavioral revolution in political science, proclaimed a bit belatedly by Truman (1955) and (with equivalent insouciance) buried prematurely by Dahl (1961), involved for many political scientists much closer interdisciplinary ties with social psychology and cultural anthropology. It was accompanied by an intensified emphasis upon scientific theory and method (Kaplan, 1964) as well as quantification in the collection and analysis of data. This fulfillment of another major facet of Merriam's vision is described by Somit and Tanenhaus (1967, ch. 12; cf. Schubert, 1968a) and was marked by the reincarnation of several harbingers who had been half a century ahead of the profession in their thinking, including both Bentley (Truman, 1951) and Wallas (Davies, 1963). But the rise of behavioralism did not signify the disappearance of either traditional or conventional political science. To the contrary, a troika replaced the former duopoly. The result was neither a dominant paradigm nor a new synthesis, but rather an extension of the preexisting condominium.

At the end of the sixties there appeared a cross-sectional report on the current state of the discipline (Eulau and March, 1969) that virtually reeked with the behavioral approach. The report stands in fascinating contrast to the Dimock report (Committee for the Advancement of Teaching, 1951), in which the word *behavior* is in no way indexed and in fact appears a total of ten times in a book of some three hundred pages. Also, at the end of the sixties came David Easton's (1969) open-armed embrace of what he called "post-behavioralism." Although this was a readily understandable gesture given both the academic-professional and public politics of the time (Schubert, 1976: 162), Easton was at the very least euphoric in his suggestion that political science was about to shuck its scientific pose for a forthright engineering stance. The American Political Science Association (APSA) never did become scientific, as we have noted, in its orientation—only behavioralism's most extreme enthusiasts, and opponents, closed ranks to proclaim *that* (Eulau, 1963; Storing, 1962; McCoy and Playford, 1967; Graham and Carey, 1972). Thus, Easton's claim that "the last revolution—behavioralism" was "completed" is as wide of the mark as his prediction that the "battle cries" of "*relevance* and *action*" (Easton's emphasis) were about to put the American Political Science Association in the vanguard of political engineering efforts to resolve "the increasing social and political crises of our time" (Easton, 1969: 1051;

and cf. Lasswell, 1963). That didn't happen either, although a greater concern for institutionalized egalitarianism in the profession and in academic life and for the public policy implications of research and teaching was certainly more evident during the seventies than before. This, however, is no more "postbehavioral" than it is "posttraditional" (Rodman, 1980: 44–45) or "postconventional." As Somit and Tanenhaus (1967: 88) point out, the APSA's Committee on Policy (under Thomas Reed) was active during the late twenties and early thirties. Furthermore, it was reinforced by Merriam's own strong interest in civic education and by the lifelong espousal of a "policy science" approach on the part of Merriam's leading student, Lasswell. Therefore, what Easton called postbehavioralism can better be understood as the coming of age of yet another long-standing though minority movement within the discipline (cf. Horowitz, 1979; and for an alternative analysis of half a dozen metaphysically defined pseudoparadigms of political science, see Haas, 1982).

There was another new development, this one *within* behavioralism (as well as, of course, within political science). It began—at least as an organized activity—at about the same time that the Caucus for a New Political Science was riding high and Easton was describing behavioralism in the past tense. The first panel on a biological approach to politics convened at the Southern Political Science Association meeting in New Orleans in November 1967. Earlier that year I published a model of academic ideology (Schubert, 1967a: 7)—one that *did* incidentally prove to be predictive (Schubert, 1967b). It distinguished among the traditional, conventional, and behavioral approaches in political science and defined biology (flanked by psychology on one side and anthropology on the other) as the *core* of the behavioral approach. A year or more earlier when the pair of articles was written, I did not know that there would *ever* be a political science panel on biopolitics (to say nothing of one that would meet as soon as the following year), so at the time this was a precocious venture in forecasting the future of the profession. In ignorance also of Easton's imminent requiem for behavioralism, my salutation to *bio*behavioralism heralded an approach in political science that by now is well established as a viable development within the profession, as attested by the convening of the panel to which this chapter was first presented under the joint auspices of the American Political Science Association and the Association for Politics and the Life Sciences.

There are several reviews of the scope and content of biobehavioral political science (Somit, 1976: Wiegele, 1979), and an extensive research literature is now available (Somit et al., 1980; Peterson et al., 1982). Biobehavioralism is also the subject of John Wahlke's APSA presidential address, of which "prebehavioralism" is his stated theme (Wahlke, 1979). His analysis rests upon a sample of articles published in the *American Political Science Review* during the decade 1968–1977. Coincidentally, a paper (Schubert, 1976: 157–161) at a small research conference where Wahlke was a commentator reported what appears to be the only relevant pilot survey along lines similar to Wahlke's subsequent empirical investigation, although the earlier sample covered only three years (1972–1974) in the middle of his. He was certainly aware of it at the time (Wahlke, 1976: 255 and 259, n. 7), but he does not mention the prototype study in his presidential address. This omission would be unremarkable except for the fact that "prebehavioralism" (a novel concept at the time of the conference paper) is defined very differently in 1976 than it is in Wahlke's 1979 presidential address. Originally, prebehavioralism was defined quite literally as traditional and conventional political science (Schubert, 1976: 163–64). Subsequently, Wahlke spoke entirely figuratively when in his analysis of the content of a decade of political behavior research, he criticized that research for its failure (1) to be "anchored in macro-level political theory" and (2) to include physiology in its conception and methodology of political psychology (Wahlke, 1979: 24). I find it easy to empathize with both criticisms. Indeed, my earlier discussion of prebehavioralism pointed out that "the implications of biological theory, for the perpetuation of much of traditional political science, are much more revolutionary than those of political behavioralism as it has been understood heretofore," and I followed this remark with an explanation of how and why this is probable (Schubert, 1976: 164). However, it is rhetorical to argue that the "real" political behavioral approach has not yet become immanent among us. Prebehavioralism as defined by Wahlke, like postbehavioralism as defined by Easton, obscures more than it reveals about the evolution of political behavioralism during the past two decades.

The one aspect of biobehavioralism to be emphasized in the discussion below is the relation of biophysics to psychobiology. At the inception of work in biopolitics, Davies (1969) directed attention to the great potential importance of brain science for political science (see also Davies, 1976; Cory, 1974; Beck, 1976b; Geigle, 1977; and Laponce, 1978), and interest in brain science has accelerated in the past two or three years (White, 1980b, 1981a, 1981b, 1982a; Peterson, 1981a, 1981b, 1982a, 1982b; Schubert, 1982c, 1983a, 1983e; Manheim, 1982; and Kitchin, 1982a). Brain science is of great potential importance to politics because it deals directly with what are rapidly becoming operationalized—not merely metaphorical or analogical—theories and processes of consciousness, memory, and decision-making in relation to both perception and emotionality (cf. Crook, 1980, especially ch. 2). Psychobiological politics deals directly with the control of human behavior, including political behavior. But precisely because the relevant theories are biophysical, their consideration requires some understanding of modern physics as well as modern biology.

Such a requirement places greater demands upon us. Nevertheless, psychobiology may proffer the best chance for political science—or at least for some political scientists—to transcend the classical as well as the antiquarian shackles that limit our ability to understand and our capacity to resolve the complex political problems that preempt our attention and concern. Living forms are more complex than inanimate matter, and the laws of physics apply to humans, but the laws of human behavior do not necessarily apply to physical interactions (cf. Margenau, 1950: 162). Our comprehension of human behavior lags by an accelerating exponential factor behind our understanding of the physical universe. Consequently our theories of politics are going to have to be more—not less—complex than the theories and methodologies that modern physics utilizes. At the very least, we must seriously consider the possibility that models of political behavior will be better informed by an awareness of what physical theory includes (and political theory doesn't) that would bring our theory closer to the facts of political life.

As a step in that direction, we can consider some of the major characteristics of relativity theory and quantum theory in relation to classical physics and each other and also in terms of what kind of social science theory modern physics has stimulated thus far. Then we can examine the physical basis of modern biology, in particular its implication for theories of consciousness and decision-making, including political decision making, and hence its implication for psychobiological politics. Any reader of Somit and Tanenhaus (1967: 113–14) should be aware that earlier presidential addresses can also serve as negative examples for us in approaching that task, for example, "Physics and Politics" by William Bennett Munro (1928). Under Munro's benign auspices was convened a conference hosted by Eulau some half a century later at which political behavioralist stalwarts critiqued biopolitical upstarts. Of course no person who was socialized as a social scientist some three and a half decades ago, and whose undergraduate education was primarily in the humanities, can profess to any profound knowledge of the physical sciences. But in the land of the blind, even a cyclops is in a position to see more and further than those who look not at all.

NEOCLASSICAL PHYSICS

Relativity Theory

Classical physics defines and describes a world that corresponds closely to the sensory perceptions that human biological evolution constrains us to accept as natural. As Ushenko (1946: xv) remarks, "the structure of the visual field is Euclidean and . . . our knowledge of this structure is intuitive." Among the principles of classical or antiquarian physics that political science continues to take for granted are those of irreversible causality, action by contact, and absolute time (Reichenbach, 1965: 15). All of these are incompatible with experimental observations that led to Einstein's special relativity theory, which in the words of Werner Heisenberg (1974: 187), "emerged from the electrodynamics of moving bodies and has led to new insights into the structure of space and time." Heisenberg (1958: 117) asserts that the most important physical consequence of the special theory concerns the equivalence of mass and energy (i.e., the inertia of energy). Einstein's statement of the theory was put in more elegant mathematical form by his own former mathematics teacher, Minkowski, as a quadratic equation in which the separation of space and time is not emphasized, leading to a four-dimensional space-time manifold or "world" (Rao, 1962: 10). It is ironic that, as we now can appreciate, the special theory is "not about those aspects of reality that are relative, but about those aspects that are *not* relative" (Zukav, 1979: 172, 180, emphasis added). It portrays a static, nonmoving picture of space and time.

In the special theory, length is not a characteristic of space, but of the bodies that occupy space, and space itself has no intrinsic metric (Salmon, 1980: 23–24). Changes of length in different manifolds (congruence) become not a question of fact but of definition. Similarly, "simultaneity is not a matter of knowledge, but of coordinative definition" (Reichenbach, 1958: 126), and as Margenau (1950: 139) has remarked, there is no "natural unit" of time and hence no way in which "purely constructional elements can be eliminated from conceptual time." Indeed Salmon (1980: 64) suggests the possibility that both space and time are *quantized* in structure, for "we do not experience time as a continuum of instants without duration, but rather, as a discrete series of specious presents" (Salmon, 1980: 42). Such experience ("history") "contains the idea of the 'arrow of time,'" but this is "a concept absent in all fundamental laws of physics that can be expressed mathematically" (Cocconi, 1970: 82). Thus classical mechanics defines a temporal order but not a direction for time. However, as Grunbaum (1973: 327) comments, "It is the essence of the relativistic account of the inanimate world as embodied in the Minkowski representation that there is change in the sense that different kinds of events can (do) occur at different times: the attributes and relations of an object associated with any given world-line may be different at different times. . . . Consequently, the total states of the world. . .are correspondingly different at different times, i.e., they change with time."

General relativity theory focuses on the relationship between gravity and inertia (Heisenberg, 1958: 121), and empirical observation demonstrates that inertial and gravitational mass are equal in the gravitational field of the earth. This contradicts both the special theory and also the Euclidian geometry of space (Reichenbach, 1965: 31). In the general theory, "Einstein has to connect the underlying physical ideas with the mathematical scheme of general geometry that had been developed by Riemann. Since the properties of space seemed to change continuously with the gravitational fields, its geometry had to be compared with the geometry on curved surfaces where the straight line of Euclidian geometry has to be replaced by the geodetical line, the line of shortest distance, and where the curvature changes continuously" (Heisenberg, 1958: 122). Einstein's general theory established a connection between the geometry of the four-dimensional space-time manifold and the distribution of masses in the world; the special theory became a limiting case of a vanishing gravitational field (Rao, 1962: 15). Thus, from the perspective of general relativity theory, "an event has many dates in the same sense in which it has many sizes and shapes . . . relatively to different standpoints" (Ushenko, 1937: 75). Because the general theory deals with vast expanses of the universe, it cannot be either proved or disproved on the basis of earthly phenomena—instead, it must be tested through astronomy. Four principal verifications (of which the first has recently been questioned) involve the perihelion of Mercury, the deflection of starlight, the gravitational "red shift" of the light from distant stars, and the still novel phenomenon of black holes.

Field Theory

A major component of Einstein's theories of relativity is electromagnetic mechanics. According to Margenau (1950: 198), "the method of science reaches its fullest development in the theory of the electromagnetic field" (for a general survey of field theory, see Sachs, 1973). Such a field is defined by two vectors, themselves the product of six scalers (just as, in quantum mechanics, there are well-established operators for six different observables); and the electromagnetic field defined by the pair of vectors is assumed to pervade the space where it is observed (or presumed) to be (Margenau, 1950: 201, 249). Because of its success in electrodynamics, the field approach (in the form of the "force field") spread to mechanics. Here, however, its use is confined to statics (Margenau, 1950: 204). "Biological fields . . . borrow . . . features from several physical types in an endeavor to account for organization and growth. One serious difficulty . . . is their failure to grasp clearly the methodological requirements involved" (Margenau, 1950: 204–5), including those relating to the unambiguous definition of the set of state variables and the laws regulating their behavior.

Unified Field Theory

It has long been part of the folklore of this century's philosophy of science that Einstein failed in the quest that preoccupied the last forty years of his life—the attempt to develop a unified theory of the gravitational and electromagnetic fields (Thomas, 1930: 761; Wheeler, 1962; Grunbaum, 1973, ch. 22). Wheeler (1962: 366–67) claims that Einstein persisted in his efforts because he wanted to derive from "some kind of classical geometry" not only electromagnetism "but also quantum theory itself." In his recollections of his own discussions with Einstein on epistemological problems in atomic physics, Niels Bohr (1958: 32–66) recounts the frequent occasions on which Einstein argued against the quantum approach and expressed unhappiness because of the "apparent lack of firmly laid down principles for the explanation of nature, in which all could agree" (Bohr, 1958: 56). Heisenberg (1974: 189) points out that "relativity theory . . . neglects those features of nature which are tied up with Planck's quantum of action; it therefore continues to presuppose that the phenomena can be objectified in the manner of classical physics." Heisenberg (1974: 5) remarks also that "Einstein's was a conservative mind. Through his years of development he had become wedded to the nineteenth century belief in progress, and his essays reflected the picture of a world which, though because of human irrationality it was exceedingly imperfect, could yet become better and better if men were ready to break with their former prejudices and put trust in reason." Five years before Einstein's death, Margenau (1950: 39, n. 1) noted that "it has become customary among physicists to regard as 'classical' the whole field of physics which antedates quantum mechanics (about 1920). Even relativity is now part of classical physics."

CLASSICAL PHYSICS AND PRECLASSICAL BIOLOGY IN POLITICAL SCIENCE

The Rashevsky Models

More than a few political scientists are familiar with the series of papers, rather tightly clustered within a period of about half a dozen years, that Nicolas Rashevsky published (1947, 1951, 1954), proposing mathematical models of social behavior, drawing in part upon his earlier work (1938) that had focused upon the biophysics of individual behavior. The author was a distinguished biophysicist, and he employs the mathematical methods of biophysics in his analyses; but only in the first part of the middle work, five chapters that discuss brain mechanism and behavior, is his subject matter biophysical. Hence for our purposes, Rashevsky's work primarily represents an example of mathematical modeling of social relationships rather than an attempt to use physics or biophysics theory to model politics. Coleman (1954) commented extensively upon the first set of papers, of which two (1947, chs. 17 and 23) do focus on political relationships. Karl Deutsch (1963: 39–41) discusses particularly the first of these two chapters (on "Interaction of Nations"), noting that the usefulness of Rashevsky's sophisticated mathematical techniques is limited by his regrettably naive assumptions about the empirical structure of the social groups that he analyzed.

Systems Theory

Landau (1961) has analyzed in detail how the paradigm of classical physics continues to dominate our constitutional polity in this country—long after such a model has proven inadequate to cope with the needs of twentieth-century physics (cf. K. Deutsch, 1963: 26–30; and Schubert, 1967c). In lieu of classical physics, Landau (1968) has discussed the differences implied by an organismic biological model that derives from antiquarian rather than classical biology (cf. K. Deutsch, 1963: 30–34). Thus, the theory that he discusses—the only kind that has influenced political science, through the respective works of Robert Merton, David Easton, Gabriel Almond, and their many followers—is actually *pre*-Darwinian, and it has certainly proven to be no more adequate than Newtonian physics as a basis for a modern theory of politics. Karl Deutsch's *Nerves of Government* (1963) sounds like it might proffer a psychobiological model of political behavior, but the nerves are only a metaphor, and the book in fact discusses a cybernetic approach based on the computer (of two decades ago) as a model for human decision making. Deutsch's model turns out to be that of a relatively sophisticated machine—yet another example of classical physics as a model for politics.

Field Theory and Politics

According to Morton Deutsch (1968: 406), "in psychology, the term 'field theory' is used primarily to designate the point of view of Kurt Lewin and his co-workers. Although the term has its origin in physics, where it is employed to refer to the conceptualization of electromagnetic phenomena in terms of fields of electromagnetic forces, field theory in psychology is not an attempt to explain psychological events in terms of physical processes." Lewin's field theory held "that events are determined by forces acting at a distance" (Lippitt, 1968: 266); but this is the theory of Aristotelian physics, certainly not that of even classical physics (Sachs, 1973: 20–24). However inconsistently, Lewin nevertheless attempted to make a literal translation of selected aspects of Einstein's theory of relativity and apply them figuratively as a metaphor for a holistic approach to social psychology. As Dorwin Cartwright, one of Lewin's leading American disciples, states, "field theory is more an approach to the scientific task than a theory about a realm of data" (Lewin, 1951: viii; and cf. Lewin's own statement, 1951: 45). Lewin conceptualizes all behavior (including acting, thinking, wishing, striving, valuing, achieving) as a differential (literally: dx/dt) of a field state in relation to time—that is to say, a locus, in the four-dimensional space-time manifold. The relevant field for an individual is her or his "life space," comprising the self and its psychological environment. The extension of this concept from the individual to group psychology is perfectly straightforward (i.e., additive). Lewin "believed strongly that a set of interdependent facts can be adequately handled conceptually only with the mathematical concept of space and the dynamic concepts of tension and force" (Lewin, 1951: xiii). Einstein's definition of simultaneity (Salmon, 1980, ch. 4, especially p. 100), which in the special theory assumes reciprocity in the speed of light between two points in a given inertial reference frame, has its counterpart (of sorts) in Lewin's principle of contemporaneity, where the mathematical time differential of physics dissolves into language about measuring behavioral change "at a given time" (Lewin, 1951: 45–46). Similarly, the topological mapping of general relativity theory is explicitly analogized (Lewin, 1951: 150–51) to what is no doubt the least accepted feature of Lewin's approach, "hodological space" (which, unlike Einstein's, is nonmetrical). Lewin ignored quantum theory, which was of course fully developed when he left Germany to come to the United States—at the same time (1933) and from the same place (the University of Berlin) as Einstein. A book describing social science applications of field theory entirely sympathetically (Mey, 1965: 47, 237) mentions the Heisenberg effect only as a problem for quantum physics, not as a problem concerning transactions between manifest change in social behavior and the latent structure of the minds of the field theorist researchers who observe it.

Mey (1965: 131, 279, n. 35) virtually ignores the criticisms of field theory by Sorokin (1964: vii), who prefaces a polemical volume with the admission that it "aims at a liberation of sociology and the social sciences from voluntary servitude to the natural sciences." Having already written off both modern physics and modern biology as unsuitable paradigms for "sociocultural space," Sorokin (1964: 97, 109–10) characterizes Lewin's approach as a

> purely analogical use of the concepts of space of physiocomathematical sciences . . . [in which] the whole operation is a mere verbal preoccupation of "ordering of some phenomena of behavior" to the "psychological field" . . . with the terms "magnitude," "force," "vector," "field," . . . which . . . do not add anything to our knowledge. . . . [S]uch a useless transcription of the terms would have some advantage and cognitive value if the transported terms could be measured adequately or were metrical, as they are in physiocomathematical sciences . . . [but] where there are no units and numbers, all the formulas and equations are void and meaningless.

Social science field theory gives lip service to the (distinctly nonverbal) field theory of electrodynamics and relativity, but its *substance*, in terms of theory and methods alike, comes from neither modern nor neoclassical physics, but rather from nineteenth-century biology in the form of Gestalt psychology.

One of the earliest and best-known uses of field theory in political science is in Quincy Wright's *Study of International Relations* (1955: 488–92, 524–28, 539–67, 586–603; and cf. McClelland, 1971). His discussion is explicitly eclectic and implicitly synthetic, but he never attempts to provide a rigorous definition of "field." Furthermore, the dimensionality of the model that he has in mind is never clarified, although virtually all of his empirical examples are two-dimensional. According to Wright (1955: 524), "a field is a system defined by time and space or by analytical co-ordinates, and by the properties, relations, and movements of the entities within it." The first of these meanings is entirely literal because Wright is talking about political geography and "conditions of actual time and space" (1955: 540). The second meaning, although more psychological, is strictly psychometric and denotes primarily factor analysis (which at that time was still a novel methodology in political science). Neither meaning has much to do with the concept of "field" in relativity theory, and although Wright (1955: 525) does remark that "Einstein's four-dimensional curved time-space may differ from Euclid's three-dimensional space," he adds, but "they are not *real* fields" (emphasis added). One wonders in what sense?

Factor analysis and other psychometric modes of multidimensional analysis were commonly employed in many fields in political science during the sixties, including judicial behavior—a development of which I am not entirely innocent (Schubert, 1961, 1962a, 1975b); but I thought then, and believe now, that the relationship between factor analysis and either Lewin's approach or Einstein's is one of orthogonality.

Rudolph Rummel is a leading contemporary exponent of the ideas of Quincy Wright regarding the use of field theory to study international relations. The author of the first comprehensive treatise by a political scientist on factor analysis (Rummel, 1970), he explicitly acknowledges his indebtedness to both Lewin and Wright (Rummel, 1975: 35–44, 60–62). Rummel's is the most thoroughgoing attempt to construct a mathematically based, operationally defined, and empirically testable field theory of politics, as presented in his reports on and appraisals of the "Dimensionality of Nations" project (Rummel, 1977, 1978, 1979; cf. Ray, 1982). However, what Rummel calls "field theory" is a statistical design for using canonical correlations to map Euclidean factorial attribute space into Euclidean factorial behavioral space, using the distance vectors of the attribute space to determine relationships in the behavioral space. Methodologically, this remains factor analysis, although it is more complex than its usual employment in political science, and it presumes linear causation from attributes to behavior. The result is no doubt a powerful analytic tool for mapping and facilitating comparative analysis of large variable-sets of data about the characteristics and behavior of the nation-states with which Rummel is concerned. However, in his opinion (Rummel, 1975: 28), 'field' cannot be operationally defined"; therefore, it is difficult to see how his own use of field theory could possibly be other than metaphorical. To use his own example, war "must be considered as part of a behavioral matrix (what I *would later call a field*)" (Rummel, 1976: 22, emphasis added). The indispensable requirement for field theory is "the notion of forces spread continuously throughout the region" (Rummel, 1975: 29), but Rummel defines "forces" as "attribute (social) distances" in factor space (Rummel, 1977: 200). The substantive content of the propositions "on social conflict" that he tests (1979: 355–75; but cf., e.g., proposition 18.1, p. 339, and proposition 18.4, p. 345; and Ray, 1982: 167) is *not* based on mathematical deductions from what he calls "field theory" because *that* mathematics is about statistical transformations between factorial spaces. Instead, these propositions are constructions subjectively related to the constellation of values that his self-taught course of the great books imparted, beginning a dozen years ago coincident with his avowed renunciation of logical positivism (Rummel, 1976; on Rummel's ideology, see Ray, 1982: 168, 178, 180–82).

A field theory of politics—however constrained in time and place its application for empirical testing may be—ought itself to specify the substantive relationships in political behavior (whether of national states or groups of individuals, and whether about war or revolution or assassination, or even about more peaceful behaviors), rather than focusing on how to process data in order to make statistical decisions about (otherwise derived) political propositions. Rummel's reversal of his own field, from hardcore scientism (before 1970) to softcore, social-science humanism (since 1975) complicates the task of interpreting the ten books that he has published within the past decade. It is clear that *he* believes that his "field theory" manifests "a quantum theory perspective" (Rummel, 1977: 483) and that his voluminous reports of complex analyses of empirically based findings provide powerful support for his theory (Rummel, 1979: 357; but cf. Rummel, 1977: 398). I remain dubious, but international politics is not a subfield of political science in which I can claim any special competence. Political scientists who can make such an assertion have thus far failed to indicate much in the way of either agreement or disagreement with the merits of Rummel's claims, either empirically or as field theory.

MODERN PHYSICS

Speaking with reference to people of the Third World, but on a subject that may conceivably have some import even for us sophisticates of the First, F. S. C. Northrop, in his introduction to Werner Heisenberg's (1958: 2–3) book on the philosophy of physics, points out that

> one cannot bring in the instruments of modern physics without sooner or later introducing its philosophical mentality, and this mentality, as it captures the scientifically trained youth, upsets the old familial and tribal moral loyalties. If unnecessary emotional conflict and social demoralization are not to result, it is important that the youth understand what is happening to them. This means that they must see their experience as the coming together of two different philosophical mentalities, that of their traditional culture and that of the new physics. Hence, the importance for everyone of understanding the philosophy of the new physics.

And so it is for political scientists who are asked to take modern physics seriously. Surely our understanding of political life is not independent of our conception of the physical world. No matter how we orient our profession toward other disciplines, our conceptions of public political life are going to be largely determined by our conceptions of the nature of the physical environment in which humans live and of the nature of humans themselves.

Quantum Physics

Newtonian classical physics deals primarily with the mesoscopic events that can be related to human sensation and experience. General relativity classicism deals primarily with megascopic phenomena (i.e., physical events in our own solar system) in their relationship with the rest of the universe. Quantum mechanics deals with submicroscopic events at the level of the infrastructure of the atom, where matter reduces into energy, and time (in the meso-macroscopic sense) becomes irrelevant—for the dual reasons that it can be neither measured nor perceived (for an excellent overview of quantum theory, see Bohm, 1980, ch. 4). In quantum mechanics, time has *neither* temporal order *nor* direction. "Time appears to be a completely macroscopic phenomenon, which cannot be traced into the microcosm" (Reichenbach, 1956: 269). The modern physical world of quantum physics presents us with images of reality that defy, and can never be confirmed by, common sense. It is little wonder that a political science that cannot yet accept nineteenth-century biology rejects twentieth-century quantum physics out of hand, as having no conceivable part to play in either our understanding or practice of politics.

More than half a century ago, it was clear "that occurrences on the atomic scale obeyed 'discontinuous' laws of quanta, that the principle of continuity prevailing in perceived microscopic work is merely simulated by an averaging process in a world which, in truth, is discontinuous by its very nature" (Rao, 1962: 26). By then, all atomic phenomena had been unified on the basis of Bohr's principle of complementarity, which emphasizes the relation between observer and observed and assumes the duality of (independent) particle and wave facets of physical reality (Heisenberg, 1958, ch. 2; Gamow, 1966; Guillemin, 1968). The principle of complementarity "is equivalent to the uncertainty principle or the principle of indeterminism which has brought about the fall of the notion of causality in physics" (Rao, 1962: 4; Bohm, 1957).

Quantum mechanics gives the same results as classical physics for the motion of ordinary objects (Margenau, 1950: 44). Indeed, "every periodic motion of classical physics, properly studied in ordinary dynamics, has as its quantum counterpart a stationary state whose study belongs to quantum statics" (Margenau, 1950: 351; and cf. Sachs, 1973). Thus, "the quantum theory gives an account of the mechanical processes in the interior of the atom, but it also incorporates Newtonian mechanics, as the limiting case in which we are able to objectify the events completely and can neglect the interaction between the object under investigation and the observer himself" (Heisenberg, 1974: 187).

Although special relativity and quantum mechanics contributed to each other's development during the early decades of this century, the general relativity standpoint has been insignificant for most purposes in quantum theory because the direct effects of gravitation in ordinary atomic phenomena are extremely small: 10^{-25} of even the weak decay interactions of elementary particles, which in turn are only 10^{-14} in electromagnetic strength as compared to the bonds in atomic nuclei (Rao, 1962: 52, 57). However, the physics of elementary particles now relies considerably upon investigations using cosmic radiation (Heisenberg, 1974: 187). Subatomic particles are energy and have no objective or independent existence of their own. They have wavelike as well as particlelike characteristics, and they may represent interactions between or among fields (Zukav, 1979: 217–20). According to Reichenbach (1956: 263), "the difference between one and two, or even three, material particles can be shown to be a matter of interpretation; that is, this difference is not an objective fact, but depends on the language used for the description. The number of material particles, therefore, is contingent upon the extension rules of language. However, the interpretations thus admitted for the language of physics differ in one essential point from all others: they require an abandonment of the order of time."

In the absence of some kind of measurement, quantum mechanics has nothing to say (Zukav, 1979: 322). The "observed" system cannot be observed until it interacts with an observing system, and only its effects on a measuring device can be observed then (Zukav, 1979: 93). Consequently, "what we perceive to be physical reality is actually our cognitive construction of it"—even light, the speed of which is the unit of measurement of relativity physics. "Without us, light does not exist" (Zukav, 1979: 105, 118).

With the use of the most precise measurement contrivable, it is still not possible in quantum physics to determine the "locus" of a particle of "matter" in the space-time manifold of classical physics. This fact leads to "the reciprocal indeterminancy of time and energy quantities" (Bohr, 1958:53). To such a moving particle two properties are attributed , position and momentum. The observer can choose which one to attempt to measure—but never both in the same observation (e.g., at the same "time"). This is because the act of measurement itself involves an interaction with the particle, which changes the particle. Hence, whichever attribute the observer selects to measure, he or she must remain uncertain about what the measurement of the other attribute would have been if it had been selected instead. Thus, the unmeasured attribute remains unobserved and indeterminate. This is the Heisenberg principle of uncertainty. There is another sense in which the predictions of quantum mechanics lead to indeterminancy. "Given a beam of electrons, quantum theory can predict the probability distribution of the electrons over a given space at a given time, but quantum theory cannot predict, even in principle, the course of a *single* electron" (Zukav, 1979: 135, emphasis added). Prediction in quantum mechanics is statistical, not particular, and high-energy particles are continuously changing anyhow. Individual events are chance happenings.

Zukav (1979: 56, 135) asserts that the uncertainty principle undermines the whole idea of a causal universe and that because objectivity—which would require the elimination of the observer—is not possible in quantum mechanics, physics has become a branch of psychology. (He suggests also the possibility that psychology is becoming a branch of physics, but for that to happen particle physicists would have to become educated as neurobiologists.) Heisenberg himself has remarked (1958: 173–74), in regard to the principle named after him, that prior to the development of quantum theory, "it was not denied that every observation had some influence on the phenomenon to be observed but it was generally assumed that by doing the experiments cautiously this influence could be made arbitrarily small. This seemed in fact a necessary condition for the ideal of objectivity which was considered as the basis of all natural science." To social scientists, this will strike a familiar chord, as one contemplates the self-satisfaction with which it is now assumed that the best antidote for investigator bias is candor in disclosure—leaving it up to the auditor to evaluate the subjective effect of the revealed bias on the reported findings (cf. Gould, 1980b: 50).

The Biophysics of the Brain

In a very influential book written during World War II while he was in exile in Ireland, Erwin Schrodinger (who had been a leading figure in the development of quantum theory) asked (1944), "Is life based on the laws of physics?" He encouraged numbers of young physicists to get involved in molecular biology, and one of the short-range consequences was the identification of the structure of DNA (Watson, 1968; and cf. Stent, 1969: 21–22, 55). The longer-range consequence has been the widespread and increasing realization, among physicists and biologists alike, of the importance of modern physics to the understanding of many biological structures and processes.

A recent article by Goldman (1980) presents the case for a unified theory of biology and physics, in terms of more than a dozen examples. Most of these originated in quantum physics, and he remarks that "there is no reason to believe that any physical process that violates the laws of physics will ever be found in an organism" (Goldman, 1980: 346). A commentator on the article (Rosen, 1980: 363–64) mentions that "many quantum physicists have felt that there must be profound relations between microphysics and life. From the earliest days of quantum theory, there has been extensive preoccupation with articulating the relation between that theory and biology." Rosen's view, which he insists is "not couched in terms of analogy, but as a causal implication of general microphysical principles," is that a phenotype comprises "a pattern generated by an *observation* of the genome, in the quantum-theoretic sense" (Rosen's emphasis).

There is an extensive literature, both research and popular, that discusses the constructional biology (Wheeler and Danielli, 1982) of models of the biophysics of the human brain. Examples of more technical models include Ingber's 1981 article "Toward a Unified Brain Theory" and a book by Edelman and Mountcastle (1978). More easily readable but equally authoritative presentations are those of Pribram (1971, 1979) and Roederer (1978; see also the section "Quantum Psychophysics" in Pelletier, 1978: 132–41). For work in this area by political scientists, see Peterson (1982a, 1982b) and my own summary of much of the recent work (Schubert, 1982a). Roederer (who is a physicist) remarks, "*Brains* have evolved as the result of adaptive interactions with the environment of living systems with locomotion capability. In this course of evolution the human brain emerged as the most organized and the most complex system in the universe as we presently know it" (1978: 426, Roederer's emphasis). Roederer states that environmental representation and prediction are the most fundamental central operations of the brain, and he presents an excellent summary of the characteristics of the operational modes of animal brains (1978: 427–28). The human brain, however, is unique in that "the human and only the human brain can perform [the recall of] images or representations, alter them, and store modified or amended versions thereof without any external input" (Roederer, 1978: 431). Roederer also points out (1978: 432) that "storage of retrievable information in the environment emerged as a most essential complement to brain operations . . . later in the form of . . . written language." Now of course, retrievable information is stored through the various modes of artificial intelligence, including computer hardware and software systems that are themselves the products of constructional biology.

In his last writing, Heisenberg (1977: 5) recalled that in 1926, when the quantum paradigm had not yet become consensual, a major problem in understanding the observational data was that "the language we used for the description of the phenomena was not quite adequate." He recalls also that while still trying to work his way through the problem, "in the despair about the futility of my attempts I remembered a discussion with Einstein and his remark: 'it is the theory which decides what can be observed.' Therefore I tried to turn around the question. Is it perhaps true that only such situations occur in nature or in the experiments which can be represented in the mathematical scheme of quantum mechanics?" This is another way of saying that perception is so dependent on conceptualization that even the most highly trained and skilled natural scientist "sees" (observes) what is expected in the language of his concept, whether mathematical or natural. Three varying perspectives of human perception (by a physicist, a biologist, and a political scientist, respectively) are presented by Bohm (1965: 185–230), Livingston (1978), and Laponce (1975, 1978). Comfort (1979) suggests what an observer *might* observe if his or her conceptions were translated (possibly through the employment of artificial intelligence) into a language (possibly mathematical and involving field theory and topology) isomorphic with what she or he assumes to be the holographic modus operandi of the human mind.

Bohm (1977: 561, 562) has commented on the two most important meanings associated with what has come to be called the Heisenberg principle. The first "showed, in effect, that the quantum needed for [the very accurate measurement of the position of an electron] would disturb the observed electron in an unpredictable and uncontrollable way, and that as a result, it is impossible to assign to the electron simultaneous values of position and momentum with greater accuracy than that specified by [their] uncertainty relations." The second showed that

> the act of observation appears to play a key part; in the sense that it not only discloses or reveals the attributes of the electron, but also in the sense that through the irreducible disturbance it helps actually to make or produce these very attributes. This means that the classical notion of a world that is essentially independent of the actions of the human observer is ultimately denied. Rather, the human being not only comes to know the world, but in doing this, he participates in an essential way.

Both meanings were incorporated in 1927 in what became known as the Copenhagen interpretation, which according to Stapp (1972: 1098) "was essentially a rejection of the presumption that nature could be understood in terms of elementary space-time realities. According to the new view, the complete description of nature at the atomic level was given by probability functions that referred, not to underlying microscopic space-time realities, but rather to the macroscopic objects of sense experience." Heisenberg himself, however, has insisted on distinguishing between "the sensual perceptions of the observer" (which he calls "positivism") and "the Copenhagen interpretation [which] regards things and processes . . . describable in terms of classical concepts . . . as the foundation of any physical interpretation" (1958: 145). So Heisenberg presumes an observer who is trained in, and who chooses to employ, the concepts of classical physics to describe observations at the level of quantum physics.

Almost four decades later came another major extension of the uncertainty principle in the form of Bell's theorem, which showed that, if the statistical predictions of quantum theory are correct, then many of our common-sense ideas about the world are profoundly mistaken. Zukav (1979: 306) comments that this "projects the 'irrational' aspects of subatomic phenomena squarely into the macroscopic domain. It says that not only do events in the realm of the very small behave in ways which are utterly different from our commonsense view of the world, but also . . . events in the world at large." Nothing is, except to the extent that seeming makes it so.

A byproduct of the Copenhagen interpretation has been a debate among philosophers of science over whether time has any physical meaning in the sense of "the present" or is exclusively a construction of human biology and culture, which reinforce each other in constraining the brain to conceptualize and therefore perceive time as a unidirectional continuum. The point is of some importance to quantum physicists because it is at the crux of what can be observed and interpreted. The matter is not made less uncomplicated by the circumstance that all physical phenomena are obeying the Second Law of Thermodynamics (entropy), and typically manifestly so. Moreover, living organisms (and by no means least, the human brain—given its last-ditch resistance, for example, to the physiological effects of starvation) characteristically contribute quite dynamically to the entropy of their environments in order to inhale, ingest, imbibe, and otherwise absorb the complex molecules that they then convert to their own needs and purposes with highly negentropic effects while life endures. At the microlevel of quantum physics, matter resolves into changing energy that has mass and momentum, but for either of these the "direction" of time is meaningless. Grunbaum (1973: 324–325) states that "the distinction between the past and the future of common sense and psychological time [with respect to which the transient now] acquires meaning has no relevance at all to the time of physical events, because it has no significance at all apart from the egocentric perspectives of a *conscious* (human) organism. . . . [so] the concept of 'now' involves features peculiar to consciousness." However, if determinism is a definitional consequence of the theory in terms of which the physical world is observed, then it too is a construct—and in this instance, a highly rational rather than merely a commonsense product of human consciousness. This implies that *neither* the past *nor* the future is physically determined from the perspective of the present (which Grunbaum insists is strictly a matter of human psychology, not of physics). Both past and future (like the present) depend on probabilities and inferences from partial data that make all time indeterminate rationally (if not psychophysiologically as well as physically).

Comfort (1980) has suggested that, if we assume alternative representations of reality that exclude a space-time-oriented observer, the understanding of reality that such representations postulate may not necessarily be confined to subatomic phenomena. It may also be useful in considering "middle-order" process models, e.g., in regard to evolutionary biology and morphogenesis. Comfort (1979b) considered the importance of non-Western philosophies and religions as the bases for such alternative representations of reality (cf. Capra, 1975, 1980; Zukav, 1979), and Comfort (1980) emphasizes Bohm's model of "implicate reality" and other direct derivatives of quantum mechanics. To examine the concept of speciation, Comfort proposes (1980: 208) a counterintuitive model derived from quantum mechanics in place of the current model of modern evolutionary genetics, which is highly acceptable intuitively. He asserts that a major justification for doing so "is the power of the demonstration in physics that reality-models are metaphors designed to be operated by, and intelligible to, a time-and-space oriented homuncular observer." Time, as he notes, "is metaphorical in particle physics, and part of the attempt to transduce phenomena to a temporal observer." Furthermore, "the relevance of brain research is in showing the artificial character of self-evident human responses to the handling of phenomena." Thus, "what Bell's theorem . . . demonstrates is the incompatibility with quantum-mechanical theory of the intuitive formalism which sees, e.g., a photon pair as separate objects; they are neither objects nor separate" (Comfort, 1980: 214). He concludes (1980: 215) that "it seems a necessary consequence of quantum physics that, as with the Newtonian model, we find out just how far it goes." A commentator on the paper (Heelan, 1980: 218) explains that "the Hilbert space model within quantum mechanics is the 'second level' 'field-type' theory, objective in that it states relations . . . that abstract from concrete events (measurement events) in the space-time of human observers."

A third paper by Comfort (1981) focuses upon Bohm's (1980) implicate-explicate model (and cf. Bohm, 1971, 1973). His concern in this paper is with the relations of Bohm's implicate to a "hyperloop" (the "observer paradox"), which Comfort variously describes as "matter thinking itself" (1981: 363), as the process "by which the brain generates 4-space by object formation from the 'plenum'" (1981: 373), and as "the idea that consciousness involves a projection of a higher multidimensional order into a series of subjective moments, exactly as particles represent a similar projection into a series of discontinuous frames" (1981: 364–65); it "implies. . .a quantization of physical time" as well as a concept of empathetic time, both of which are united through the continuity of the implicate. Comfort continues:

> The field which Bohm terms the "holomovement" is of high order [political scientists: read high *dis*order] but non-random (because the quantization of manifest phenomena is . . . a graphic representation of this subjacent non-local structure. . . . One effect of this idea is to transpose the kind of non-locality and interconnectedness which Wheeler has postulated at a micro-level ("the quantum foam") into a far more general property of Nature, which could be looked for macroscopically. Manifest phenomena are related to the holomovement by a process which Bohm calls explication. Since manifest phenomena are identified by their regularities—which are what we measure—it follows that any implicate exhibits regularities which are expressed in space-time.

A heterogeneous group of measurables (location, mass, spin, number, probability) constitutes the "observables." These represent "a tabulation of basic phenomena as they appear in the 'real' (explicate) world" (Comfort, 1981: 366). The resulting projected reality bears a close resemblance to a hologram of a special kind, a reverse one, because "in our model . . . a pattern is translated into virtual objects, as when a hologram is viewed." Bohm's concept of "holo*movement* implies motion or change. . . . [movement, change, and the like are themselves] transductions to a consciousness—*change has no meaning except to a positional observer*" (1981: 366, Comfort's emphasis). "If anything is moving it has to be ourselves" (Comfort, 1981: 367). "Our brain reduces reality to a Boolean logic, probably by reason of its hardwired structure." One consequence is that "since all input takes a finite time for a mental encoding, there are no objective 'nows,' and we operate exclusively on a nondisobservable memory tape."

"A manifest world," says Comfort (1981: 369), "is a hands-on-model in which organisms can operate without performing cumbersome calculations before doing so—for the relatively slow neurochemical mode used in brains, the algebra necessary . . . would be transcomputable. This is precisely what Kant meant by *a prioris*—we are hard-wired to see a hands-on world composed of centroids and yes-no approximations. . . . The conclusion is overwhelming that time is [an] . . . algorithm, that the subjective model has fundamental validity different from that of *t* in the Minkowski figure. . . . How about . . . positional self? Very probably that too is *a priori.*"

"There is no difficulty," Comfort concludes (1981: 370–71), "in talking scientific sense about explication in relation to electrons and similar inferential structures; the exercise consists in developing an algebra and then seeing if it fits, or predicts, observables. Moreover, particle events, though inferential, are in a sense continuous with the normal, optical world, and display the same local consistency." As we experience "mind," it is "at least transduced by, and possibly wholly generated by, brain," but the brain's "hardware" has been uniquely generated epigenetically by the relevant body's lifetime of experience. So the bottom-line question is whether consciousness is "as primary and as fundamental as 'matter,'" and this question "no longer makes scientific nonsense, because it no longer runs counter to the prevalent understanding of physics."

What is in the brain is the implicate; "mind" itself "*is* the process of explication"; and what is observed is the explicate (Comfort, 1980: 372, Comfort's emphasis). It follows that "without an observer, there is effectively no explication and there are no phenomena. . . . The conclusion of a Bohmian model is. . .that we should recognize. . ." phenomenal reality to be just as subjective as mind itself.

How modern psychobiology—the physics of the brain—operationalizes the much more primitive concept of "psychology" entertained by the physicist discoverers of "the observer paradox" has been well summarized by Morowitz (1980: 16; see also the article's exceptionally apt set of three illustrations by Victor Juhasz, pp. 12, 15, and 16).

> First, the human mind, including consciousness and reflective thought, can be explained by activities of the central nervous system, which, in turn, can be reduced to the biological structure and function of that physiological system. Second biological phenomena at all levels

> can be totally understood in terms of atomic physics, that is, through the action and interaction of the component atoms of carbon, nitrogen, oxygen, and so forth. Third and last, atomic physics, which is now understood most fully by means of quantum mechanics, must be formulated with the mind as a primitive component of the system. We have thus, in separate steps, gone around an epistemological circle—from the mind, back to the mind.

QUANTIZING POLITICAL THEORY

What are the implications for political theory of the acceptance and use in the study of politics of the paradigm of modern physics? Obviously we need not consider those implications for politics itself because our worlds—terrestrial and extraterrestrial alike—are replete with both material and psychic phenomena that are the products and by-products of modern physics. So the only question that remains is when political science is going to begin to study its subject matter in terms isomorphic with those used by us (humans, not us political scientists) to define politics.

Decision Making

The modern physics of psychobiology focuses dead center on what has long been a major concern of political scientists: how and why both individual and group decisions get made as they do. As quantum physicists and biologists have realized for four decades, and as we discussed in detail in the preceding section of the present chapter, the new paradigm of modern physics speaks not only to biology. Bell's theorem tolls for us political scientists, too. So likewise do Heisenberg's uncertainty principle, the Copenhagen interpretation, and Bohm's model of the explication of reality (see especially Bohm, 1980, ch. 3). The paradoxes of observing and becoming challenge our accepted epistemology of politics in its most fundamental respects. The new paradigm teaches us that such leading postulates as rationality in political decision making and objectivity in the scientific method are themselves tautologies of psychobiological construction. So is our favorite methodological postulate of causality—that by sufficiently refined analysis and reduction we can determine linear (if parallel) action sequences, partitioning pieces of the action among independent and intermediate variables that determine change in one or more dependent variables. Aristotelian direct action is replaced by Bohmian action at a distance because of the basic interconnectedness of all matter (cf. Sachs, 1973).

The deterministic political thinking derived from the classical physical paradigm continues to dominate not only the American Constitution at all levels (Landau, 1961) but also the politics of Western societies generally. However, the new paradigm signifies that if our political world (like all of our other worlds) inescapably represents a highly indeterminate complex of events regarding which great and continuing uncertainty is bound to continue (Bohm, 1957), then we ought to be constructing models of politics and political behavior in which chance plays a major part in explaining what is happening politically and why (Aubert, 1959; Schubert, 1975b: 330) instead of using models that restrict chance to the residual terms of our regression equations. There is, at the very least, no reason to assume that models making explicit room for chance effects in political processes and outcomes will do less well than the causative models on which we now typically rely, even though only too frequently such models fail to do better than the null hypothesis of nothing-but-chance.

The Psychobiological Paradox

Once the paradigm of modern physics becomes more generalized in the training and thinking of political scientists, the political explications of the quantum model of reality will extend over a much broader domain of political inquiry than the core subject of decision-making theory. This is so because such characteristics of reality as indeterminancy, irrationality, chance variation, diffuse transactionality, and the artificiality of time as a measure of change are elements of physical and social—not only psychological—human environments. Since a beginning already has been made in political psychobiology, it is appropriate that we undertake a brief review of that work in concluding our consideration of the future paradigm of political science.

Our first example derives from a lecture, "Of Clouds and Clocks," delivered by Karl Popper (1965), himself a sometime contributor to the development of quantum physics. The clocks are of course symbols of classical physics, and the clouds (as in "cloud chamber") signify quantum physics (cf. Margenau's remark [1950: 43, n. 1] that it was Schrodinger who first pictured the electron as a cloud). Popper echoed what is by now for us a familiar theme, saying (1965: 10) that "behaviorist 'laws' are not, like those of Newtonian physics, differential equations and . . . every attempt to introduce such differential equations would lead beyond behaviorism into physiology, and thus ultimately into physics; so it would lead us back to the problem of *physical determinism*" (Popper's emphasis). Popper proclaimed himself to be a physical indeterminist, although he added that "indeterminism is not enough" (1965: 13) because "men, and perhaps animals, can be 'influenced' or 'controlled' by such things as aims, or purposes, or rules, or agreements. This then is our central problem" (1965: 15). Popper expressed his personal dislike of psychobiological theories in which the central nervous system acts like an electronic amplifier with "quantum jumps." Popper calls such theories "tiny baby theories," saying that "they seem to me to be almost as unattractive as tiny babies" (1965: 17)—a remark reminiscent of the similar but perhaps more notorious views of the late W. C. Fields—*mais chaqu'un à son goût*. Apart from his feeling about neonates, however, Popper is criticizing here the psychobiologically naive views of an earlier generation of quantum physicists, who entertained too literal a mechanistic model of brain operations. The lecture is reprinted as chapter 6 of Popper's *Objective Knowledge* (1972), and the thrust of that book's argument strongly favors a biologically based, evolutionary theory of human perception and consciousness, strongly reinforcing the theories we reviewed in the immediately preceding sections of this chapter (e.g., Popper, 1972: 145–46).

> Sense organs, such as the eye, are prepared to react to certain selected environmental events—to those events which they "expect," and *only* to those events. Like theories (and prejudices) they will in general be blind to others: to those which they do not understand, which they cannot interpret (because they do not correspond to any specific problem which the organism is trying to solve).
>
> Classical epistemology which takes our sense perceptions as "given," as the "data" from which our theories have to be constructed by some process of induction, can only be described as pre-Darwinian. It fails to take account of the fact that the alleged data are in fact adaptive reactions, and therefore interpretations which incorporate theories and prejudices and which, like theories, are impregnated with conjectural expectations; that there can be no pure perception, no pure datum; exactly as there can be no pure observational language, since all languages are impregnated with theories and myths. Just as our eyes are blind to the unforeseen or unexpected, so our languages are unable to describe it (though our languages can grow—as can our sense organs, endosomatically as well as exosomatically).
>
> This consideration of the fact that theories or expectations are built into our very sense organs shows that the epistemology of induction breaks down even before taking its first step. . . . Sense organs incorporate the equivalent of primitive and uncritically accepted theories, which are less widely tested than scientific theories.

Almond and Genco (1978) take off from Popper's "Clouds and Clocks" chapter, although they ignore the remainder of his book, as well as the entire corpus of research literature in modern physics and psychobiology alike, to set forth an argument against "hard science"—which they equate with quantum physics! Such an identification seems possible only in the light of their failure to examine either what they think is "hard science" or the "hard- wiring" of the mind. Instead, they come up with a plug for a rejuvenation of the "policy-science" approach in political science.

A recent article by White (1982a) also uses Popper's chapter on clouds and clocks as a springboard for repudiating physics as a relevant paradigm for political science. However, unlike Almond and Genco, who make no distinction between the "clocks" and the "clouds" (although that *is* the core of Popper's own argument), White does make it clear that what he rejects is classical physics. Curiously, however, White then proceeds to ignore Popper's use and discussion of quantum physics. Thus "clouds" embellish the title of White's paper, but not its content, except briefly and obliquely on page 23. Here, White interjects the caveat that "classical physics is taken as a prototype for the contemporary physical sciences, when in reality it may only represent the physics of a bygone century." This assertion, as we are by now quite aware, is mistaken on several counts. Classical physics remains highly relevant to most of what is done in the physical sciences as well as elsewhere in human life today, and relativity theory certainly represents the physics of *this* century. Yet it is considered to be a component of classical physics—at least by physicists. White then cites (though to a later edition) Popper (1972), who follows the classical physicist and pragmatic philosopher C. S. Pierce rather than any of the many quantum physicists who have propounded the indeterminist perspective. As Morowitz (1980: 15) points out, "the implications of the developing paradigm greatly surprised early quantum physicists and led them to study epistemology and the philosophy of science. Never before in scientific history . . . had all of the leading contributors produced books and papers expounding the philosophical and humanistic meaning of their results."

White does quote the Italian quantum physicist Cocconi (1970: 87), but for a statement of what patently represents not a quantum-theory but rather an evolutionary perspective. "The immutable laws of physics could become as 'ephemeral' as those of organic life, immutable only for observations limited in space and time, and even more exotic, the evolution of these laws would depend on history, a history that has followed a path that, to a great extent, must have been determined by chance." In context, however, Cocconi had prefaced the quoted sentence with the statement that "it is appealing to think that in the realm of high energies, situations could develop similar to those possible for molecules. . . . [But] the only justification today for this kind of science fiction is the observation of a great variety of endothermal high-energy reactions." The sentence immediately following the one quoted by White and completing the paragraph is: "Seen from this point of view, even the Heisenberg uncertainty principle could be considered a temporary consequence of laws establishing themselves in an undeterminable manner." And Cocconi quickly adds, "I realize that I have carried the argument to its extreme consequences and beyond, into the metaphysical sphere." Cocconi's main interest was whether the high-energy phenomena of quantum physics could help to explain how life once began on this earth.

Consciousness

The facets of consciousness in which political scientists have tended to be interested relate to such matters as linguistic phenomenology and analyses of language that demonstrate in detail the extent and modes of observer bias as an inescapable consequence of the use of language to communicate. As Michael Shapiro (1981: 55) remarks, "part of the impact of the phenomenological perspective is owed to the philosophy of mind it offers, one which conceives of consciousness as an active, meaning constituting process"—and one which, at least to that extent, is in considerable agreement with the role assigned to both mind and language by quantum physics (see Bohm, 1980, ch. 7). A few political scientists have even begun to consider the effect of language upon thinking from the perspectives of the psychology of language and linguistic relativity (Landau, 1965) and from the perspective of biolinguistics (Kitchin, 1982a; Schubert, 1983a; and cf. TenHouten and Kaplan, 1973, chs. 4 and 5; and Rogers, TenHouten, Kaplan, and Gardner, 1977). Nancy Hartsock, a Marxist-feminist, claims (1979: 57) that "not only is our theory implicit in our conception of the world, but our conception of the world is itself a political choice." This is a profound half-truth. Our conception of the world surely is a *psychobiologically constrained* political choice (cf. Gould, 1980b: 49–51), and it is the transactional relationship between the psychobiological and the political that we must study if we are to understand political choice.

White (1982a) discusses theories of consciousness drawn from contemporary psychobiology, although his attraction is to dualist theories of brain function such as the one that Popper coauthored with Eccles (1977). I have discussed elsewhere (Schubert, 1983a) the holistic but materialistic theory of consciousness that is predominant in contemporary psychobiology (Morowitz, 1980: 12) with particular regard to how this relates to human perception and to the thinking process with which the observer paradox of quantum physics is concerned. I have also discussed what such a theory implies for political science.

Steven Peterson (1982b) focuses even more explicitly on the question of neurophysiology and rationality in political thinking and comes to conclusions highly supportive of the psychobiological perspective that has been propounded here—that human perception is tuned to macroscopic physical events, but the human brain operates at the level of microscopic (particle) physics, where the uncertainty principle operates pervasively. For the life sciences, the relevant "fields" are electrodynamic, to be sure, but in a fully materialistic sense they are *in the human brain*, not in any social or political or other external reality.

20
Life-Science Politics

This book is about theory: the kind of changes in political theory implied by the life-science paradigm. The chapters above frequently also discuss both methodology and empirical data; but the recurring and most consistent focus of this work is theoretical.

It begins with consideration of what a life-science paradigm involves and how it differs from the well-established humanities and social-science orientation of political science, and therefore of political theory. The political behavioral-science revolution that began within a year or two of the end of World War II was a move towards the *social* sciences, including their quantification, which contrasted so sharply with the methodology of traditional (pre World War II) political science; but the empirical data base remained anchored in human culture. It therefore failed to include virtually anything about human physiology or any other biological substrates of political behavior. Life science, to the contrary, starts with the presumption that, as animals, humans must develop and behave in ways that in many critical respects cannot be unlike those of other mammals generally (and of other primates in particular). The task of a life-science approach to politics is to identify the principal components of life-science theory, methods, and empirical knowledge about animal behavior, which conventional political science has ignored; then it must show how that information necessarily transforms an exclusively culturally determined understanding of *both* human politics *and* political science.

There are three major facets of such a life-science approach: ethology (how and why animals behave as they do); ecology (how environmental definitions of niches provide the stimulation and stress that proffer the opportunities and constraints to which animals respond in their behavior); and evolution (theories of how generalized and persistent changes in ecology result in reciprocal changes, first in animal behavior and then in animal physiology). Recently Gould (1985: 37) remarked that over 175 years ago, about half a century before Darwin's major work appeared, Lamarck recognized "that change of behavior must precede alteration of form. An animal enters a new environment with its old form unsuited to other styles of life."

We have examined evolutionary theory from many points of view. First we considered alternative models of the two orders of mammals that humans are deemed to resemble most: other primates and carnivores. Humans are closest genetically to other primates (especially to chimpanzees). But human behavior is closer to that of some carnivores (both large felids and the much smaller extant feral canids). Nevertheless *Homo sapiens* has always been omnivorous, not carnivorous; and the more sophisticated our comprehension of hominid evolution becomes, the greater the importance attached to hominid gathering of plants and scavenging of animals for food. Our species is unique in its *combination* of adapted behaviors, but those adaptations phyletically must be measured in the millions of years of geological time epochs, rather than the few thousand years for which historical reconstruction has been possible. As exemplified in chapter 6 (above), these truly ancient evolutionary considerations raise a very different set of questions about human political behavior than the contemporary ways in which politics is conceptualized by either politicians, political scientists, or by other commentators on politics (such as journalists and historians).

A second approach to evolutionary theory is to analyze the similarities and differences between biological and cultural evolution. Darwin and Mendel were contemporaries; but we now understand (as Darwin did not and could not know—given the seclusion of Mendel's life and the crypticity of his chosen mode of publication) that the mechanism of biological inheritance is genetic, as between biological parents and their offspring. Genes (like viruses) are complex nucleic acid compounds of DNA molecular aggregations that can act to regulate cell growth and production only in vivo; the information that they recombine through chromosomal meiosis and transmit through mitosis *defines* biological inheritance. Culture includes all nongenetic information that is transmitted between and among human conspecifics, plus certain information that humans transmit to some other (usually domesticated) animals and to some (especially cybernetic) machines. Genetic evolution is restricted (for humans) to the constraints of sexual reproduction and of female mammalian production in utero during gestation (notwithstanding certain novel biogenetic technologies not discussed in this book; see Blank, 1981, 1984); culture, on the other hand, can be and frequently is broadcast by dispersion, using methods that permit virtually instantaneous exchange of information between and among humans almost (in principle) anywhere in the world.

Gould states (1980a: 83–84) that Lamarckism is "the mode of 'inheritance' for . . . human cultural evolution. *Homo sapiens* arose at least 50,000 years ago, and we have not a shred of evidence for any genetic improvement since then. . . . All that we have accomplished, for better or for worse, is a result of cultural evolution.

And we have done it at rates unmatched by orders of magnitude in all the previous history of life. . . . [The present] crux in the earth's history has been reached because Lamarckian processes have finally been unleashed upon it. Human cultural evolution, in strong opposition to our biological history, is Lamarckian in character." (Gould also acknowledges Lamarck as a "great French biologist" and "a very fine scientist" for his important contributions to *biological* evolutionary theory, for which Lamarck has not been credited, and which are unrelated to the one typically associated with his name.)

Cultural communicators need not be (and frequently are not) in any sort of genetic degree of relationship—beyond conspecificity, of course; neither need they be in any face-to-face relationship, nor indeed members of closely contiguous generations. The respective life periods of communicators may be separated by thousands of years. Nevertheless much culture *is* transmitted biologically via the thought and memory processes of living humans, although much more of culture probably is stored in vitro and recalled from such storage for subsequent retransmission. Furthermore, genetic evolution is expressed epigenetically, and only the genes per se get sexually reproduced in meiosis—*not* the epigenetic expression through combination and recombination of genes in response to somatic and environmental evocation. Eldredge and Tattersall (1982: 177) point out that the cultural "style of transmission is 'Lamarckian' because a cultural 'item' may be acquired (either learned or invented) in the course of one's lifetime and transmitted" to one's *parents*, for example—that is, the previous, not the next, generation. Thus environmental change leads directly and can lead immediately to a change in cultural response that can be imitated, repeated ("cloned"), revised, or whatever. The processes of cultural production and reproduction have no relationship whatsoever to biological processes of either meiosis or (so far as we know) mitosis. Therefore it is ridiculous for sociobiologist (and some other geneticist) theorists (as discussed and criticized in part 4) to tout highly mathematicized models of "coevolution" in which they endeavor to constrain processes of cultural evolution by the same constraints of mammalian sexual reproduction and production that limit genetic evolution.

Sociobiology, a reactionary version of gradualist neo-Darwinian natural selection theory that stresses genetic determination as the direct cause of human decisional choices (among other modes of behavior), is criticized in part 3 not for its conservatism but because of its inapplicability to such a spectacularly open-to-learning and adaptable species as humans. On the other hand, chapter 11 advocates a conservative stance in opposition to radical proposals for direct technological intervention in human genetic evolution, on the ground that that constitutes a method of ultramodern postindustrial cultures for domesticating our own species, with consequences for its future adaptability that can neither be predicted nor understood. The research on evolutionary extinction emphasizes that the best protection against the elimination of a genus or family of animals exists if there is a diverse array of monophyletic clades (of species, genera, and other higher taxa) so that within that array of heterogeneity the probabilities increase that one or more of the included species will be capable of quantum speciation into a sufficiently modified genetic form that will allow successful adaptation to a drastically changing niche.

Such a solution holds as long as the scope and intensity of environmental change remain within the bounds of punctuational ("background") extinction. The same line of argument applies at the level of species gene pools; but even for a species with such extraordinary plasticity as *Homo sapiens*, the absence of any other monophyletic taxa below the level of the superfamily *Hominoidea* (see Eldredge and Tattersall, 1982: 26, 129–30) diminishes the survival odds for extant *Hominidae* (i.e., *Homo sapiens sapiens*). It is in that context of probabilities that the claimed benefits of biogenetic engineering, as applied to human genes, must be appraised. Given what has long been known about mutations (accidental, random gene reconstructions) and domestication side-effects, it is not merely possible but highly probable that the effects of human gene manipulation in the form of genetic engineering (planned mutations) will be catastrophic—indeed, they will certainly also be *epigenetically* catastrophic.

We should also recall that neither individual animals nor their species (or higher taxa) have any defense against catastrophes, other than luck. Natural catastrophes are not predictable and therefore their effects are random. But cultural catastrophes are quite a different matter. We can predict well enough what chlorofluorocarbonization of earth's atmosphere will do to the ozone layer; and we have had enough experience now with domestication (i.e., 10 ± 2 thousand years) to be able to estimate the extent to which we are likely to improve our species adaptation by accelerating to the max our experiments in conspecific domestication.

Political and other social scientists are most likely to be misled by "biological dogmas and mythologies" (below) if they remain ignorant of how biologists themselves respond to and counteract the proponents of such dogmas and myths—of which many of the most attractive (to those who are biologically illiterate) deal explicitly with pretensions about the human mind and cultures, as well as about genes (e.g., Dawkins, 1976; Lumsden and Wilson, 1981; but cf. Lewontin et al., 1984). Discussions of genetic engineering (among social scientists and humanists) typically focus upon what is ethical, in relation to what is technically feasible, in manipulating changes in the information biochemically encoded in the existing human genome (Blank, 1981, 1984). But one of the biologists who has made an effort to interact professionally with political scientists, behavioral geneticist Benson Ginsburg, has explicated (1978: 13–14) a very different kind of social genetic alternative instead of genetic engineering.

> The concept of the genomic repertoire, once operationally understood, holds promise for the effective utilization of our genetic reserves, consisting of genes that we possess physically, but that do not enter into the formation of our phenotype. Many of these undoubtedly represent superior capacities to those we manifest, and a fuller understanding of how to detect and activate these seems . . . to offer greater promise for so-called genetic engineering at

> the human level than the introduction of exogenous DNA. To understand and evaluate our social institutions and goals and to develop methods for achieving these is the realm of political science. This is a discipline that deals with human units and human behavior. To be misled by biological dogmas and mythologies poses serious pitfalls. Our knowledge of biological man is growing rapidly, and the confluence of our disciplines is inevitable, since what we are and can become depends equally on understanding our immutabilities and our modifiabilities, and on applying this understanding to the optimizing of the human condition.

Bleier (1985: 38) agrees, stating that "biology and culture together have provided us with a brain that has enormous potential for flexibility, learning, and creativity. Rather than biology, it is the culture we have created that limits, by its institutions, ideologies, and expectations, the full expression of that potential. One of the most important political problems we need to face is what role public policy, as the authoritative enforcer of cultural values, will play in the development or limitation of that potential."

Global catastrophes lead to mass extinctions (such as the cometary impact sixty-five million years ago that extinguished the dinosaurs, and thus directly cleared the way for mammalian expansion and radiation and indirectly made possible the evolution of *Homos* two million years ago and the recent radiation and contemporary global expansion of humans). But the implications of any such global catastrophe are strictly negative for our biological survival as a species. Nevertheless the implications of catastrophe *theory* may be highly positive as applied to such cultural threats to *Homo sapiens* survival as nuclear war; global poisoning of the air, waters, and land surfaces by our industrial and personal wastes; and the destruction of the biota that define the human psychological environment. The sensory deprivations imposed by our technological progress—the artificial climates, vistas, ingestions, imbibitions, atmospheres, sights, sounds, smells, tastes, touches, and remembered images that preempt our consciousness and unconsciousness alike—need to be naturalized drastically and soon, especially for those (on this criterion) maximally abused and deprived humans who inhabit the "richest" and most industrial/technologized countries of Western Europe, North America, and easternmost Asia. This is especially true if the catastrophic pace of technological innovation (Green, 1986) can be modeled by an exponential curve even approaching the approximately vertical line by means of which Pettersson (1978: 202) measures the rate of acceleration in change in present human technologized culture. Only revolutionary political action— which I mean literally as well as figuratively—is likely to escalate political change in both time and space enough even to arrest, let alone attempt to reverse, present trends in regard to nuclear winter and the wasting of what remains of the natural environment.

A major facet of evolutionary theory concerns the developmental change discussed in part 4. The dominant position, within professional evolutionary theory, of neo-Darwinian gradualism has come into confrontation with increasing dissent during the past half century. The principal dissidents include Sewall Wright, who died in 1988 at age 98, and who, throughout the latter half of his most fruitful life as one of the leading geneticists of the twentieth century, urged the importance of genetic drift and group selection for small, isolated populations of colonists in novel and challenging environments. Another dissenter was the late Conrad Hal Waddington, whose research and theories of epigenesis described the transactional effects between changing environments in the evocation of gene expression and behavioral positive feedback upon *both* an animal's environment and expressed genome. Valerius Geist emphasized the combination of Wright's theories with those of Waddington, in application to the development of the human species in the stressing but bountiful periglacial environments of Eurasia during the glaciation and interglacials of the past one hundred thousand years. Together these theories explain how and why the quantum speciation, recently proposed by punctuationalist evolutionary theorists, takes place. In barely a decade and a half, punctuated equilibria theory has generated considerably wider and more consistent criticism of modern-synthesis gradualism, than any or all of Wright, Waddington, and Geist—each working and acting independently—had been able to attract. The time for a Kuhnian paradigm change in evolutionary theory may be at hand.

Psychobiological theory directs attention away from the mind (and/or soul) to the human brain; how the brain is constructed, and why and how the phylogeny of all animals for some purposes, but especially, that of other mammals, defines much of the structure that makes possible the autonomic (visceral and muscular) control functions of the spinal cord and the emotional responses of the lower to mid-brain; but the self-consciously rational calculations and the language production of the human neocortex are specific to the human species. Thus chapter 17 asserts that an understanding of how the rational/linguistic interacts with the emotional, and the left hemisphere with the right, is essential to understand how individual humans make decisions on political questions.

Our present political theories reflect the tendencies toward polarization and dichotomization that come readily to mind, to the type of mind that was naturally selected for human males. Up to twelve thousand years ago, those males lived in habitats where their task was to cooperate together in small groups of related individuals, in confrontation with large animals and an often hostile natural environment. To encounter a strange human male was to confront the most dangerous enemy of all. The ethnocentrism and group paranoia, the fear of strangers, the reliance upon physical strength to both achieve and justify social superiority, all may have made good sense for males confined to life in hunting bands. Today, however, the only wild animals, either prey or predators, that most humans can conceivably confront are other humans. This expands enormously the possibilities for defining other humans as enemies. Certainly, however, the primordial attitudes and behaviors describe well the relationships among the "major powers" (e.g., the U.S.A., the U.S.S.R., the People's Republic of China; cf. Pettman, 1981); the other principal industrialized countries; and the Third-World aspirants who emulate to the extent possible the role models supplied by the "firsts" and "seconds." Equivalent bifurcations often are conceptualized to catalyze social divisions among classes or subcultures within a political society; and the "enemy within" usually is deemed to be treated as far more dangerous than the uitlander. Nor should we forget that the specific hormonal catalyst that provokes relatively extreme lateralization in the male brain—and consequently also in male thinking and behavior—is testosterone, which is responsible for male aggressiveness as well as male reproductive behavior per se. The attitudes and associated behaviors relating to both aggressiveness and sexual potency in primate males are transactionalized in the brain in complex ways that we are just beginning to understand (cf. J. C. Davies, 1980; McGuire, 1982; Schubert, 1982b): the male attitudes and behavior that define on a global basis the major problems of both international and domestic politics are all mixed up with male aggressiveness and sexuality as well as with the bifurcation in structure and function of the male brain. One thing that we obviously need is a less male (e.g., lateralized) theory of politics; and more generally, we need a politics characterized by a maximum (*not*, as at present, a minimum) of political androgeny (see Schubert, 1987).

All human perception is developmentally epigenetic; and the range of politically relevant perceptual differences among most people is probably both just as important and as extreme as are the politically conceptual differences among those same people. Conception relies much more on individual ontogenetic experience (present culture) than on hard-wiring of the brain (i.e., phyletic experience, including that of cultures past). But the transactional relationships between perceptual and conceptual processes (as in Laponce's theory of how the hard-wiring of the *per*ception of three-dimensional "natural" space strongly influences the *con*ception of political relationships and symbolism) are what political scientists must learn to understand political decision making.

Chapter 19 proposes that beyond the life sciences there are questions of relativity and indeterminism that are culturally important to more realistic modelling of political decision making. To deal with such matters, political scientists need also to acquire some minimal literacy in modern physics.

References
Name Index
Subject Index

References

Abercrombie, Nicholas, C. J. Hickman, and M. L. Johnson (1980). *The Penguin Dictionary of Biology*. 7th ed. New York: Penguin Books.

Abramovitch, Rona (1980). "Attention Structures in Hierarchically Organized Groups." In Donald R. Omark, Fred F. Strayer, and Daniel G. Freedman, eds., *Dominance Relations*, 381–96. New York: Garland.

Abramovitch, Rona, and Fred F. Strayer (1978). "Preschool Social Organization: Spacing and Attentional Behaviors." In J. L. Krames, P. Pliner, and T. Alloway, eds., *Aggression, Dominance, and Individual Spacing*. New York: Plenum.

Adorno, Theodore W., Else Frenkel-Brunswick, D. J. Levinson, and R. N. Sanford (1950). *The Authoritarian Personality*. New York: Harper.

Adrian, Charles (1969). "Implications for Political Science and Public Policy of Recent Ethological Research." Paper presented at the Second International Sinological Conference. Taipei, Taiwan: The China Academy.

Albin, Rochelle Semmel (1981). "Biopolitics: Odd Hybrid or a Synthesis?" *New York Times*, Aug. 23: E–7.

Alcock, John (1983). *Animal Behavior: An Evolutionary Approach*. 4th ed. Sunderland, MA: Sinauer (1975, 1st; 1979, 2nd).

Alexander, Richard D. (1974). "The Evolution of Social Behavior." *Annual Review of Ecology and Systematics*, 5: 324–83.

——— (1979a). *Darwinism and Human Affairs*. Seattle, WA: University of Washington Press.

——— (1979b). "Evolution and Culture." In N. A. Chagnon and W. Irons, eds., *Evolutionary Biology and Human Social Behavior*, 59–78. North Scituate, MA: Duxbury Press.

Allison, A. C. (1971). "Polymorphism and Natural Selection in Human Populations." In Carl Jay Bajema, ed., *Natural Selection in Human Populations: The Measurement of Ongoing Genetic Evolution in Contemporary Societies*, 166–90. New York: Wiley.

Alloway, Thomas, Patricia Pliner, and Lester Krames, eds. (1977). *Attachment Behavior*. New York: Plenum.

Almond, Gabriel, and Stephen Genco (1978). "Clouds, Clocks, and the Study of Politics." *World Politics*, 29: 489–523.

Altmann, S. A. (1962). "A Field Study of the Sociobiology of Rhesus Monkeys, *Macaca Mulatta*." *Annals of the New York Academy of Sciences*, 102: 338–435.

——— (1978). "The Politics of Sociobiology." Paper presented at the Annual Meeting of the Animal Behavior Society. Seattle.

Andrew, R. J. (1963). "The Origin and Evolution of the Calls and Facial Expressions of the Primates." *Behavior*, 20: 1–109.

——— (1972). "The Information Potentially Available in Mammal Displays." In Robert A. Hinde, ed., *Non-verbal Communication*, 179–206. New York: Cambridge University Press.

Ann Arbor Science for the People Editorial Collective (1977). *Biology as a Social Weapon*. Minneapolis: Burgess.

Annett, Marian (1980). "Sex Differences in Laterality—Meaningfulness and Reliability." *Behavioral and Brain Sciences*, 3: 227–28.

Ardrey, Robert (1961) *African Genesis*. New York: Atheneum.

——— (1970). *The Social Contract*. New York: Dell.

——— (1976). *The Hunting Hypothesis*. London: Fontana.

Argyle, Michael (1972). "Non-verbal Communication in Human Social Interaction." In R. A. Hinde, ed., *Non-verbal Communication*, 243- 69. New York: Cambridge University Press.

——— (1975). *Bodily Communication*. London: Methuen.

Aubert, Vilhelm (1959). "Chance in Social Affairs." *Inquiry*, 2: 1–24.

Auel, Jean M. (1980). *The Clan of the Cave Bear*. New York: Crown.

——— (1982). *The Valley of the Horses*. New York: Crown.

——— (1985). *The Mammoth Hunters*. New York: Crown.

Axelrod, Robert (1981). "The Emergence of Cooperation among Egoists." *American Political Science Review*, 75: 306–18.

——— (1984). *The Evolution of Cooperation*. New York: Basic Books.

Axelrod, Robert, and William D. Hamilton (1981). "The Evolution of Cooperation." *Science*, 211: 1390–96.

Baer, Darius, and Donald L. McEachron (1982). "A Review of Selected Sociobiological Principles: Application to Hominid Evolution: I. The Development of Group Social Structure." *Journal of Social and Biological Structures*, 5: 69–90.

Baer, Denise L. (1980). "Disentangling Gender Differences: An Inquiry into Biological and Learning Based Explanation." Paper presented at the Annual Meeting of the Midwest Political Science Association. Chicago.

Baerends, Gerhard (1975). "On Evolution of the Conflict Hypothesis as an Explanatory Principle for the Evolution of Displays." In C. Beer and A. Manning, eds., *Function and Evolution in Behavior: Essays in Honour of Professor Niko Tinbergen*. Oxford: Oxford University.

——— (1976a). "On Drive, Conflict and Instinct, and the Functional Organization of Behavior." In M. A. Corner and D. F. Swaab, eds., *Perspectives in Brain Research, Progress in Brain Research*. Vol. 45. Amsterdam: Elsevier/North-Holland Biomedical Press.

——— (1976b). "The Functional Organization of Behavior." *Animal Behavior*, 24: 726–38.

Baldwin, John D., and Janice I. Baldwin, (1977a). "The Role of Learning Phenomena in the Ontogeny of Exploration and Play." In Chevalier-Skolnikoff and F. E. Poirier, eds., *Primate Bio-Social Development*, 343–406. New York: Garland.

——— (1977b). "Reinforcement Theories of Exploration, Play, Creativity, and Psycho-Social Growth." Paper presented at the Annual Meeting of the Animal Behavior Society. Pennsylvania State University.

——— (1979). "The Phylogenetic and Ontogenetic Variables that Shape Behavior and Social Organization." In I. S. Bernstein and E. O. Smith, eds., *Primate Ecology and Human Origins*, 89–116. New York: Garland.

——— (1981). *Beyond Sociobiology*. New York: Elsevier.

Ball, Donald W. (1973). "The Biological Bases of Human Society." In Jack D. Douglas, ed., *Introduction to Sociology*, 118–38. New York: The Free Press.

Barash, David, et al. (1984) Symposium review of *Promethean Fire*. *Politics and the Life Sciences*, 2: 213–24.

Barkow, Jerome H. (1977). "Human Ethology and Intra-individual Systems." *Social Science Information*, 16: 133–45.

——— (1978). "Evolution and Sexuality," *Human Ethology Newsletter*, 23: 9–13.

Barlow, George W., and James Silverberg, eds. (1980). *Sociobiology: Beyond Nature/Nurture?* Boulder, CO: Westview.

Barner-Barry, Carol (1977). "An Observational Study of Authority in a Preschool Peer Group." *Political Methodology*, 4: 415–49.

——— (1978). "The Biological Correlates of Power and Authority: Dominance and Attention Structure." Paper presented at the Meeting of the American Political Science Association. New York City.

——— (1979). "The Utility of Attention Structure Theory and the Problem of Human Diversity." Paper presented at the World Congress of International Political Science Association. Moscow.

——— (1981). "Longitudinal Observational Research and the Study of Basic Forms of Political Socialization." In M. Watts, ed., *Biopolitics: Ethological and Physiological Approaches* 51–60. San Francisco: Jossey-Bass.

——— (1982). "An Ethological Study of Leadership Succession." *Ethology and Sociobiology*, 3: 199–207.

Barnes, S. B. (1968). "Paradigms--Scientific and Social." *Man*, 4: 94–102.

Barsley, Michael (1970). *Left-Handed Man in a Right-Handed World*. London: Pitman.

Barwick, Judith M. (1971). *Psychology of Women: A Study of Biocultural Conflicts*. New York: Harper and Row.

Bay, Christian (1965). "Politics and Pseudopolitics: A Critical Evaluation of Some Behavioral Literature." *American Political Science Review*, 59: 39–51.

Beccaria, Cesare (1963). *On Crime and Punishment*. Translated, with an introduction, by Henry Paolucci. Indianapolis: Bobbs-Merrill.

Beck, Henry (1975). "Ethological Considerations on the Problem of Political Order." *Political Anthropology*, 1: 109–35.

——— (1976a). "Attentional Struggles and Silencing Strategies in a Human Political Conflict: The Case of the Vietnam Moratoria." In Michael R. A. Chance and Ray R. Larsen, eds., *The Social Structure of Attention*. New York: Wiley.

——— (1976b). "Neuropsychological Servosystems, Consciousness and the Problem of Embodiment. *Behavioral Science*, 21: 139–60.

Beckwith, Jonathan, et al. (1975). "Against Sociobiology." *New York Review of Books*, Nov. 13.

Beer, Colin G. (1973). "Species-Typical Behavior and Ethology." In Donald A. Dewsbury and Dorothy A. Rethlingshafer, eds., *Comparative Psychology: A Modern Survey*. New York: McGraw-Hill, 21–78.

Bekoff, Marc (1974). "Social Play-Soliciting by Infant Canids." *American Zoologist*, 14: 323–40.

——— (1977). "Socialization in Mammals with an Emphasis on Nonprimates." In Suzanne Chevalier-Skolnikoff and Frank Poirier, eds., *Primate Bio-Social Development*, 603–36. New York: Garland.

Bennison, Gray (1983). "Historicism vs. History: The Comparative Method." *Journal of Social and Biological Structures*, 6: 193–206.

Bentley, Arthur F. (1908). *The Process of Government: A Study of Social Pressures*. Bloomington, IN: Principia Press.

Berns, Walter (1963). "Law and Behavioral Science." *Law and Contemporary Problems*, 28: 185–212.

Bernstein, Irwin S. (1981). "Dominance: The Baby and Behavior." *Behavioral and Brain Sciences*, 4: 419–57.

Bernstein, Irwin S., and Euclid O. Smith. (1979). *Primate Ecology and Human Origins*. New York: Garland.

Binford, Lewis (1971). "Post-Pleistocene Adaptations." In S. Struever, ed., *Prehistoric Agriculture*. Garden City, NY: Natural History Press.

Birdsell, Joseph B. (1972). *Human Evolution: An Introduction to the New Physical Anthropology*. Chicago: Rand-McNally.

Birdwhistell, R. L. (1952). *Introduction to Kinesics*. Louisville, KY: University of Louisville Press.

Bischof, Norbert (1975). "Comparative Ethology of Incest Avoidance." In Robin Fox, ed., *Biological Anthropology*, 37–67. New York: Wiley.

Black, Hugo (1969). *A Constitutional Faith*. New York: Knopf.

Black, Hugo, Jr. (1975). *My Father: A Remembrance*. New York: Random House.

Blank, Robert H. (1981). *The Political Implications of Human Genetic Technology*. Boulder, CO: Westview.

——— (1982). "Biomedical Issues and Political Science: A Time for Action." Paper presented at the Meeting of the Western Political Science Association.

——— (1984). *Redefining Human Life: Reproductive Technologies and Social Policy*. Boulder, CO: Westview.

Blaustein, Albert P., and Roy M. Mersky (1978). *The First One Hundred Justices*. Hamden, CT: Shoe String Press.

Bleier, Ruth (1985). "Biology and Women's Policy: A View from the Biological Sciences." In Virginia Sapiro, ed., *Women, Biology, and Public Policy*, 19–40. Beverly Hills, CA: SAGE.

Blurton-Jones, Nicolas G. (1967). "An Ethological Study of Some Aspects of Social Behaviour of Children in Nursery School." In D. Morris, ed., *Primate Ethology*. London: Weidenfeld and Nicolson.

——— (1972). "Characteristics of Ethological Studies of Human Behavior." In N. G. Blurton-Jones, ed., *Ethological Studies of Child Behavior*, 3–33. Cambridge: Cambridge University Press.

——— (1975). "Ethology, Anthropology, and Childhood." In Robin Fox, ed. *Biosocial Anthropology*, 69–92. New York: Wiley.

——— (1976). "Growing Points in Human Ethology: Another Link Between Ethology and the Social Sciences?" In P. P. G. Bateson and R. A. Hinde, eds., *Growing Points in Ethology*, 427–50. New York: Cambridge University Press.

Blurton-Jones, Nicolas G., and Melvin J. Konner (1973). "Sex Differences in Behaviour of London and Bushmen Children." In R. P. Michael and J. H. Crook, eds., *Comparative Ecology and Behaviour of Primates*. London: Academic Press.

Blurton-Jones, Nicolas G., and R. Sibly (1978). "Testing Adaptiveness of Culturally Determined Behaviour: Do Bushman Women Maximize Their Reproductive Success by Spacing Births Widely and Foraging Seldom?" In V. Reynolds and N. Blurton-Jones, eds., *Human Behaviour and Adaptation* 135–58 London: Taylor and Francis.

Boas, Franz (1927). *Primitive Art*. Oslo: H. Aschehoug.

Bock, Kenneth (1980). *Human Nature and History: A Response to Sociology*. New York: Columbia University Press.

Boggess, Jane (1979). "Troop Male Membership Changes and Infant Killing in Langurs." *Folia Primatologica*, 32: 65–107.

Bohm, David (1957). *Causality and Chance in Modern Physics*. London: Routledge and Kegan Paul.

——— (1965). *The Special Theory of Relativity*. New York: Benjamin.

——— (1971). "Quantum Theory as an Indication of a New Order in Physics, Part A." *Foundations of Physics*, 1: 359–81.

——— (1973). "Quantum Theory as an Indication of a New Order in Physics, Part B." *Foundations of Physics*, 3: 139–68.

——— (1977). "Heisenberg's Contributions to Physics." In William Price and Seymour Chissick, eds., *The Uncertainty Principle and Quantum Mechanics*, 559–63. New York: Wiley.

——— (1980). *Wholeness and the Implicate Order*. London: Routledge and Kegan Paul.

Bohr, Niels (1958). *Atomic Physics and Human Knowledge*. New York: Wiley.

Boklage, Charles (1980). "The Sinistral Blastocyst: An Embriologic Perspective on the Development of Brain-Function Asymmetries." In Jeannine Herron, ed., *Neuropsychology of LeftHandedness*. New York: Academic Press.

Bonner, John T. (1980). *The Evolution of Culture in Animals*. Princeton: Princeton University Press.

Boorman, Scott A., and Paul R. Levitt (1972). "Group Selection on the Boundary of a Stable Population." *Proceedings of the National Academy of Sciences*, 69: 2711–13.

Borgia, Gerald (1980). "Human Aggression as a Biological Adaptation." In Joan Lockard, ed., *The Evolution of Human Social Behavior*, 165–93. New York: Elsevier.

Boulding, Elise (1977). *Women in the Twentieth Century World*. New York: SAGE/John Wiley.

Boulding, Kenneth E. (1972). "Economics as a Not Very Biological Science." In John A. Benke, ed., *Challenging Biological Problems*. New York: Oxford University Press.

Bowlby, John (1958). "The Nature of a Child's Tie to His Mother." *International Journal of Psycho-Analysis*, 39: 1–23.

Boyd, Robert, and Peter Richerson (1985). *Culture and the Evolutionary Process*. Chicago: University of Chicago Press.

Brace, C. Loring (1979). "Biological Parameters and Pleistocene Hominid Life-Ways." In I. Bernstein and E. O. Smith, eds., *Primate Ecology and Human Origins*, 263–89. New York: Garland.

Bradshaw, John L. (1980). "Sex and Side: A Double Dichotomy Interacts." *Behavioral and Brain Sciences*, 3: 229–30.

Bradshaw, John L., and N. C. Nettleton (1981). "The Nature of Hemispheric Specialization in Man." *Behavioral and Brain Sciences*, 4: 51–91.

Braidwood, Robert J. (1975). *Prehistoric Man*. Glenview, IL: Scott, Foresman.

Brannigan, Christopher R., and David A. Humphries (1972). "Human Non-verbal Behaviour, a Means of Communication." In Nicholas G. Blurton-Jones, ed., *Ethological Studies of Child Behaviour*, 37–64. Cambridge: Cambridge University Press. "The Breast and the Bottle" (1981). *Newsweek*, Jun. 1: 54–55.

Breines, Wini, Margaret Cerullo, and Judith Stacey (1978). "Social Biology, Family Studies, and Anti-feminist Backlash." *Feminist Studies*, 4: 43–67.

Bromley, Dennis B. (1974). *The Psychology of Human Ageing*. 2d ed. Princeton: Princeton University Press.

Brothwell, Don, ed. (1977). *Biosocial Man*. London: Institute of Biology.

Brown, Jethro (1920). "Law and Evolution." *Yale Law Journal*, 29: 394–400.

Brown, Lester R. (1974). *In the Human Interest: A Strategy to Stabilize World Population*. New York: Norton.

Bruce, Virginia (1983). "Rock Throwing as a Determining Event in Human Development." *Human Ethology Newsletter*, 4: 5–16.

Bruner, Jerome S., Allison Jolly, and Kathy Sylva, eds. (1976). *Play: Its Role in Development and Evolution*. New York: Basic Books.

Burhoe, Ralph W. (1979). "Religion's Role in Human Evolution: The Missing Link between Ape-Man's Selfish Genes and Civilized Altruism," *Zygon*, 14: 135–62.

Burian, Richard M. (1978). "Methodological Critique of Sociobiology." In Arthur L. Caplan, ed., *The Sociobiology Debate*, 376–95. New York: Harper and Row.

Burke, Kenneth (1937). *Attitudes toward History*, Vol. 2. New York: New Republic Press.

——— (1945). *A Grammar of Motives*. New York: Prentice-Hall.

Butzer, Karl W. (1971). "Agricultural Origins in the Near East as a Geographical Problem." In S. Struever, ed., *Prehistoric Agriculture*, 209–35. Garden City, NY: Garland.

Bygott, J. D. (1972). "Cannibalism among Wild Chimpanzees." *Nature*, 238: 410–11.

Caldwell, Lynton K. (1964). "Biopolitics: Science, Ethics, and Public Policy." *The Yale Review*, 54: 1–16.

——— (1966). "Problems of Applied Ecology: Perceptions, Institutions, Methods, and Operational Tools." *Bioscience*, 16: 424–527.

——— (1980). "Biology and Bureaucracy: The Coming Confrontation." *Public Administration Review*, 40: 1–12.

Callan, Hilary (1970). *Ethology and Society: An Anthropological View*. New York: Oxford University Press.

Calvin, William H. (1981). "The Throwing Theory for Language Origins." *Human Ethology Newsletter*, 314: 17–22.

——— (1982). "Did Throwing Stones Shape Hominid Brain Evolution?" *Ethology and Sociobiology*, 3: 115–24.

Campbell, Bernard G. (1966). *Human Evolution: An Introduction to Man's Adaptations.* Chicago: Aldine.

——— (1979). "Ecological Factors and Social Organization in Human Evolution." In I. Bernstein and E. O. Smith, eds., *Primate Ecology and Human Origins.* New York: Garland.

Campbell, Donald T. (1972). "On the Genetics of Altruism and the Counter-Hedonic Components in Human Culture." *Journal of Social Issues*, 28: 21–37.

——— (1975). "On the Conflicts between Biological and Social Evolution, and between Psychology and Modern Tradition." *American Psychologist*, 30: 1103–26.

Caplan, Arthur L., ed. (1978a). *The Sociobiology Debate.* New York: Harper and Row.

——— (1978b). "Testability, Disreputability, and the Structure of the Modern Synthetic Theory of Evolution." *Erkenntnis*, 13: 261–78.

Caplan, Arthur L., et al. (1982). (Symposium review of *Genes, Mind, and Culture.*) *Behavioral and Brain Sciences*, 5: 8–31.

Capra, Fritof (1975). *The Tao of Physics.* London: Fontana/Collins.

——— (1980). "The New Physics: Implications for Psychology." *American Theosophist*, 68: 114–20.

Cardozo, Benjamin (1921). *The Nature of the Judicial Process.* New Haven, CT: Yale University Press.

Carneiro, Robert L. (1970). "A Theory of the Origin of the State." *Science*, 159: 733–38.

Cattani, Richard J. (1981). "Monitor Poll of Political Scientists." *The Christian Science Monitor*, Sept. 11: 3; Sept. 15: 6; Sept. 16: 7.

Cavallaro, Sahli (1978). "An Ethological Study of the Social Structure of Attention in a Preschool Mainstream Classroom." Paper presented at the meeting of the Animal Behavior Society. New Orleans.

Cavalli-Sforza, Luigi L., and M. W. Feldman (1981). *Cultural Transmission and Evolution: A Quantitative Approach.* Princeton: Princeton University Press.

Chagnon, Napoleon, and William Irons, eds. (1979). *Evolutionary Biology and Human Social Behavior.* North Scituate, MA: Duxbury.

Chance, Michael R. A., and Clifford J. Jolly (1970). *Social Groups of Monkeys, Apes and Men.* New York: E. P. Dutton.

Chance, Michael R. A., and Ray R. Larsen, eds. (1976). *The Structure of Attention.* New York: Wiley.

Chapple, Elliot D. (1970). *Culture and Biological Man: Explorations in Behavioral Anthropology.* New York: Holt, Rinehart, and Winston.

Charlesworth, William R. (1978). "Ethology: Understanding the Other Half of Intelligence." *Social Science Information*, 17: 231–77.

——— (1982). "The Epigenetic Connection between Genes and Culture: Environment to the Rescue." *Behavioral and Brain Sciences*, 5: 9–10.

Chase, Allan (1971). *The Biological Imperatives: Health, Politics, and Human Survival.* Baltimore: Penguin.

Chase, H. Peter, and David R. Metcalf (1975). "Undernutrition and Brain Development: Clinical, Biochemical, and Experimental Encephalographic Studies." In James W. Prescott, Merrill S. Read, and David B. Coursin, eds., *Brain Function and Malnutrition.* New York: Wiley.

Cherfes, Jeremy, and John Gribbin (1981a). "The Molecular Making of Mankind." *New Scientist*, 91: 518–21.

——— (1981b). "Descent of Man--or Ascent of Ape?" *New Scientist*, 91: 592–95.

Chevalier-Skolnikoff, Suzanne, and Frank E. Poirier, eds. (1977). *Primate Bio-Social Development: Biological, Social, and Ecological Determinants.* New York: Garland.

Chomsky, Noam (1975). *Reflections on Language.* New York: Pantheon.

Cigler, Allan J., and Burdett A. Loomis (1983). *Interest Group Politics.* Washington, DC: CQ Press.

Clapham, Wentworth B., Jr. (1973). *Natural Ecosystems.* New York: Macmillan.

Clinton, Richard L. (1973). *Population and Politics: New Directions in Political Science Research* Lexington, MA: D.C. Heath.

Clinton, Richard L., Willian Flash, and R. Kenneth Godwin, eds. (1972). *Political Science in Population Studies.* Lexington, MA: D.C. Heath.

Clinton, Richard L., and R. Kenneth Godwin, eds. (1972). *Research in the Politics of Population.* Lexington, MA: D.C. Heath.

Cloak, F. T., Jr. (1975). "Is a Culture Ethology Possible?" *Human Ecology,* 3: 161–82.

Cocconi, G. (1970). "The Role of Complexity in Nature." In M. Conversi, ed., *The Evolution of Particle Physics,* 81–88. New York: Academic Press.

Cohen, Felix (1935). "Transcendental Nonsense and the Functional Approach." *Columbia Law Review,* 35: 809–49.

Cohen, L. Jonathan (1981). "Can Human Rationality Be Experimentally Demonstrated?" *Behavioral and Brain Sciences,* 4: 317–70.

Cohen, Mark Nathan (1977). *The Food Crisis in Prehistory: Overpopulation and the Origins of Agriculture.* New Haven, CT: Yale University Press.

Cohen, Mark Nathan, Roy S. Malpaso, and Harold G. Klein, eds. (1980). *Biosocial Mechanisms of Population Regulation.* New Haven, CT: Yale University Press.

Coleman, James S. (1954). "An Expository Analysis of Some of Rashevsky's Social Behavior Models." In Paul F. Lazasfeld, ed., *Mathematical Thinking in the Social Sciences,* 105–55. Glenco, IL: The Free Press.

Comfort, Alex (1979a). *The Biology of Senescence.* 3d ed. New York: Elsevier.

——— (1979b). "The Cartesian Observer Revisited: Ontological Implications of the Homuncular Illusion." *Journal of Social and Biological Structures,* 2: 211–23.

——— (1980). "Demonic and Historical Models in Biology." *Journal of Social and Biological Structures,* 3: 207–15.

——— (1981). "The Implications of an Implicate," *Journal of Social and Biological Structures,* 4: 363–74.

Committee for the Advancement of Teaching, American Political Science Association (1951). *Goals for Political Science.* New York: William Sloane.

Comrey, Andrew, and John Newmeyer (1965). "Measurement of Radicalism—Conservatism." *Journal of Social Psychology,* 67: 357–69.

Congressional Quarterly (1979). *Guide to the U.S. Supreme Court.* Washington DC: Congressional Quarterly.

Corballis, Michael (1978). "Brain Twisters and Hand Wringers." *Behavioral and Brain Sciences,* 1: 331–36.

Corballis, Michael, and Ivan Beale (1976). *The Psychology of Left and Right.* Hillsdale, NJ: Lawrence Erlbaum Associates.

Corballis, Michael, and Michael Morgan (1978). "On the Biological Basis of Human Laterality: I. Evidence for a Maturational Left-Right Gradient," *Behavioral and Brain Sciences,* 1:261–69.

Corning, Peter A. (1970). "Theory of Evolution as a Paradigm for the Study of Political Phenomena." Ph. D. Diss. New York University.

——— (1971). "The Biological Bases of Behavior and Some Implications for Political Science." *World Politics*, 23: 321–70.

——— (1973). "Politics and Evolutionary Process." In T. Dobzhansky, et al., eds., *Evolutionary Biology*, vol. 7, ch. 4, 253–94. New York: Appleton-Century-Crofts.

——— (1977). "Human Nature Redivivus." In J. Roland Pennock and John W. Chapman, eds., *Human Nature and Politics*. New York: New York University Press.

——— (1980). "'Ethopolitics' and Political Science." *Human Ethology Newsletter*, 31: 12–13.

——— (1983). *The Synergism Hypothesis: A Theory of Progressive Evolution*. New York: McGraw-Hill.

Corning, Peter A., S. Kessler, and R. Kakihana (1977). "Three Kinds of Aggressive Behavior in Laboratory Mice." *Behavior Genetics*, 7: 51–52.

Corning, Peter A., Joseph Losco, and Thomes C. Wiegele (1981). "Political Science and the Life Sciences." *PS*, 14: 590–94.

Corson, Samuel A., and Elizabeth O'Leary Corson, eds. (1980). *Ethological and Nonverbal Communication in Mental Health: An Inter-disciplinary Biopsychosocial Exploration*. New York: Pergamon.

Cory, Gerald A., Jr. (1974). "The Biopsychological Basis of Political Socialization and Political Culture." Ph.D. Diss. Stanford University.

Crombie, Alistair (1961). "Quantification in Medieval Physics." In Harry Woolf, ed., *Quantification: A History of the Meaning of Measurement in the Natural and Social Sciences*, 13–30. Indianapolis: Bobbs-Merrill.

Crook, John Hurrell (1975). "Problems of Inference in the Comparison of Animal and Human Organization." *Social Science Information*, 14: 89–112.

——— (1980). *The Evolution of Human Consciousness*. Oxford: Oxford University Press.

Curio, Eberhard (1976). *The Ecology of Predation*. New York: Springer-Verlag.

Curtin, Richard, and Phyllis Dolhinow (1978). "Primate Social Behavior in a Changing World." *American Scientist*, 66: 468–75.

Cutler, Neal E. (1973). "Aging and Generations in Politics: The Conflict of Explanations and Inference." In Allen R. Wilcox, ed., *Public Opinion and Political Behavior*. Chicago: Aldine.

——— (1975). "Toward a Generational Conception of Political Socialization." In David C. Schwartz and Sandra K. Schwartz, eds., *New Directions in Political Socialization*, 254–88. New York: The Free Press.

——— (1977). "Demographic, Social-Psychological, and Political Factors in the Politics of Aging: A Foundation for Research in 'Political Gerontology.'" *American Political Science Review*, 71: 1011–25.

Cutler, Neal G., and Vern L. Bengtson (1974). "Age and Political Alienation: Maturation, Generation and Period Effects." *The Annals of the American Academy of Political and Social Science*, 4: 160–74.

Cutler, Neal G., and John R. Schmidhauser (1975). "Age and Political Behavior." In Diana S. Woodruff and James E. Birren, eds., *Aging: Scientific Perspectives and Social Issues*, 374–406. New York: Van Nostrand.

Dahl, Robert (1961). "The Behavioral Approach." *American Political Science Review*, 55: 763–72.

Daly, Martin, and Margo Wilson (1978). "Functional Significance of the Psychology of Men and Women." *Human Ethology Newsletter*, 23: 6–8.

Danelski, David J. (1966). "The Japanese Supreme Court: An Exploratory Study." Paper presented at the Annual Meeting of the American Political Science Association. New York.

——— (1969). "The Supreme Court of Japan: An Exploratory Study." In Glendon Schubert and D. Danelski, eds., *Comparative Judicial Behavior*, 121–56. New York: Oxford University Press.

Danielli, James, et al. (1982). "Constructional Biology." *Journal of Social and Biological Structures*, 5:11–14.

d'Aquili, Eugene G. (1978). "The Neurobiological Bases of Myth and Concepts of Deity." *Zygon*, 13: 257–75.

d'Aquili, Eugene G., Charles D. Laughlin, Jr., and John McManus (1979). *The Spectrum of Ritual: A Biogenetic Structural Analysis*. New York: Columbia University Press.

Darlington, Cyril D. (1969). *The Evolution of Man and Society*. New York: Simon and Schuster.

——— (1978). *The Little Universe of Man*. London: George Allen and Unwin.

Darwin, Charles (1871). *The Descent of Man and Selection in Relation to Sex*. London: J. Murray; New York: D. Appleton.

——— (1965 [1872]). *The Expression of Emotions in Man and Animals*. Chicago: University of Chicago Press.

Davidson, Julian, and Richard Davidson, eds. (1980). *The Psychobiology of Human Consciousness*. New York: Plenum.

Davies, James C. (1963). *Human Nature in Politics: The Dynamics of Political Behavior*. New York: Wiley.

——— (1969). "The Psychobiology of Political Behavior: Some Provocative Developments." Paper presented at the Annual Meeting of the Western Political Science Association. Honolulu.

——— (1976). "Ions of Emotions and Political Behavior: A Proto-Theory." In Albert Somit, ed., *Biology and Politics*. The Hague: Mouton.

——— (1977). "The Priority of Human Needs and the Stages of Political Development." In J. R. Pennock and J. W. Chapman, eds., *Human Nature in Politics*. New York: New York University Press.

——— (1980). "Biological Perspectives of Human Conflict." In Ted Robert Gurr, ed., *Handbook of Political Conflict: Theory and Research*, 19–68. New York: The Free Press.

Davies, Paul (1980). *Other Worlds*. New York: Simon and Schuster.

——— (1982). *The Edge of Infinity*. New York: Simon and Schuster.

Davis, Patrick D. C., and A. A. Dent (1968). *Animals That Changed the World*. New York: Crowell-Collier/Macmillan.

Dawkins, Richard (1976). *The Selfish Gene*. New York: Oxford University Press.

Dawkins, Richard, and J. R. Krebs (1978). "Animal Signals: Information or Manipulation?" In J. R. Krebs and N. B. Davies, eds., *Behavioral Ecology: An Evolutionary Approach*. Sunderland, MA: Sinauer.

Dawson, John (1977). "An Anthropological Perspective on the Evolution and Lateralization of the Brain." *Annals of the New York Academy of Sciences*, 299: 424–47.

Day, Clarence (1936 [1920]). *This Simian World*. New York: Knopf.

Dearden, John (1974). "Sex-Linked Differences of Political Behavior." *Social Science Information*, 13: 19–46.

de Luce, Judith, and Hugh T. Wilder (1983). *Language in Primates*. New York: Springer-Verlag.

Demarest, William J. (1977). "Incest Avoidance among Human and Non-human Primates." In Suzanne Chevalier-Skolnikoff and Frank E. Poirier, eds., *Primate Bio-Social Development*, 323–42. New York: Garland.

Denenberg, Victor (1981). "Hemispheric Laterality in Animals and the Effects of Early Experience." *Behavioral and Brain Sciences*, 4: 1–49.

de Nicolas, Antonio T. (1982). "Audial and Literary Cultures: Bhasavad Gita as a Case Study," 4: 1–49.

Deutsch, Karl W. (1963). *The Nerves of Government*. New York: The Free Press.

Deutsch, Karl, and Thomas Edsall (1972). "The Meritocracy State." *Society*, 9: 71–79.

Deutsch, Morton (1968). "Field Theory." *International Encyclopedia of the Social Sciences*, 5: 406–17.

DeVore, Irven, ed. (1965). *Primate Behavior: Field Studies of Monkeys and Apes*. New York: Holt, Rinehart and Winston.

de Waal, Frans B. M. (1978). Exploitative and Familiarity-Dependent Support Strategies in a Colony of Semi-Free Living Chimpanzee." *Behaviour*, 66: 268–312.

——— (1982). *Chimpanzee Politics: Power and Sex among Apes*. London: Jonathan Cape.

de Waal, Frans B .M., and Jan A. R. A. M. van Hooff (1981). "Side Directed Communication and Agonistic Interaction in Chimpanzee." *Behavior*, 77: 164–98.

Dimond, Stuart (1977). "Concluding Remarks: The Diversity of the Human Brain." *Annals of the New York Academy of Sciences*, 299: 490–501.

Dobzhansky, Theodore D. (1937). *Genetics and the Origins of Species*. New York: Columbia University Press.

Dreyfus, Hubert L. (1972). *What Computers Can't Do: A Critique of Artificial Reason*. New York: Harper and Row.

Dreyfus, Hubert L., and Stuart E. Dreyfus (1986). *Mind over Machine: The Power of Human Intuition and Expertise in the Era of the Computer*. New York: The Free Press.

Dror, Y. (1984). "Policymaking as Fuzzy Gambling." Paper presented at the Annual Meeting of the American Political Science Association. Washington, DC.

Dunbar, M. J. (1960). "The Evolution of Stability in Marine Environments: Natural Selection at the Level of the Ecosystem." *American Naturalist*, 94: 129–36.

Dunne, Gerald T. (1977). *Hugo Black and the Judicial Revolution*. New York: Simon and Schuster.

Durham, William H. (1976a). "The Adaptive Significance of Cultural Behavior." *Human Ecology*, 4: 89–121.

——— (1976b). "Resource Competition and Human Aggression, Part 1: A Review of Primitive War." *Quarterly Review of Biology*, 51: 385–415.

——— (1979). "Toward a Coevolutionary Theory of Human Biology and Culture." In N. A. Chagnon and W. Irons, eds., *Evolutionary Biology and Human Social Behavior*, 39–58. North Scituate, MA: Duxbury Press.

Easton, David (1969). "The New Revolution in Political Science." *American Political Science Review*, 63: 1051–61.

——— (1973). "System Analysis and Its Classical Critics." *Political Science Reviewer*, 3: 269–301.

Ebbesson, Sven O. E. (1984). "Evolution and Ontogeny of Neural Circuits." *Behavioral and Brain Sciences*, 7: 321–66.

Eccles, John (1977). "Evolution of the Brain in Relation to the Development of the Self-Conscious Mind." *Annals of the New York Academy of Sciences*, 299: 161–79.

Edelman, Gerald, and Vernon Mountcastle (1978). *The Mindful Brain*. Cambridge: MIT Press.

Ehrman, Lee, and P. A. Parsons (1976). *The Genetics of Behavior*. Sunderland, MA: Sinauer.

Eibl-Eibesfeldt, Irenaus (1972). "Similarities and Differences Between Cultures in Expressive Movements." In R.A. Hinde, ed., *Nonverbal Communication*. Cambridge: Cambridge University Press.

——— (1974). *Love and Hate: The Natural History of Behavior Patterns*. New York: Schocken.

——— (1975). *Ethology: The Biology of Behavior*. 2d ed. New York: Holt, Rinehart and Winston.

——— (1979). "Human Ethology: Concepts and Implications for the Sciences of Man." *Behavioral and Brain Sciences*, 2: 1–26.

——— (1980). "Too Many Jumping on the Bandwagon of Sociobiology." *Human Ethology Newsletter*, 29: 7–10.

Ekman, Paul and W. V. Friesen (1975). *Unmasking the Face*. Englewood Cliffs, NJ: Prentice-Hall.

Eldredge, Niles (1985a). *Time Frames: The Rethinking of Darwinian Evolution and the Theory of Punctuated Equilibria*. New York: Simon and Schuster.

——— (1985b). *Unfinished Business*. New York: Oxford University Press.

Eldredge, Niles, and Stephen Jay Gould (1972). "Punctuated Equilibria: An Alternative to Phyletic Gradualism." In Thomas J. M. Schopf, ed., *Models in Paleobiology*, 82–115. San Francisco: Freeman and Cooper.

Eldredge, Niles, and Ian Tattersal (1982). *The Myths of Human Evolution*. New York: Columbia University Press.

Elliott, David K. (1986). *Dynamics of Extinction*. New York: Wiley.

"Enigmas of Evolution" (1982). *Newsweek*, Mar. 29: 44–49.

Eshel, Ilan (1972). "On the Neighbor Effect and the Evolution of Altruistic Traits." *Theoretical Population Biology*, 3: 258–77.

Etheredge, Lloyd (1979). "Hardball Politics: A Model." *Political Psychology*, 1: 3–26.

Eulau, Heinz (1963). *The Behavioral Persuasion in Politics*. New York: Random House.

Eulau, Heinz, and James G. March, eds. (1969). *Political Science*. Englewood Cliffs, NJ: Prentice-Hall.

Eysenck, Hans J. (1954). *The Psychology of Politics*. London: Routledge and Kegan Paul.

——— (1967). *The Biological Basis of Personality*. Springfield, IL: Charles C. Thomas.

Eysenck, Hans J., and Glenn Wilson, eds. (1978). *The Psychological Basis of Ideology*. Lancaster, England: MTP Press.

Fairman, Charles (1938). "The Retirement of Federal Judges." *Harvard Law Review*, 51: 406–40.

Falger, V. S. E. (1978). "Ethologie en Politicologie: Tegenstelling of Aanvulling?" *Politica*, 13: 3–47.

Festinger, Leon, Henry W. Riecken, and Stanley Schachter (1956). *When Prophecy Fails: A Social and Pychological Study of a Modern Group That Predicted the Destruction of the World*. Minneapolis: University of Minnesota Press.

Filskov, Susan, and Thomas Boll, eds. (1981). *Handbook of Clinical Neuropsychology*. New York: Wiley.

Fincher, Jack (1977). *Sinister People: The Looking-Glass World of the Left-Hander*. New York: Putnam.

Findlay, C. Scott, and Charles J. Lumsden (1988). "The Creative Mind: Toward an Evolutionary Theory of Discovery and Innovation." *Journal of Social and Biological Structures*, 11(1): 3–55.

Firestone, Shulamith (1970). *The Dialectic of Sex: The Case for Feminist Revolution*. New York: William Morrow.

Fisher, Elizabeth (1979). *Woman's Creation: Sexual Evolution and the Shaping of Society*. New York: McGraw-Hill.

Fisher, Helen E. (1982). *The Sex Contract*. New York: William Morrow.

Fisher, Ronald A. (1930). *The Genetical Theory of Natural Selection*. Oxford: Clarendon Press.

Flannery, Kent (1972). "The Cultural Evolution of Civilization." *Annual Review of Ecology and Systematics*, 3: 339–426.

Fossey, Dian (1983). *Gorillas in the Mist*. Boston: Houghton Mifflin.

Foucault, Michel (1965). *Madness and Civilization*. New York: Pantheon.

Fox, Michael W., ed. (1975). *The Wild Canids*. Princeton: Van Nostrand-Reinhold.

Frank, Jerome (1930). *Law and the Modern Mind*. New York: Coward-McCann.

——— (1931). "Are Judges Human?" *University of Pennsylvania Law Review*, 80: 17–53.

Frank, Pat (1947). *Mr. Adam*. London: Victor Gollantz.

Frankel, Mark S. (1974). "Political Responses to Controversial Issues in the Development of Biomedical Technologies." Paper presented at the Annual Meeting of the American Political Science Association.

Frankovic, Kathleen A. (1982). "Sex and Politics—New Alignments, Old Issues." *PS*, 15: 439–47.

Freedman, Daniel G. (1979). *Human Sociobiology: A Holistic Approach*. New York: The Free Press.

Freeman, L. G. (1981). "The Fat of the Land: Notes on Paleolithic Diet in Iberia." In Robert Harding and Geza Teleki, eds., *Omnivorous Primate: Gathering and Hunting in Human Evolution*, 104–65. New York: Columbia University Press.

Fried, Morton H. (1967). *The Evolution of Political Society*. New York: Random House.

Fries, James F., and Lawrence M. Crapo (1981). *Vitality and Aging*. San Francisco: W. H. Freeman.

Futuyama, Douglas J. (1979). *Evolutionary Biology*. Sunderland, MA: Sinauer.

Gadbois, George H., Jr. (1969). "The Selection, Background Characteristics, and Voting Behavior of Indian Supreme Court Judges, 1950–1959." In Glendon Schubert and David Danelski, eds., *Comparative Judicial Behavior*, 221–56. New York: Oxford University Press.

Gamow, George (1966). *Thirty Years That Shook Physics*. Garden City, NY: Doubleday.

Gardner, Martin (1964). *The Ambidextrous Universe*. New York: Basic Books.

Gazzaniga, Michael, and Joseph LeDoux (1978). *The Integrated Mind*. New York: Plenum.

Geber, Marcelle (1958). "The Psycho-motor Development of African Children in the First Year, and the Influence of Maternal Behavior," *Journal of Social Psychology*, 47: 185–95.

Geber, Marcelle, and R. F. A. Dean (1957). "Gesell Tests on African Children." *Pediatrics*, 20: 1055–65.

Geertz, Clifford (1973). *The Interpretation of Cultures*. New York: Basic Books.

Geigle, Ray A. (1977). "Theoretical Considerations of Psychobiological Adaptation and Political Response Predisposition." Paper presented at the Annual Meeting of the American Political Science Association.

Geist, Valerius (1978). *Life Strategy, Human Evolution, Environmental Design: Toward a Biological Theory of Health.* New York: Springer-Verlag.

Gianos, Phillip L. (1982). *Political Behavior: Metaphors and Models of American Politics.* Pacific Palisades, CA: Palisades Publishers.

Gibson, Kathleen (1977). "Brain Structure and Intelligence in Macaques and Human Infants from a Piagetian Perspective." In S. Chevalier-Skolnikoff and F. E. Poirier, eds., *Primate Bio-Social Development.* New York: Garland.

Gilligan, Carol (1982). *In a Different Voice.* Cambridge: Harvard University Press.

Gilman, Charlotte (1911). *The Man-made World; or Our Androcentric Culture.* New York: Charlton Co.

Gilman, Stuart C., Robert L. Simon, and Stephen L. Zegura (1979). "Evolution, Ethics, and Equality." Paper presented at the 75th Annual Meeting of the American Political Science Association. Washington, DC.

Ginsburg, Benson E. (1978). "What Will Students in Political Science Have to Know about Biology to Understand the New Dimensions of Their Discipline and to Advance the Frontiers of Knowledge?" Paper presented at the Annual Meeting of the American Political Science Association. New York City.

Glamser, Francis D. (1974). "The Importance of Age to Conservative Opinions: A Multivariate Analysis." *Journal of Gerontology,* 29: 549–54.

Glenn, Noval D. (1974). "Aging and Conservatism." *Annals of the American Academy of Political and Social Science,* 4: 176–86.

Globus, Charles, Grover Maxwell, and Irwin Savodnik, eds. (1976). *Consciousness and the Brain.* New York: Plenum.

Goff, John S. (1960). "Old Age and the Supreme Court." *American Journal of Legal History,* 4: 95–106.

Goffman, Erving (1961). *Asylums.* Garden City, NY: Anchor.

Golding, William (1955). *The Inheritors.* New York: Harcourt, Brace and World.

Goldman, Sheldon (1975). "Voting Behavior on the US Courts of Appeals Revisited." *American Political Science Review,* 69: 491–506.

——— (1982). *Constitutional Law and Supreme Court Decision-Making.* New York: Harper and Row.

Goldman, Stanford (1980). "A Unified Theory of Biology and Physics." *Journal of Social and Biological Structures,* 3: 331–60.

Goldsmith, Donald, ed. (1985). *Nemesis: The Death-Star and Other Theories of Mars Extinction.* New York: Walker.

Goodall, Jane (1977). "Infant Killing and Cannibalism in Free-Living Chimpanzees." *Folia Primatologica,* 28: 259–82.

——— (1986). *The Chimpanzees of Gombe: Patterns of Behavior.* Cambridge, MA: Belknap Press.

Goodall, Jane, et al. (1979). "Intercommunity Interactions in the Chimpanzee Population of the Gombe National Park." In David Hamburg and Elizabeth R. McCown, eds., *The Great Apes.* Menlo Park, CA: Benjamin/Cummings.

Goodman, Ellen (1982). "Woman's Vote Really Counts Now." *Honolulu Advertiser,* May 21, p. A–18, col. 3–5.

Gould, Stephen Jay (1977a). *Ontogeny and Phylogeny.* Cambridge: Harvard University Press.

——— (1977b). *Ever since Darwin.* New York: Norton.

——— (1980a). *The Panda's Thumb.* New York: Norton.

——— (1980b). "The Evolutionary Biology of Constraint." *Daedalus,* 109: 39–52.

——— (1981). *The Mismeasure of Man.* New York: Norton.

——— (1983). *Hen's Teeth and Horse's Toes.* New York: Norton.

——— (1984). "Toward the Vindication of Punctuational Change." In W. A. Berggren and John A. Van Couvering, eds., *Catastrophes and Earth's History,* 9–34. Princeton: Princeton University Press.

——— (1985). *The Flamingo's Smile.* New York: Norton.

Gould, Stephen Jay, and Niles Eldredge (1977). "Punctuated Equilibria: The Tempo and Mode of Evolution Reconsidered." *Paleobiology,* 3: 115–51.

Gow, David John (1981). "The Cognitive Organization of Political Attitudes and Their Psycho-Physiological Bases." Ph.D. Diss. University of Hawaii-Manoa, Honolulu.

Graber, Doris (1982a). "Strategies for Processing Political Information." Paper presented at the Meeting of the Midwest Political Science Association. Milwaukee.

——— (1982b). "Have I Heard This Before and Is It Worth Knowing?: Variations in Political Information Processing." Paper presented at the Annual Meeting of the American Political Science Association. Denver.

Grady, K. E. (1977). "The Belief in Sex Differences." Paper presented at the Meeting of the Eastern Psychological Association. Boston.

Graham, George, and George Carey, eds. (1972). *The PostBehavioral Era.* New York: David McKay.

Grant, Ewan C. (1969). "Human Facial Expression." *Man,* 4: 525–36.

Grant, J. A. C. (1965). "Felix Frankfurter: A Dissenting Opinion." *UCLA Law Review,* 12: 1013–42.

Graubard, Mark (1985). "The Biological Foundation of Culture." *Journal of Social and Biological Structures,* 8: 109–24.

Gray, Bennison (1983). "Historicism vs. History: The Comparative Method." *Journal of Social and Biological Structures,* 6: 193–206.

Gray, Jeffrey A. (1982a). *The Neurophysiology of Anxiety: An Enquiry into the Functions of the Septo-hippocampal System.* Oxford: Oxford University Press.

——— (1982b). "Precis of the Neurophysiology of Anxiety: An Enquiry into the Functions of the Septo-hippocampal System." *Behavioral and Brain Sciences,* 5: 469–534.

Green, Halcott P. (1986). *Power and Evolution: The Disequilibrium Hypothesis.* Columbia, SC: Institute of International Studies, University of South Carolina.

Greenberg, L. M. (ed.) (1982). *Evolution, Extinction, and Catastrophism.* Wynnewood, PA: Kronos.

Greenwood, Davydd, and Willian A. Stini (1977). *Nature, Culture, and Human History.* New York: Harper and Row.

Gregory, Michael S., Anita Silvers, and Diane Sutch, eds. (1978). *Sociobiology and Human Nature.* San Francisco: Jossey-Bass.

Griffin, Donald R. (1978). "Prospects for a Cognitive Ethology." *Behavioral and Brain Sciences,* 1: 527–38.

——— (1981). *The Question of Animal Awareness: Evolutionary Continuity of Mental Experience.* 2d ed. New York: Rockefeller University Press.

——— (1982). *Animal Mind—Human Mind.* New York: Springer-Verlag.

——— (1984). *Animal Thinking.* Cambridge: Harvard University Press.

Grings, W. W., and M. E. Dawson (1978). *Emotions and Bodily Responses: A Psychophysiological Approach.* New York: Academic Press.

Groos, Karl (1898). *The Play of Animals.* New York: Appleton.

Grunbaum, Adolf (1973). *Philosophical Problems of Space and Time.* 2d ed. Boston: Reidel.

Gruter, Margaret, and Roger D. Masters, eds. (1986). *Ostracism as a Social and Biological Phenomenon.* New York: Elsevier.

Gualtieri, Thomas, and Robert E. Hicks (1985). "An Immunoreactive Theory of Selective Male Affliction." *Behavioral and Brain Sciences,* 8: 427–40.

Guillemin, Victor (1968). *The Story of Quantum Mechanics.* New York: Scribner.

Gustafsson, Jan-Eric (1985). "Measuring and Interpreting *g.*" *Behavioral and Brain Sciences,* 8: 231–32.

Guthrie, Stewart (1980). "A Cognitive Theory of Religion." *Current Anthropology,* 21:181–94.

Haas, Michael (1969). "Toward the Study of Biopolitics: A Cross-Sectional Analysis of Mortality Rates." *Behavioral Science,* 14:257–80.

——— (1982). "Paradigms of Explanation in Political Science:Metaphysical Underpinnings." Paper presented at the Meeting of the American Political Science Association. Denver.

Haddon, Alfred Cort (1895). *Evolution in Art.* London: W. Scott.

Haddow, Anna (1939). *Political Science in American Colleges and Universities, 1836–1900.* New York: D. Appleton-Century.

Hager, Joseph C., and Paul Ekman (1979). "Long Distance Transmission of Facial Affect Signals." *Ethology and Sociobiology,* 1: 77–82.

Hailman, Jack P. (1982). "Creationism." *Academe,* 68: 6–26.

——— (1982). "Creation Stories." *BioScience,* 32: 129–30.

Haldane, John B. S. (1932). *The Causes of Evolution.* London:Longmans, Green.

Hall, Edward T. (1959). *The Silent Language.* New York: Doubleday.

Hall, Roberta L. (1982). *Sexual Dimorphism in Homo Sapiens: A Question of Size.* New York: Praeger.

——— , ed. (1985). *Male-Female Differences: A Biocultural Perspective.* New York: Praeger.

Hall, Roberta L., and Henry S. Sharp, eds. (1978). *Wolf and Man: Evolution in Parallel.* New York: Academic Press.

Halliday, R. J. (1971). "Social Darwinism." *Victorian Studies,* 14: 389–405.

Hamburg, David, and Elizabeth R. McGowan (1979). *The Great Apes.* Reading, MA: Addison-Wesley.

Hamilton, William J. (1964). "The Genetic Theory of Social Behavior: I and II." *Journal of Theoretical Biology,* 7: 1–51.

——— (1975). "Innate Social Aptitudes of Man: An Approach from Evolutionary Genetics." In Robin Fox, ed., *Biosocial Anthropology.* New York: Wiley.

Handberg, Roger, Jr. (1976). "Decision-Making in a Natural Court, 1916–1921." *American Politics Quarterly,* 4: 357–78.

Handberg, Roger, Jr., and William S. Maddox (1981). "Public Controversy and Science Policy: Genetic Engineering, Genetic Research, and Public Support." Paper presented at the Annual Meeting of the American Political Science Association. New York City.

Handler, Philip, ed. (1970). *Biology and the Future of Man.* New York: Oxford University Press.

Hannon, Bruce (1985). "World Shogun." *Journal of Social and Biological Structures*, 8: 329–31.

Harding, Robert (1975). "Meat-Eating and Hunting in Baboons." In Russell H. Tuttle, ed., *Socioecology and Psychology of Primates*, 245–57. The Hague: Mouton.

Harding, Robert, and Geza Teleki, eds. (1981). *Omnivorous Primates: Gathering and Hunting in Human Evolution*. New York: Columbia University Press.

Harlow, Harry F., and Margaret K. Harlow (1962). "Social Deprivation in Monkeys." *Scientific American*, 207: 136–46.

Harnad, Steven, and Horst Steklis (1976). "Comment on Split-Brain Research and the Culture-and-Cognition Paradox." *Current Anthropology*, 17: 302–22.

Harrington, F. H., and L. D. Mech (1978). "Wolf Vocalization." In R. L. Hall and H. S. Sharp, eds., *Wolf and Man*. New York: Academic Press.

Harris, Lauren Julius (1978). "Sex Differences in Spatial Ability: Possible Environmental, Genetic, and Neurological Factors." In Marcel Kinsbourne, ed., *Asymmetrical Function of the Brain*. New York: Cambridge University Press.

——— (1980). "Which Hand Is the 'Eye' of the Brain?—A New Look at Old Question." In Jeannine Herron, ed., *Neuropsychology of Left-Handedness*. New York: Academic Press.

Harris, Marvin (1977). *Cannibals and Kings: The Origins of Culture*. New York: Random House.

Hartl, Daniel L. (1982). "A Too Simple View of Population Genetics." *Behavioral and Brain Sciences*, 5: 13–14.

Hartsock, Nancy (1979). "Feminist Theory and the Development of Revolutionary Strategy." In Zillah Eisenstein, ed., *Capitalist Patriarchy and the Case for Socialist Feminism*. New York: Monthly Review Press.

Hass, Hans (1972). *The Human Animal: The Mystery of Man's Behavior*. New York: Dell.

Hediger, Heini (1964). *Wild Animals in Captivity: An Outline of the Biology of Zoological Gardens*. New York: Dell.

——— (1968). *The Psychology and Behavior of Animals in Zoos and Circuses*. New York: Dover.

Heelan, Patrick (1980). "Comments on 'A Unified Theory of Biology and Physics' by S. Goldman." *Journal of Social and Biological Sciences*, 3: 361–62.

Heisenberg, Werner (1958). *Physics and Philosophy*. New York: Harper.

——— (1974). *Across the Frontier*. New York: Harper.

——— (1977). "Remarks on the Origin of the Relations of Uncertainty." In William Price and Seymour Chrissick, eds., *The Uncertainty Principle and Foundations of Quantum Mechanics*, 3–6. New York: John Wiley.

Henley, Nancy (1977). *Body Politics: Power, Sex and Nonverbal Communication*. Englewood Cliffs, NJ: Prentice-Hall.

Herrnstein, Richard F. (1973). *IQ in the Meritocracy*. Boston: Little, Brown.

Herron, Jeannine (1980). *Neuropsychology of Lefthandedness*. New York: Academic Press.

Hershey, Marjorie R. (1977). "The Politics of Androgyny? Sex Roles and Attitudes toward Women in Politics." *American Politics Quarterly*, 5: 261–87.

Heymer, Armin (1978). *The Ethological Dictionary: In English, French, and German*. New York: Garland.

Hildebrand, James L. (1968). "Soviet International Law: An Exemplar for Optimal Decision Theory Analysis." *Case Western Reserve Law Review*, 20: 141–250.

Hill, Jack (1978). "The Origin of Sociocultural Evolution." *Journal of Social and Biological Structures*, 1: 377–86.

——— (1984). "Human Altruism and Sociocultural Fitness." *Journal of Social and Biological Structures*, 7: 17–35.

Hill, Sandra (1983). "The Grass Is Much Greener Now for this Peace Activist." United Press International dispatch (from Bonn. West Germany), in the *Sunday Star Bulletin and Advertiser* (Honolulu, HI), May 1, p. A–29.

Hinde, Robert A. (1966). *Animal Behavior: A Synthesis of Ethology and Comparative Psychology*. New York: McGraw-Hill.

——— (1970). *Animal Behavior: A Synthesis of Ethology and Comparative Psychology*. Rev. ed. New York: McGraw-Hill.

——— , ed. (1972). *Non-verbal Communication*. Cambridge: Cambridge University Press.

——— (1974). *Biological Bases of Human Social Behavior*. New York: McGraw-Hill.

Hines, Samuel M., Jr. (1979). "Evolutionary Epistemology and Political Knowledge." In Maria Falco, ed., *Through the Looking Glass: Epistemology and the Study of Political Inquiry*, 329–65. Washington, DC: University Press of America.

——— (1982a). "Biopolitics and the Evolution of Inquiry in Political Science." *Politics and Life Sciences*, 1: 5–16.

——— (1982b). "Ordering Political Space: A Biobehavioral Approach with Special Reference to the Work of Jean Laponce." Paper presented at the Meeting of the Western Political Science Association.

Hirsch, Harry N. (1981). *The Enigma of Felix Frankfurter*. New York: Basic Books.

Hofer, Myron A. (1981). *The Roots of Human Behavior: An Introduction to the Psychobiology of Early Development*. San Francisco: Freeman.

Hofstadter, Richard (1944). *Social Darwinism in American Thought*. Philadelphia: University of Pennsylvania Press.

Hopple, Gerald (1980). *Political Psychology and Biopolitics: Assessing and Predicting Elite Behavior in Foreign Policy Crises*. Boulder, CO: Westview.

Horowitz, Irving Louis (1979). "Paradigms of Political Psychology." *Political Psychology*, 1: 99–103.

Hrdy, Sarah Blaffer (1974). "Male-Female Competition and Infanticide among the Langurs (*Presbytis Entellus*) of Abu, Rajasthan." *Folia Primatologica*, 22: 19–58.

——— (1977a). "Infanticide as a Primate Reproductive Strategy." *American Scientist*, 65: 40–49.

——— (1977b). *The Langurs of Abu*. Cambridge, MA: Harvard University Press.

——— (1979). "Infanticide among Animals: A Review, Classification, and Examination of the Implications for the Reproductive Strategies of Females." *Ethology and Sociobiology*, 1: 13–40.

——— (1981). *The Female That Never Evolved*. Cambridge, MA: Harvard University Press.

Hubbard, Ruth (1981). "The Emperor Doesn't Wear Clothes: The Impact of Feminism on Biology." In Dale Spender, ed., *Men's Studies Modified*, 213–35. New York: Oxford University Press.

Hull, David L. (1978). "Altruism in Science: A Sociobiological Model of Co-operative Behavior among Scientists." *Animal Behavior*, 26: 685–97.

Hummel, Ralph (1973). "The Psychology of Charismatic Leaders." Paper presented at the World Congress of the International Political Science Association. Montreal.

——— (1974). "Freud's Totem Theory as Complement to Max Weber's Theory of Charisma." *Psychological Reports*, 35: 683–66.

——— (1975). "Psychology of Charismatic Followers." *Psychological Reports*, 37: 759–70.

Hummel, Ralph, and Robert Isaak (1980). *Politics for Human Beings*. N. Scituate, MA: Duxbury.

Hutt, Corinne (1966). "Exploration and Play in Children." *Symposia of the Zoological Society of London*, 18: 61–81.

Huxley, Julian S. (1942). *Evolution: The Modern Synthesis*. London: Allen and Unwin.

Ingber, Lester (1981). "Toward a Unified Brain Theory." *Journal of Social and Biological Structures*, 4: 211–24.

Inter-University Consortium for Political and Social Research (1979). *Guide to Resources and Services, 1979–1980*. Ann Arbor: Institute for Social Research, University of Michigan.

Itani, Junichiro (n. d.). *Intraspecific Killing among Nonhuman Primates*. Kyoto University Laboratory of Human Evolution Studies.

Itani, Junichiro, and A. Nishimura (1973). "The Study of Infrahuman Culture in Japan: A Review." In E. W. Menzel, ed., *Precultural Primate Behavior*. Basel: S. Karger.

Iverson, Leslie L. (1979). "The Chemistry of the Brain." *Scientific American*, 241: 146, 148.

Izard, Carroll (1980). "The Emergence of Emotions and the Development of Consciousness in Infancy." In Julian Davidson and Richard Davidson, eds., *The Psychobiology of Consciousness*. New York: Plenum.

Jahn, Robert G., ed. (1981). *The Role of Consciousness in the Physical World*. Boulder, CO: Westview.

Jantsch, Erich, and Conrad H. Waddington, eds. (1976). *Evolution and Consciousness: Human Systems in Transition*. Reading, MA: Addison-Wesley.

Jaros, Dean (1972). "Biochemical Desocialization: Depressants and Political Behavior." *Midwest Journal of Political Science*, 16: 1–28.

——— (1973). *Socialization to Politics*. New York: Praeger.

Jaros, Dean, and Lawrence V. Grant (1974). *Political Behavior: Choices and Perspectives*. New York: St. Martin's Press.

Jastrow, Robert, and Malcolm H. Thompson (1974). *Astronomy: Fundamentals and Frontiers*. 2d ed. New York: Wiley.

Jay, Antony (1971). *Corporation Man: Who He Is, What He Does, Why His Ancient Tribal Impulses Dominate the Life of the Modern Corporation*. New York: Random House.

Jaynes, Julian (1976). *The Origin of Consciousness in the Breakdown of the Bicameral Mind*. Boston: Houghton Mifflin.

Jensen, Arthur R. (1980). "Bias in Mental Testing." *Behavioral and Brain Sciences*, 3: 325–33.

——— (1985). "The Nature of the Black-White Difference on Various Psychometric Tests: Spearman's Hypothesis." *Behavioral and Brain Sciences*, 8: 193–219.

Jerison, Harry J. (1977). "Evolution of the Brain." In M. C. Whittrock, ed., *The Human Brain*. Englewood Cliffs, NJ: Prentice-Hall.

John, E. Roy (1980). "Multipotentiality: A Statistical Theory of Brain Function: Evidence and Implications." In J. Davidson and R. Davidson, eds., *The Psychobiology of Consciousness*. New York: Plenum.

Johnson, Ronald C., and Craig T. Nagoshi (1985). "Do We Know Enough about g to Be Able to Speak of Black-White Differences." *Behavioral and Brain Sciences*, 8: 232–33.

Jones, Alexander, ed. (1968). *The Jerusalem Bible: Reader's Edition.* Garden City, NY: Doubleday.

Jones, Diane C. (1983). "Power Structures and Perceptions of Power Holder in Same-Sex Groups of Young Children." *Women and Politics*, 4: 147–64.

"Judicial Disability and the Good Behavior Clause" (1976). Anonymous. *Yale Law Journal*, 85: 706–20.

Kaplan, Abraham (1964). *The Conduct of Inquiry: Methodology for Behavioral Science.* San Francisco: Chandler.

Kapp, William (1950). *The Public Costs of Private Enterprise.* Cambridge, MA: Harvard University Press.

Karl, Barry D. (1974). *Charles E. Merriam and the Study of Politics.* Chicago: University of Chicago Press.

Katchadourian, Herant A., ed. (1979). *Human Sexuality: A Comparative and Developmental Perspective.* Berkeley and Los Angeles: University of California Press.

Katz, Neil H., and David C. List (1981). "Seabrook: A Profile of Anti-nuclear Activists, June 1978." *Peace and Change*, 7: 59–68.

Kaufman, Herbert (1985). *Time, Chance, and Organizations: Natural Selection in a Perilous Environment.* Chatham, NJ: Chatham House.

Kavanagh, Dennis (1983). *Political Science and Political Behavior.* London: George Allen and Unwin.

Kawanaka, K. (1981). "Infanticide and Cannibalism in Chimpanzees, with Special Reference to the Newly Observed Case in the Mahale Mountains." *African Studies Monographs*, 1: 69–99.

Keith, Arthur (1949). *A New Theory of Human Evolution.* New York: Philosophical Library.

Kent, Ernest W. (1981). *The Brain of Men and Machines.* Peterborough, NH: Byte/McGraw-Hill.

Key, Mary R., ed. (1979). *The Relationship of Verbal and Nonverbal Communication.* The Hague: Mouton.

Key, Mary R., and D. Preziosi (1982). *Nonverbal Communication Today: Current Research.* The Hague: Mouton.

Kimble, Daniel P. (1973). *Psychology as a Biological Science.* Pacific Palisades, CA: Goodyear.

Kimura, M., and T. Ohta (1971). "On the Role of Molecular Evolution." *Journal of Molecular Evolution*, 1: 1–17.

King, Glen E. (1976). "Society and Territory in Human Evolution." *Journal of Human Evolution*, 5: 323–32.

——— (1980). "Alternative Uses of Primates and Carnivores in the Reconstruction of Early Hominid Behavior." *Ethology and Sociobiology*, 1: 99–109.

King, William L. (1950). *Melville Weston Fuller: Chief Justice of the United States, 1888–1910.* Chicago: University of Chicago Press.

Kinsbourne, Marcel (1978). "Evolution of Language in Relation to Lateral Action." In M. Kinsbourne, ed., *Asymmetrical Functions of the Brain.* New York: Cambridge University Press.

Kirkpatrick, Samuel A., and William Lyons (1976). "Age-Related Effects and Contending

Impacts on the Vote: A Cohort Analysis of US Presidential Election." Paper presented at the Tenth World Congress of the International Political Science Association. Edinburgh.

Kitchin, William (1982a). "Hemispheric Lateralization and Political Communication." Paper presented at the Annual Meeting of the American Political Science Association.

——— (1982b). "The Split Brain and Presidential Behavior: The Cognitive Styles of President Reagan." Paper presented at the Meeting of the Northeastern Political Science Association.

Klinghammer, Erich, ed. (1978). *The Wolf: Behavior and Ecology.* New York: Garland.

Kocel, Katherine (1980). "Age-Related Changes in Cognitive Abilities and Hemispheric Specialization." In Jeannie Herron, eds., *Neuropsychology of Left-Handedness.* New York: Academic Press.

Konner, Melvin (1982). *The Tangled Wing: Biological Constraints on the Human Spirit.* New York: Holt, Rinehart and Winston.

Kort, Fred (1983). "An Evolutionary-Neurobiological Explanation of Political Behavior and the Lumsden-Wilson 'Thousand Year Rule.' " *Journal of Social and Biological Structures,* 6: 219–30.

——— (1986). "Considerations for a Biological Basis of Civil Rights and Liberties." *Journal of Social and Biological Structures,* 9: 37–52.

Kortlandt, Adriaan (1972). *New Perspectives on Ape and Human Evolution.* Amsterdam: Department of Animal Psychology and Ethology, University of Amsterdam.

Krebs, J. R., and N. B. Davies, eds. (1978). *Behavioural Ecology: An Evolutionary Approach.* Sunderland, MA: Sinauer.

Kuhn, Thomas S. (1970). *The Structure of Scientific Revolution.* 2d ed.; 1st ed. 1962. Chicago: University of Chicago Press.

Kurten, Bjorn (1980). *Dance of the Tiger: A Novel of the Ice Age.* New York: Pantheon.

——— (1986). *Singletusk.* New York: Pantheon.

Kurth, G. (1976). "Neencephalization, Hominization, and Behavior." *Journal of Human Behavior,* 5: 501–9.

Lamb, Charles M. (1976). "Exploring the Conservatism of Federal Appeals Court Judges." *Indiana Law Journal,* 51: 257–79.

Lancaster, Jane B. (1975). *Primate Behavior and the Emergence of Human Culture.* New York: Holt, Rinehart and Winston.

Landau, Martin (1961). "On the Use of Metaphor in Political Analysis." *Social Research,* 28: 331–53.

——— (1965). "Due Process of Inquiry." *American Behavioral Scientist,* 9/2: 4–10.

——— (1968). "On the Use of Functional Analysis in American Political Science." *Social Research,* 35: 48–75.

——— (1972). *Political Theory and Political Science.* New York: Macmillan.

Lane, Robert (1962). *Political Ideology: Why the American Common Man Believes What he Does.* New York: Free Press of Glencoe.

Laponce, Jean A. (1970). "Dieu: à droite ou à gauche?" *Canadian Journal of Political Science* 3: 257–74.

——— (1972a). "In Search of the Stable Elements of the Left-Right Landscape." *Comparative Politics,* 4: 455–75.

——— (1972b). "The Use of Visual Space to Measure Ideology." In J. Laponce and Paul Smoker, eds., *Experimentation and Simulation in Political Science*. Toronto: University of Toronto Press.

——— (1974). "Of Gods, Devils, Monsters, and One-Eyed Variables." *Canadian Journal of Political Science*, 7: 199–209.

——— (1975). "Spatial Archetypes and Political Perception." *American Political Science Review*, 69: 11–20.

——— (1976). "The Left-Hander and Politics." In Albert Somit, ed., *Biology and Politics: Recent Explorations*, 45–57. Paris: Mouton.

——— (1978). "Relating Biological, Physical and Political Phenomena: The Case of Up and Down." *Social Science Information*, 17: 385–97.

——— (1981). *Left and Right: The Topography of Political Perceptions*. Toronto: University of Toronto Press.

——— (1987). "Relating Physiological, Physical, and Political Phenomena: Center and Centrality." *International Political Science Review*, 8: 175–82.

Larsen, Ray (1976). "Charisma: A Reinterpretation." In Michael R. A. Chance and Ray Larsen, eds., *The Social Structure of Attention*, 253–72. New York: Wiley.

Lasswell, Harold D. (1930). *Psychopathology and Politics*. Chicago: University of Chicago Press.

——— (1936). *Who Gets What, When, How*. New York: McGraw-Hill.

——— (1948). *Power and Personality*. New York: Norton.

——— (1963). *The Future of Political Science*. New York: Atherton.

Laughlin, Charles D., Jr., and Ivan Brady, eds. (1978). *Extinction and Survival in Human Populations*. New York: Columbia University Press.

Laughlin, Charles D., Jr., and Eugene G. d'Aquili (1974). *Biogenetic Structuralism*. New York: Columbia University Press.

Layzer, David (1978). "Altruism and Natural Selection." *Journal of Social and Biological Structures*, 1: 297–305.

Leavitt, Donald (1974). "Changing Issues, Ideological and Political Influences on the US Supreme Court, 1893–1945." Paper presented at the 70th Annual Meeting of the American Political Science Association. Chicago.

Lee, Richard B. (1972). "!Kung Spatial Organization: An Ecological and Historical Perspective." *Human Ecology*, 1: 125–47.

Lee, Richard B., and Irwin DeVore, eds. (1968). *Man the Hunter*. Chicago: Aldine.

——— , eds. (1976). *Kalahari Hunter-Gatherers: Studies of the !Kung San and Their Neighbors*. Cambridge, MA: Harvard University Press.

Lehman, Harvey C. (1953). *Age and Achievement*. Princeton: Princeton University Press.

Lehner, Philip (1979). *Handbook of Ethological Methods*. New York: Garland.

LeMay, M. (1977). "Asymmetries of the Skill and Handedness." *Journal of the Neurological Sciences*, 32: 243–53.

Lenneberg, Eric H. (1967). *Biological Foundations of Language*. New York: Wiley.

Lenski, Gerhard (1970). *Human Societies: A Macrolevel Introduction to Sociology*. New York: McGraw-Hill.

——— (1972). "A Letter to the Editor." *Contemporary Sociology*, 1: 306.

Lerner, Gerda (1979). *The Majority Finds Its Past*. New York: Oxford University Press.

Lerner, I. Michael (1968). *Heredity, Evolution, and Society*. San Francisco: W. H. Freeman.

Leventhal, Howard, and Elizabeth Sharp (1965). "Facial Expressions as Indicators of Distress." In Carroll E. Izard and Silvan S. Tomkins, eds., *Affect, Cognition, and Personality: Empirical Studies*. New York: Springer.

Levine, Adeline G. (1982). *Love Canal: Science, Politics and People*. Lexington, MA: Lexington Books.

Levins, R. (1970). "Extinction." In M. Gerstenhaber, ed., *Some Mathematical Questions in Biology*, 77–107. Providence, RI: American Mathematical Society.

Levinson, Boris M. (1980). "The Child and His Pet: A World of Non-Verbal Communication." In S. Corson and E.O. Corson, eds., *Ethology and Nonverbal Communication in Mental Health*. New York: Pergamon.

Levy, Jerre (1980). "Varieties of Human Brain Organization and the Human Social System." *Zygon*, 15: 351–75.

Levy, Jerre, and J. M. Levy (1978). "Human Lateralization from Head to Foot: Sex-Related Factors." *Science*, 200: 1291–92.

Levy, Jerre, and Ruben Gur (1980). "Individual Differences in Psycho-neurological Organization." In J. Herron, ed., *Neuropsychology of Left-Handedness*. New York: Academic Press.

Lewin, Kurt (1951). *Field Theory in Social Science*. New York: Harper.

Lewin, Roger (1987). *Bones of Contention: Controversies in the Search for Human Origins*. New York: Simon and Schuster.

Lewontin, Richard C. (1974). *The Genetic Basis of Evolutionary Change*. New York: Columbia University Press.

——— (1977). "Biological Determinism as a Social Weapon." In Ann Arbor Science for the People Editorial Collective, ed., *Biology as a Social Weapon*, 6–20. Minneapolis: Burgess.

——— (1978). "Adaptation," *Scientific American*, 239: 157–69.

Lewontin, Richard C., S. Rose, and L. J. Kamin (1984). *Not in Our Gene: Biology, Ideology and Human Nature*. New York: Pantheon.

Lex, Barbara (1978). "Neurological Bases of Revitalization Movements," *Zygon*, 13: 276–312.

Libet, Benjamin (1985). "Unconscious Cerebral Initiative and the Role of Conscious Will in Voluntary Action." *Behavioral and Brain Sciences*, 8: 529–39.

Lippitt, Ronald (1968). "Lewin, Kurt." *International Encyclopedia of the Social Sciences*, 9: 266–70.

Livingston, Robert B. (1978). *Sensory Processing, Perception, and Behavior*. New York: Raven Press.

——— (1980). "Frontiers of Neurosciences." Lectures at National Science Foundation/Chautauqua Course, Oregon Graduate Center. Beaverton, OR.

——— (1981). "Frontiers of Neurosciences." Lectures at National Science Foundation/Chautauqua Course, Oregon Graduate Center. Beaverton, OR.

Lockard, Joan S., ed. (1980). *The Evolution of Human Social Behavior*. New York: Elsevier.

Lockard, Joan S., et al. (1978). "Human Postural Signals: Stance, Weight-Shifts and Social Distance as Intention Movements to Depart." *Animal Behaviour*, 26: 219–24.

Lockwood, Randall (1979). "Dominance in Wolves: Useful Construct or Bad Habit?" in E. Klinghammer, ed., *The Behavior and Ecology of Wolves*, 225–44. New York: Garland.

Lodge, Milton, and John Wahlke (1982). "Politics, Apoliticals, and the Political Information." *International Political Science Review*, 3: 130–50.

Lorenz, Konrad (1966). *On Aggression.* New York: Harcourt Brace.

Losco, Joseph and Donna Day Baird (1982). "The Impact of Sociobiology on Political Science." *American Behavioral Scientist*, 25: 335–60.

Lovett-Doust, L., and J. Lovett-Doust (1985). "Gender Chauvinism and the Division of Labor in Humans." *Perspectives in Biology and Medicine*, 28: 526–42.

Loy, James (1975). "The Descent of Dominance in Macaca: Insights into the Structure of Human Societies." In H. Tuttle, ed., *Socioecology and Psychology of Primates*, 153–80 The Hague: Mouton.

Lumsden, Charles J. (1983). "Cultural Evolution and the Development of Tabula Rasa." *Journal of Social and Biological Structures*, 6: 101–14.

Lumsden, Charles J., and Edward O. Wilson (1981). *Genes, Mind, and Culture: The Coevolutionary Process.* Cambridge, MA: Harvard University Press.

——— (1982a). "Precis of Genes, Mind, and Culture." *Behavioral and Brain Sciences*, 5: 1–17.

——— (1982b). "Authors' Response." *Behavioral and Brain Sciences*, 5: 31–35.

——— (1983). *Promethean Fire: Reflections on the Origin of Mind.* Cambridge: Harvard University Press.

——— (1984). "Promethean Fire: Reflections on the Origin of Mind." *Politics and Life Sciences*, 2: 213.

——— (1985). "The Relation between Biological and Cultural Evolution." *Journal of Social and Biological Sciences* 8: 343–59.

Lyons, J. (1972). "Human Language." In Robert A. Hinde, ed., *Non-verbal Communication.* New York: Cambridge University Press.

MacArthur, Robert H. (1972). *Geographical Ecology: Patterns in the Distributions of Species.* New York: Harper and Row.

MacArthur, Robert H., and Edward O. Wilson (1967). *The Theory of Island Biogeography.* Princeton: Princeton University Press.

Maccoby, Eleanor E., ed. (1966). *The Development of Sex Differences.* Stanford: Stanford University Press.

MacLean, Paul D. (1958). "The Limbic System with Respect to Self-Preservation and the Preservation of the Species." *Journal of Nervous Mental Disease*, 127: 1–11.

——— (1973). "A Triune Concept of the Brain and Behavior." In T. Boag and D. Campbell, eds., *The Hicks Memorial Lectures*, 6–66. Toronto: University of Toronto Press.

Macphail, Eulan M. (1985). "Comparative Studies of Animal Intelligence: Is Spearman's *g* Really Hull's *D*?" *Behavioral and Brain Sciences*, 8: 234–35.

Madsen, Douglas (1985). "A Biochemical Property Relating to Power-Seeking in Humans." *American Political Science Review*, 79: 448–57.

——— (1986). "Power Seekers Are Different: Further Biochemical Evidence." *American Political Science Review*, 80: 261–69.

Mainardi, Danilo (1980). "Tradition and the Social Transmission of Behavior in Animals," in G. W. Barlow and J. Silverberg, eds., *Sociobology: Beyond Nature/Nurture?* 227–55. Boulder, CO: Westview.

Manheim, Jarol B. (1982). *The Politics Within: A Primer in Political Attitudes and Behavior.* 2d ed. New York: Longman.

Manicas, Peter T. (1982). "The Human Sciences: A Radical Separation of Psychology and the

Social Science." In Paul F. Secord, ed., *Explaining Human Behavior: Consciousness, Human Action, and Social Structure*, 155–73. Beverly Hills: Sage.
——— (1983). "Reduction, Epigenesis, and Explanation." *Journal of the Theory of Social Behavior*, 13: 331–54.
Mann, Thomas (1938). "The Blood of the Walsungs." In *Stories of Three Decades*, 297–319. New York: Knopf.
Marais, Eugene N. (1967 [1934]). *The Soul of the Ape*. New York: Atheneum.
——— (1937). *The Soul of the White Ant*. New York: Dodd Mead.
Margenau, Henry (1950). *The Nature of Physical Reality*. New York: McGraw-Hill.
Markl, Hubert (1982). "The Power of Reduction and the Limits of Compressibility." *Behavioral and Brain Sciences*, 5: 18–19.
Marler, Peter and William J. Hamilton, III (1966). *Mechanisms of Animal Behavior*. New York: Wiley.
Marshack, Alexander (1972). *The Roots of Civilization*. New York: McGraw-Hill.
Martin, Paul S., and R. G. Klein, eds. (1984). *Quarternary Extinctions: A Prehistoric Revolution*. Tucson, AZ: University of Arizona Press.
Maruyama, Magoroh (1980a). "Summary of Tsunoda's Seven Experimental Methods." *Journal of Social and Biological Structures*, 3: 267–71.
——— (1980b). "Comments on Tsunoda's Book." *Journal of Social and Biological Structures*, 3: 273–76.
Marx, Karl (1970). "A Contribution to the Critique of Hegel's "Philosophy of Right" Introduction." In Joseph O'Malley, ed., *Critique of Hegel's "Philosophy of Right."* Cambridge: Cambridge University Press.
Maslow, Abraham H. (1954). *Motivation and Personality*. New York: Harper and Row.
Mason, William A. (1965). "The Social Development of Apes and Monkeys." In I. DeVore, ed., *Primate Behavior: Field Studies of Monkeys and Apes*. New York: Holt, Rinehart and Winston.
Masters, Roger D. (1967). "La Redécouverte de la Nature Humaine." *Critique*, 245: 857–76.
——— (1968). *The Political Philosophy of Rousseau*. Princeton: Princeton University Press.
——— (1970). "Genes, Language, and Evolution," *Semiotica*, 2: 295–320.
——— (1973a). "Functional Approaches to Analogical Comparisons between Species." *Social Science Information*, 12/4: 7–28.
——— (1973b). "On Comparing Humans—and Human Politics—with Animal Behavior." Paper presented to the Ninth World Congress of the International Political Science Association. Montreal.
——— (1974). "Ethological Comparisons between Animal and Social Behavior." Paper presented at the Conference on the Relations between Biological and Social Theory, American Academy of Art and Sciences. Brookline, MA.
——— (1975). "Political Behavior as a Biological Phenomenon." *Social Science Information*, 14: 7–63.
——— (1976). "The Impact of Ethology on Political Science." In Albert Somit, ed., *Biology and Politics: Recent Explorations*, 197–233. Paris: Mouton.
——— (1977). "Human Nature, Nature, and Political Thought." In J. Roland Pennock and John W. Chapman, eds., *Human Nature in Politics*, 69–110. New York: New York University Press.

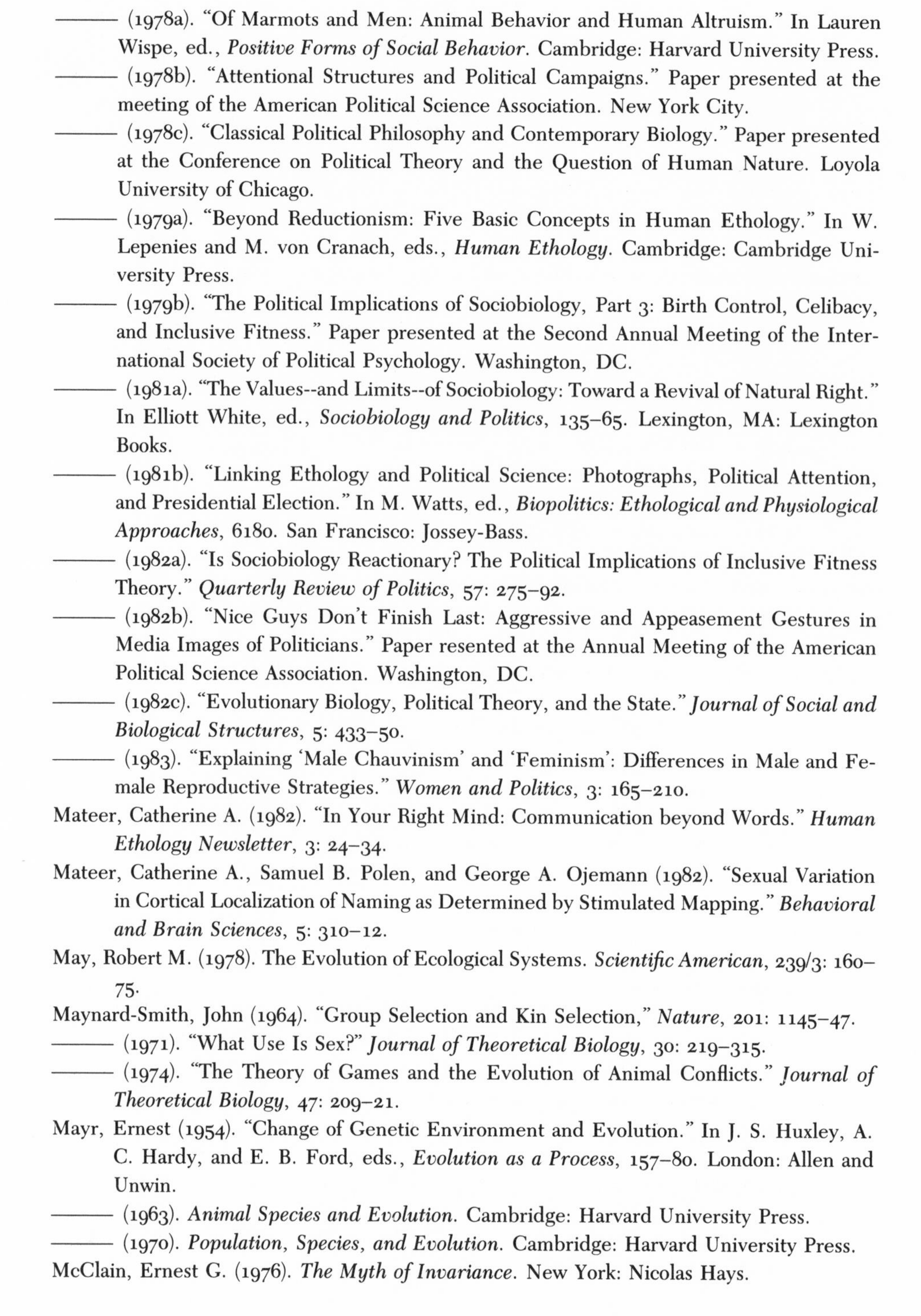

——— (1978a). "Of Marmots and Men: Animal Behavior and Human Altruism." In Lauren Wispe, ed., *Positive Forms of Social Behavior*. Cambridge: Harvard University Press.

——— (1978b). "Attentional Structures and Political Campaigns." Paper presented at the meeting of the American Political Science Association. New York City.

——— (1978c). "Classical Political Philosophy and Contemporary Biology." Paper presented at the Conference on Political Theory and the Question of Human Nature. Loyola University of Chicago.

——— (1979a). "Beyond Reductionism: Five Basic Concepts in Human Ethology." In W. Lepenies and M. von Cranach, eds., *Human Ethology*. Cambridge: Cambridge University Press.

——— (1979b). "The Political Implications of Sociobiology, Part 3: Birth Control, Celibacy, and Inclusive Fitness." Paper presented at the Second Annual Meeting of the International Society of Political Psychology. Washington, DC.

——— (1981a). "The Values--and Limits--of Sociobiology: Toward a Revival of Natural Right." In Elliott White, ed., *Sociobiology and Politics*, 135–65. Lexington, MA: Lexington Books.

——— (1981b). "Linking Ethology and Political Science: Photographs, Political Attention, and Presidential Election." In M. Watts, ed., *Biopolitics: Ethological and Physiological Approaches*, 6180. San Francisco: Jossey-Bass.

——— (1982a). "Is Sociobiology Reactionary? The Political Implications of Inclusive Fitness Theory." *Quarterly Review of Politics*, 57: 275–92.

——— (1982b). "Nice Guys Don't Finish Last: Aggressive and Appeasement Gestures in Media Images of Politicians." Paper resented at the Annual Meeting of the American Political Science Association. Washington, DC.

——— (1982c). "Evolutionary Biology, Political Theory, and the State." *Journal of Social and Biological Structures*, 5: 433–50.

——— (1983). "Explaining 'Male Chauvinism' and 'Feminism': Differences in Male and Female Reproductive Strategies." *Women and Politics*, 3: 165–210.

Mateer, Catherine A. (1982). "In Your Right Mind: Communication beyond Words." *Human Ethology Newsletter*, 3: 24–34.

Mateer, Catherine A., Samuel B. Polen, and George A. Ojemann (1982). "Sexual Variation in Cortical Localization of Naming as Determined by Stimulated Mapping." *Behavioral and Brain Sciences*, 5: 310–12.

May, Robert M. (1978). The Evolution of Ecological Systems. *Scientific American*, 239/3: 160–75.

Maynard-Smith, John (1964). "Group Selection and Kin Selection," *Nature*, 201: 1145–47.

——— (1971). "What Use Is Sex?" *Journal of Theoretical Biology*, 30: 219–315.

——— (1974). "The Theory of Games and the Evolution of Animal Conflicts." *Journal of Theoretical Biology*, 47: 209–21.

Mayr, Ernest (1954). "Change of Genetic Environment and Evolution." In J. S. Huxley, A. C. Hardy, and E. B. Ford, eds., *Evolution as a Process*, 157–80. London: Allen and Unwin.

——— (1963). *Animal Species and Evolution*. Cambridge: Harvard University Press.

——— (1970). *Population, Species, and Evolution*. Cambridge: Harvard University Press.

McClain, Ernest G. (1976). *The Myth of Invariance*. New York: Nicolas Hays.

——— (1982). "Structure in the Ancient Wisdom Literature: The Holy Mountain." *Journal of Social and Biological Structures*, 5: 233–48.

McClelland, Charles (1971). "Field Theory and System Theory in International Relations." In Albert Lepawsky, Edward Beuhrig, and Harold D. Lasswell, eds., *The Search for World Order*. New York: Appleton-Century-Crofts.

McClosky, Herbert (1958). "Conservatism and Personality." *American Political Science Review*, 52: 27–45.

McCoy, Charles A., and John Playford, eds. (1967). *Apolitical Parties: A Critique of Behavioralism*. New York: Crowell.

McEacheron, Donald L. (1984). "Hypothesis and Explanation in Human Evolution." *Journal of Social and Biological Structures*, 7: 9–15.

McEacheron, Donald L., and Darius Baer (1982). "A Review of Selected Sociobiological Principles: The Effects of Intergroup Conflict." *Journal of Social and Biological Structures*, 5: 121–39.

McGlone, Jeannette (1980). "Sex Differences in Human Brain Asymmetry: A Critical Survey." *Behavioral and Brain Sciences*, 3: 215–27.

McGrew, William C. (1972). *An Ethological Study of Children's Behavior*. London: Academic Press.

——— (1977). "Socialization and Object Manipulation of Wild Chimpanzees." In S. Chevalier-Skolnikoff and F. E. Poirier, eds., *Primate Bio-Social Development*. New York: Garland.

McGuigan, F. J. (1978). "Imagery and Thinking: Covert Functioning of the Motor System." In Gary Schwartz and David Shapiro, eds., *Consciousness and Self-Regulation*, vol. 2. New York: Plenum.

McGuinness, Diane (1979). "Was Darwin Conscious of His Mother?" In *The Re-evaluation of Existing Values and the Search for Absolute Value*, 725–35. New York: Intercultural Foundation.

——— (1980). "Strategies, Demands, and Lateralized Sex Differences." *Behavioral and Brain Sciences*, 3: 244.

——— (1981). "Auditory and Motor Aspects of Language Development in Males and Females." In A. Ansara, et al., eds., *The Significance of Sexual Differences in Dyslexia*. Towson: The Orton Society.

McGuinness, Diane, and Karl Pribram (1978). "The Origins of Sensory Bias in the Development of Gender Differences in Perception and Cognition." In M. Bortner, ed., *Cognitive Growth and Development: Essays in Memory of Herbert G. Birch*, 3–56. New York: Bruner/Mazel.

McGuire, Michael T. (1982). "Social Dominance Relationships in Male Vervet Monkeys: A Possible Model for the Study of Dominance Relationships in Human Political Systems." *International Political Science Review*, 3: 11–32.

McKelvey, Richard D., and Peter C. Ordeshook (1976). "Symmetric Spatial Games without Majority Rule Equilibria." *American Political Science Review*, 70: 1172–84.

Means, R. (1967). "Sociology, Biology, and the Analysis of Social Problems." *Eugenics Quarterly*, 15: 200–12.

Mecacci, Luciano (1981). "Brain and Socio-Cultural Environment." *Journal of Social and Biological Structures*, 4: 319–27.

Medawar, Peter B. (1976). "Does Ethology Throw any Light on Human Behaviour?" In P. P.

G. Bateson and R. A. Hinde, eds., *Growing Points in Ethology*. New York: Cambridge University Press.

——— (1981). "Stretch Genes." *New York Review of Books*, Jul. 16: 45–48.

Menzel, Emil W. (1972). "Spontaneous Invention of Ladders in a Group of Young Chimpanzees." *Folia Primatologica*, 17: 87–106.

——— (1973). "Leadership and Communication in Young Chimpanzees." In E. W. Menzel, ed., *Precultural Primate Behavior*. Basel: S. Karger.

——— (1978). "Implications of Chimpanzee Language-Training Experiments for Primate Field Research--and Vice Versa." In D. J. Chivers and J. Hebert, eds., *Recent Advances in Primatology: Vol. 1, Behavior*. London: Academic Press.

Menzel, Emil W., and N. K. Johnson (1976). "Communication and Cognitive Organization in Humans and Other Animals." In Steven R. Harnard, Holst D. Steklis, and Jane Lancaster, eds., *Origins and Evolution of Language and Speech: Annals of the New York Academy of Sciences*, 280: 131–42.

Merriam, Charles E. (1921). "The Present State of the Study of Politics," *American Political Science Review*, 15: 173–85.

——— (1925 [1970]). *New Aspects of Politics*. 3d ed. Chicago: University of Chicago Press.

Mersky, Roy M. (1978). *Louis Dembitz Brandeis, 1856–1941: A Biography* (New Haven: Yale Law School).

Messinger, Harley B., and Edward S. Rogers (1967). "Human Ecology: Toward a Holistic Method." *Milbank Memorial Fund Quarterly*, 45: 25–42.

Mey, Harold (1965). *Field-Theory: A Study of Its Applications in the Social Sciences*. New York: St. Martin's Press.

Miller, Eugene F. (1971). "David Easton's Political Theory." *Political Science Reviewer*, 1: 184–235.

Mirabile, C. S., Jr., G. C. Glueck, and C. F. Stroebel (1976). "Spatial Orientation, Cognitive Process and Cerebral Specialization." *Psychiatric Journal of the University of Ottawa*, 1: 99–104.

Mittwoch, Ursula (1978). "Charges in the Direction of the Lateral Growth Gradient in Human Development: Left to Right and Right to Left," *Behavioral and Brain Sciences*, 1: 306–07.

Monckeberg, Fernando (1976). "The Effect of Malnutrition on Physical Growth and Development." In James W. Prescott, Merrill S. Read, and David B. Coursin, eds., *Brain Function and Malnutrition*. New York: Wiley.

Monod, Jacques (1971). *Chance and Necessity: An Essay on the Natural Philosophy of Modern Biology*. New York: Knopf.

Moran, Greg, and John Fentress (1979). "A Search for Order in Wolf Social Behavior." In E. Klinghammer, ed., *Behavior and Ecology of Wolves*, 245–83. New York: Garland.

Morgan, Charles J. (1979). "Eskimo Hunting Groups, Social Kinship, and the Possibility of Kin Selection in Humans." *Ethology and Sociobiology*, 1: 83–86.

Morgan, Elaine (1984). *The Aquatic Ape*. NY: Stein and Day.

Morgan, Michael (1978). "Genetic Models of Asymmetry Should Be Asymmetrical." *Behavioral and Brain Sciences*, 1: 325–31.

Morgan, Michael J., and Michael C. Corballis (1978). "On the Biological Basis of Human Later-

ality: The Mechanisms of Inheritance." *Behavioral and Brain Sciences*, 1: 270–77.
Morowitz, Harold J. (1980). "Rediscovering the Mind." *Psychology Today*, 14: 12–18.
Morris, Desmond (1967). *The Naked Ape: A Zoologist's Study of the Human Animal*. New York: McGraw-Hill.
——— (1969). *The Human Zoo*. London: Jonathan Cape.
——— (1977). *Manwatching: A Field Guide to Human Behavior*. London: Jonathan Cape.
Morris, Desmond, Peter Collett, Peter Marsh, and Marie O'Shaughnessy (1979). *Gestures: Their Origins and Distributions*. London: Jonathan Cape.
Mortenson, F. Joseph (1975). *Animal Behavior: Theory and Research*. Monterey, CA: Brook/Cole.
Munro, William Bennett (1928). "Physics and Politics--An Old Analogy Revisited" *American Political Science Review*, 22: 1–11.
Murphy, Bruce Allen (1982). *The Brandeis/Frankfurter Connection: The Secret Political Activities of Two Supreme Court Justices*. New York, Oxford University Press.
Nash, John (1970). *Developmental Psychology: A Psychobiological Approach*. Englewood Cliffs, NJ: Prentice-Hall.
Newman, Oscar (1972). *Defensible Space: Crime Prevention through Urban Design*. New York: Macmillan.
Newman, Robert C., and Herman J. Eckelmann, Jr. (1977). *Genesis One and the Origin of the Earth*. Downers Grove, IL: Inter-Varsity Press.
Nishida, Toshisada (1980). "On Inter-Unit-Group Aggression and Intra-Group Cannibalism among Wild Chimpanzees." *Human Ethology Newsletter*, 31: 21–24.
Nitecki, Matthew H., ed. (1984). *Extinction*. Chicago: University of Chicago Press.
Noe, Ronald, Frans B. M. de Waal, and Jan A. R. A. M. van Hooff (1980). "Types of Dominance in a Chimpanzee Colony." *Folia Primatologica*, 34: 90–110.
Nye, Russel B. (1951). *Midwestern Progressive Politics: A Historical Study of Its Origins and Development, 1870–1950*. East Lansing: Michigan State College Press.
Odum, Eugene P. (1971). *Fundamentals of Ecology*. Philadephia: W. B. Saunders.
Odum, Howard T. (1971). *Environment, Power, and Society*. New York: Wiley.
O'Keefe, John, and Lynn Nadel (1979). "Precis of O'Keefe and Nadel's 'The Hippocampus as a Cognitive Map.'" *Behavioral and Brain Sciences*, 2: 487–533.
Olton, David S., James T. Becker, and Gail E. Handelmann (1977). "Hippocampus, Space, and Memory." *Behavioral and Brain Sciences*, 2: 313–65.
O'Malley, J. (ed.) (1970). *Critique of Hegel's "Philosophy of Right."* Cambridge: Cambridge University Press.
Omark, Donald R., and M. S. Edelman (1975). "A Comparison of Status Hierarchies in Young Children: An Ethological Approach." *Social Science Information*, 14: 87–107.
Omark, Donald R., M. L. Fiedler, and R. S. Marvin (1976). "Dominance Hierarchies: Observational Techniques Applied to the Study of Children at Play." *Instructional Science*, 5: 403–23.
Omark, Donald R., Fred F. Strayer, and Daniel G. Freedman (1980). *Dominance Relations: An Ethological View of Human Conflict and Social Interaction*. New York: Garland.
O'Neil, Robert M. (1982). "Creationism, Curriculum, and the Constitution." *Academe*, 68: 21–26.

Ornstein, Robert E. (1972). *The Psychology of Consciousness*. San Francisco: W. H. Freeman.
Orwell, George (1949). *Nineteen Eighty-Four*. New York: Harcourt, Brace.
Overton, William R. (1982). "Rev. Bill McLean vs. Arkansas Board of Education, *Academe*, 68(2): 27–36.
Pagels, Heinz R. (1982). *The Cosmic Code: Quantum Physics as the Language of Nature*. New York: Simon and Schuster.
Paige, Karen Ericksen, and Jeffery M. Paige (1981). *The Politics of Reproductive Ritual*. Berkeley and Los Angeles: University of California.
Panksepp, Jaak (1982). "Toward a General Psychobiological Theory of Emotions." *Behavioral and Brain Sciences*, 5: 407–67.
Parker, Sue Taylor, and Kathleen Rita Gibson (1979). "A Developmental Model of the Evolution of Language and Intelligence in Early Hominids." *Behavioral and Brain Sciences*, 2: 367–407.
Parsons, Jacquelynne, ed. (1980). *The Psychobiology of Sex Differences and Sex Roles*. New York: Hemisphere/McGraw-Hill.
Paxson, Frederic L. (1929). *Recent History of the United States, 1865–1929*. Boston: Houghton Mifflin.
Pelletier, Kenneth (1978). *Toward a Science of Consciousness*. New York: Delta.
Peters, Roger (1978a). "Scent-Marking in Wolves." In R. L. Hall and H. S. Sharp, eds., *Wolf and Man: Evolution in Parallel*, 133–47. New York: Academic Press.
——— (1978b). "Communication, Cognitive Mapping, and Strategy in Wolves and Hominids." In Roberta L. Hall and Henry S. Sharp, eds., *Wolf and Man: Evolution in Parallel*, 95–107. New York: Academic Press.
——— (1979). "Mental Maps in Wolf Territoriality." In Erich Klinghammer, ed., *The Behavior and Ecology of Wolves*, 119–54. New York: Garland.
Peters, Roger, and David Mech (1975). "Behavioral and Intellectual Adaptations of Selected Mammalian Predators to the Problem of Hunting Large Animals." In R. H. Tuttle, ed., *Socioecology and Psychology of Primates*, 279–300. The Hague: Mouton.
Peterson, Steven A. (1973). "The Effects of Physiological Variables upon Student Protest Behavior." Paper presented at the Ninth World Congress of the International Political Science Association Montreal.
——— (1979). "The Biopolitics of Political Behavior: Past and Future." Paper presented at the Second Annual Meeting of the International Society of Political Psychology. Washington, DC.
——— (1981a). "Cognitive Development, Biology, and Political Socialization." Paper presented at the Annual Meeting of the International Society of Political Psychology. Mannheim, Germany.
——— (1981b). "Sociobiology and Ideas-Become-Real: Case Study and Assessment." *Journal of Social and Biological Structures*, 4: 125–43.
——— (1982a). "The Brain and Hypostatizing." Paper presented at the Meeting of the International Society of Political Psychology.
——— (1982b). "Neurophysiology and Rationality in Political Thinking." Paper presented at the Annual Meeting of the American Politiical Science Association. Denver.
——— (1983a). "The Psychobiology of Hypostatizing." *Micropolitics*, 2: 423–51.
——— (1983b). "Biology and Political Socialization: A Cognitive Development Link?" *Political Psychology*, 4: 265–88.

——— (1985). "Neurophysiology, Cognition, and Political Thinking." *Political Psychology*, 6: 495–518.

——— (1986). "Why Policies Don't Work: A Biocognitive Perspective." In Elliott White and Joseph Losco (eds.), *Biology and Bureaucracy: Public Administration and Public Policy from the Perspective of Evolutionary, Genetic and Neurobiological Theory*, 447–502. Lanham, MD: University Press of America.

Peterson, Steven A., and Robert Lawson (1982a). "Cognitive Psychology and Predicting Political Attitudes and Behavior." Paper presented at the Meeting of the Midwest Political Science Association.

——— (1982b). "Cognitive Psychology and the Study of Politics." Paper presented at the Annual Meeting of the American Political Science Association. Denver.

Peterson, Steven A., and Albert Somit (1980). "Cost-Benefit Analysis, Shifting Coalitions, and Human Evolution." *Human Ethology Newsletter*, 31: 16–18.

——— (1982). "Cognitive Development and Childhood Political Socialization: Questions about the Primacy Principle." *American Behavioral Scientist*, 25: 313–33.

——— (1983a). "Biology and Political Violence: An Assessment." Paper presented at the meeting of the International Society of Political Psychology. Oxford, England.

——— (1983b). "Punctuated Equilibria, Sociobiology, and Politics." Paper presented the Southern Political Science Association. Birmingham, AL.

Peterson, Steven A., Albert Somit, and Barbara Brown (1983). "Biopolitics in 1982." *Politics and the Life Sciences*, 2: 76–80.

Peterson, Steven A., Albert Somit, and Robert Slagter.

——— (1982). "Biopolitics: 1980–1981 Update." *Politics and the Life Sciences*, 1: 52–57.

Pettersson, Max (1978). "Acceleration in Evolution before Human Times." *Journal of Social and Biological Structures*, 1: 201–6.

Pettman, Ralph (1975). *Human Behavior and World Politics*. New York: St. Martin's Press.

——— (1981). *Biopolitics and International Values*. New York: Pergamon.

Pfeiffer, John E. (1977). *Emergence of Society: A History of the Establishment*. New York: McGraw-Hill.

Piaget, Jean (1970). *Structuralism*. New York: Basic Books.

Piaget, Jean, and Barbara Inhelder (1969). *The Psychology of the Child*. New York: Basic Books.

Pilbeam, David (1972). *The Ascent of Man*. New York: Macmillan.

Pitcairn, Thomas K., and Margaret Schleidt (1976). "Dance and Decision: An Analysis of a Courtship Dance of the Medlpa, New Guinea." *Behavior*, 58: 298–316.

Plutchik, Robert (1980). *Emotion: A Psychoevolutionary Synthesis*. New York: Harper and Row.

——— (1982). "Genes, Mind, and Emotion." *Behavioral and Brain Sciences*, 5: 21–22.

Poirier, Frank E., A. Bellisari, and L. Haines (1977). "Functions of Primate Play Behavior." Paper presented at the Meeting of the Animal Behavior Society. Pennsylvania State University.

Poortinga, Ype H. (1985). "Empirical Evidence of Bias in Choice Reaction Time Experiments." *Behavioral and Brain Sciences*, 8: 236–37.

Pope, Kenneth, and Jerome Singer (1978). "Regulation of the Stream of Consciousness: Toward a Theory of Ongoing Thought." In Gary Schwartz and David Shapiro, eds., *Consciousness and Self-Regulation*. Vol. 2. New York: Plenum.

Popper, Karl (1965). *Of Clouds and Clocks.* St. Louis, MO: Washington University Press.

——— (1972). *Objective Knowledge.* Oxford: Oxford University Press.

——— (1977). *The Self and Its Brain.* New York: Springer-Verlag.

Porter, Joshua R., and W. M. S. Russell, eds. (1978). *Animals in Folklore.* Totowa, NJ: Rowman and Littlefield.

Pranger, Robert (1967). "Ethology and Politics: The Work of Konrad Lorentz." Paper presented at the Annual Meeting of the Southern Political Science Association. New Orleans.

Premack, David, and Guy Woodruff (1978). "Does the Chimpanzee Have a Theory of Mind?" *Behavioral and Brain Sciences*, 1: 515–26.

Presthus, Robert (1977). "Some Conditions of Comparative Analysis in Canada and the United States." In Robert Presthus, ed., *Crossnational Perspectives: United States and Canada.* Leiden: Brill.

Pribram, Karl H. (1971). *The Language of the Brain.* Englewood Cliffs, NJ: Prentice-Hall.

——— (1978). "Consciousness and Neurophysiology." *Federation Proceedings*, 37/9: 2271–74.

——— (1979). "Behaviourism, Phenomenology, and Holism in Psychology." *Journal of Social and Biological Structures*, 2: 65–72.

——— (1980a). "The Role of Analogy in Transcending Limits in the Brain Sciences." *Daedalus*, 109: 19–38.

——— (1980b). "Mind, Brain, and Consciousness: The Organization of Competence and Conduct." In J. Davidson and R. Davidson, eds., *The Psychobiology of Consciousness.* New York: Plenum.

Pribram, Karl, and Diane McGuinness (1975). "Arousal, Activation, and Effort in the Control of Attention." *Psychological Review*, 82: 116–49.

Pritchett, C. Herman (1948). *The Roosevelt Court: A Study in Judicial Politics and Values.* New York: Macmillan.

Pryor, Gordon (1975). "Malnutrition and the 'Critical Period' Hypothesis." In J. W. Prescott, M. S. Reed, and D. B. Coursin, eds., *Brain Function and Malnutrition.* New York: Wiley.

Puccetti, Roland, and Robert W. Dykes, eds. (1978). "Sensory Cortex and the Mind-Brain Problem." *Behavioral and Brain Sciences*, 1: 337–76.

Pulliam, H. Ronald, and Christopher Dunford (1980). *Programmed to Learn: An Essay on the Evolution of Culture.* New York: Columbia University Press.

Pyke, Graham H. (1978). "Are Animals Efficient Harvesters?" *Animal Behavior*, 26: 241–50.

Ra, Chong Phil (1986). "Toward a Biosociopolitical Transactional Model of Cognitive Psychology." Paper presented at the Annual Meeting of the American Political Science Association. Washington, DC.

Rabbitt, P. M. A. (1985). "Oh g Dr. Jensen, or g-ing up Cognitive Psychology?" *Behavioral and Brain Sciences*, 8: 238–39.

Rainier, H. S. H., III, Prince of Monaco, and Geoffrey H. Bourne, eds. (1977). *Primate Conservation.* New York: Academic Press.

Rajecki, D. W., Michael E. Lamb, and Pauline Obmascher (1978). "Toward a General Theory of Infantile Attachment: A Comparative Review of Aspects of the Social Bond." *Behavioral and Brain Sciences*, 1: 417–64.

Rao, B. S. Madhava (1962). *Principles of Relativity and Quantum Mechanics.* India: Annamalai University.

Rashevsky, Nicolas (1938). *Mathematical Biophysics*. Chicago: University of Chicago Press.
——— (1947). *Mathematical Theory of Human Relations*. Bloomington, IN: Principia Press.
——— (1951). *Mathematical Biology of Social Behavior*. Chicago: University of Chicago Press.
——— (1954). "Two Models: Imitative Behavior and Distribution of Status." In Paul F. Lazarsfeld, ed., *Mathematical Thinking in the Social Sciences*, 67–104. New York: The Free Press.
Raskin, Allen, and Lissy F. Jarvik, eds. (1979). *Psychiatric Symptoms and Cognitive Loss in the Elderly: Evolution and Assessment Techniques*. New York: Wiley.
Ray, James Lee (1982). "Understanding Rummel." *Journal of Conflict Resolution*, 26: 161–87.
Ray, R. M. (1978). "The Evolution of Ecological Systems." *Scientific American*, 239(3): 160–75.
Reed, Charles A. (1971). "Animal Domestication in the Prehistoric Near East." In Stuart Struever, ed., *Prehistoric Agriculture*, 423–50. Garden City, NY: National History Press.
Reichenbach, Hans (1956). *The Direction of Time*. Los Angeles: University of California Press.
——— (1958). *The Philosophy of Space and Time*. New York: Dover.
——— (1965). *The Theory of Relativity and A Priori Knowledge*. Los Angeles: University of California Press.
Restivo, Sal (1983). "Remarks on the Structure of Ancient Wisdom." *Journal of Social and Biological Structures*, 6: 169–71.
Reynolds, Peter (1981). *On the Evolution of Human Behavior: The Argument from Animals to Man*. Berkeley and Los Angeles: University of California Press.
Reynolds, Vernon (1976). *The Biology of Human Action*. San Francisco: W. H. Freeman.
——— (1984). "Primate Social Thinking." Paper presented at the Tenth Congress of the International Primatological Society. Nairobi.
Richerson, Peter J., and Robert Boyd (1978). "A Dual Inheritance Model of the Human Evolutionary Process: I. Basic Postulates and a Simple Model." *Journal of Scoial and Biological Structures*, 1: 127–54.
Riesman, David, N. Glazer, and R. Denny (1951). *The Lonely Crowd: A Study of the Changing American Character*. New Haven, CT: Yale University Press.
Riker, William H. (1962). *The Theory of Political Coalitions*. New Haven, CT: Yale University Press.
Riker, William H., and Peter C. Ordeshook (1973). *An Introduction to Positive Political Theory*. Englewood Cliffs, NJ: Prentice-Hall.
Riley, Matilda White, ed. (1979). *Aging from Birth to Death: Interdisciplinary Perspectives*. Boulder, CO: Westview.
Riley, Matilda White, Ronald P. Abeles, and Michael S. Teitelbaum, eds. (1982). *Aging from Birth to Death, Vol. 2. Sociotemporal Perspectives*. Boulder, CO: Westview.
Roberts, Morley (1938). *Bio-Politics: An Essay in the Physiology and Pathology and Politics of the Social and Somatic Organism*. London: J. M. Dent & Sons.
——— (1941). *The Behavior of Nations*. London: J. M. Dent & Sons.
Robertson, Alec (1977). *Biographical Memoirs of the Royal Society*. Vol. 23, 575–622. London: The Royal Society.
Robinson, John P., Jerrold G. Rusk, and Kendra G. Head (1968). *Measures of Political*

Attitudes. Ann Arbor: Survey Research Center, Institute for Social Research, University of Michigan.

——— (1969). *Measures of Social Psychological Attitudes.* Ann Arbor: Survey Research Center, Institute for Social Research, University of Michigan.

Rodman, John (1980). "Paradigm Change in Political Science: An Ethological Perspective." *American Behavioral Scientist*, 24: 49–78.

Roederer, Juan (1976). "Functions of the Human Brain: An Interdisciplinary Introduction to Neuropsychology for Students of Science and Humanities, Part I." Lecture notes for "Physics of the Brain." Department of Physics and Astronomy University of Denver, CO.

——— (1978). "On the Relationship between Human Brain Functions and the Foundations of Physics, Science, and Technology." *Foundations of Physics*, 8: 423–38.

——— (1979a). "Human Brain Functions and the Foundations of Science." *Endeavour, New Series*, 3: 99–103.

——— (1979b). "The Perception of Music by the Human Brain." *Humanities Association Review*, 10: 11–23.

——— (1979c). *Physics and Psychophysics of Music.* 2d ed. New York: Springer-Verlag.

Roell, A. (1978). "Social Behavior of the Jackdaw: *Corvus Monedula*, in Relation to Its Niche." *Behavior*, 64: 1–124.

Roelofs, Joan (1979). "The Supreme Court and Corporate Capitalism: An Iconoclastic View of the Warren Court in the Shadow of Critical Theory." *Telos*, 39: 94–112.

Rogers, Linda, Warren TenHouten, Charles D. Kaplan, and Martin Gardiner (1977). "Hemispheric Specialization of Language: An EEG Study of Bilingual Hopi Indian Children." *International Journal of Neuroscience*, 8: 1–6.

Roitblat, H. L. (1982). "The Meaning of Representation in Animal Memory." *Behavioral and Brain Sciences*, 5: 353–406.

Rokeach, Milton (1960). *The Open and Closed Mind.* New York: Basic Books.

——— (1968). *Beliefs, Attitudes, and Values.* San Francisco: Jossey-Bass.

Rosen, Robert (1980). "Comments on 'A Unified Theory of Biology and Physics,' by S. Goldman." *Journal of Social and Biological Structures*, 3: 363–65.

Rosenbaum, Walter A. (1975). *Political Culture.* New York: Praeger.

Ross, Barry E. (1978). "Food Taboos, Diet, and Hunting Strategy: The Adaptation to Animals in Amazon Cultural Ecology." *Current Anthropology*, 19: 1–36.

Rossi, Alice S. (1977). "A Biosocial Perspective on Parenting." *Daedalus*, 106: 1–31.

Rubenstein, Robert, and Harold D. Lasswell (1966). *The Sharing of Power in a Psychiatric Hospital.* New Haven, CT: Yale University Press.

Rumbaugh, Duane M., ed. (1977). *Language Learning by a Chimpanzee: The LANA Project.* New York: Academic Press.

Rummel, Rudolph J. (1970). *Applied Factor Analysis.* Evanston, IL: Northwestern University Press.

——— (1975). *The Dynamic Psychological Field: Understanding Conflict and War*, Vol. 1. New York: Wiley.

——— (1976). "The Roots of Faith." In James N. Rosenau, ed., *In Search of Global Patterns*, 10–30. New York: The Free Press.

——— (1977). *Field Theory Evolving.* Beverly Hills, CA: Sage.

——— (1978). *National Attributes and Behavior.* Beverly Hills, CA: Sage.

——— (1979). *War, Power, Peace: Understanding Conflict and War.* Vol. 4. Beverly Hills, CA: Sage.

Ruse, Michael (1979). *Sociobiology: Sense or Nonsense?* Boston: D. Reidel.

Rushton, J. Phillipe (1985). "Differential K Theory and Group Differences in Intelligence." *Behavioral and Brain Sciences*, 8: 239–40.

Russell, Bertrand (1918). *Mysticism and Logic.* New York: W. W. Norton.

Ruyle, Eugene E. (1973). "Genetic and Cultural Pools: Some Suggestions for a Unified Theory of Biocultural Evolution. *Human Ecology*, 1: 201–15.

Ruyle, E. E. N., F. T. Cloak, L. B. Slobodkin, and W. H. Durham (1977). "The Adaptive Significance of Cultural Behavior: Comments and Reply." *Human Ecology*, 5: 49–68.

Ryan, John Paul, and C. Neal Tate (1981). *The Supreme Court in American Politics: Policy through Law.* SETUPS No. 5. Rev. ed. Washington, DC: American Political Science Association and Inter-University Consortium for Political and Social Research.

Ryder, Norman B. (1965). "The Cohort as a Concept in the Study of Social Change." *American Sociological Review*, 30: 843–61.

Sacher, George A. (1978). "Longevity and Aging in Vertebrate Evolution." *BioScience*, 28: 497–501.

Sachs, Mendel (1973). *The Field Concept in Contemporary Science* Springfield, IL: Charles C. Thomas.

Sackheim, Harold, and Ruben Gur (1978). "Self-Deception, Self-Confrontation, and Consciousness." In G. Schwartz and D. Shapiro, eds., *Consciousness and Self-Regulation.* Vol. 2. New York: Plenum.

Sacksteder, Richard (1982). "Mathematics and Numerology." *Journal of Social and Biological Structures*, 5: 252–53.

Sagan, Carl (1977). *The Dragons of Eden: Speculations on the Evolution of Human Intelligence.* New York: Random House.

Sahlins, Marshall D. (1976). *The Use and Abuse of Biology.* Ann Arbor: University of Michigan Press.

Salmon, Wesley (1980). *Space, Time, and Motion.* Minneapolis: University of Minnesota Press.

Sambursky, Samuel (1975). *Physical Thought from the Presocratics to the Quantum Physics.* New York: PICA Press.

Sasanuma, Sumiko (1980). "Do Japanese Show Sex Differences in Brain Asymmetry? Supplementary Findings." *Behavioral and Brain Sciences*, 3: 247–48.

Savage, E. Sue, and Duane M. Rumbaugh (1977). "Communication, Language, and LANA: A Perspective." In D.M. Rumbaugh, ed., *Language Learning by a Chimpanzee: The LANA Project.* New York: Academic Press.

Savage-Rumbaugh, E. Sue, Duane Rumbaugh, and Sally Boysen (1978). "Linguistically Mediated Tool Use and Exchange by Chimpanzee (*Pan Troglodytes*)." *Behavioral and Brain Sciences*, 1: 539–54.

Schacter, Stanley (1964). "The Interaction of Cognitive and Physiological Determinants of Emotional State." In L. Berkowitz, ed., *Advances in Experimental Social Psychology*, Vol. 1, 49–80. New York: Academic Press.

Schacter, Stanley, and Singer, J. E. (1962). "Cognitive, Social, and Physiological Determinants of Emotional States." *Psychological Review*, 69: 379–99.

Schaller, George B., and Gordon R. Lowther (1969). "The Relevance of Carnivore Behavior to the Study of Early Hominids." *South West Journal of Anthropology*, 25: 307–41.

Scher, Jordan M. (1962). *Theories of the Mind.* New York: The Free Press.

Schmidhauser, John R. (1962a). "Age and Judicial Behavior: American Higher Appellate Judges." In Wilma Donahue and Clark Tibbits, eds., *Politics of Age*, 101–16. Ann Arbor: Division of Gerontology, University of Michigan.

——— (1962b). "When and Why Justices Leave the Supreme Court." In Wilma Donahue and Clark Tibbits. eds., *Politics of Age*, 117–31. Ann Arbor: Division of Gerontology, University of Michigan.

Schrodinger, Erwin (1944). *What Is Life?* Cambridge: Cambridge University Press.

Schubert, Glendon (1960). *Constitutional Politics: The Political Behavior of Supreme Court Justices and the Constitutional Policies That They Make.* New York: Holt, Rinehart and Winston.

——— (1961). "A Psychometric Model of the Supreme Court." *American Behavioral Scientists*, 5: 14–18.

——— (1962a). "The 1960–61 Term of the Supreme Court: A Psychological Analysis." *American Political Science Review*, 56: 90107.

——— (1962b). "A Solution to the Indeterminate Factorial Resolution of Thurstone and Degan's Study of the Supreme Court." *Behavioral Science*, 7: 448–58.

——— (1965a). *Judicial Policy Making: The Political Role of the Courts.* Chicago: Scott, Foresman.

——— (1965b). *The Judicial Mind: The Attitudes and Ideologies of Supreme Court Justices, 1946–1963.* Evanston, IL: Northwestern University Press.

——— (1966). "The Future of Public Law." *George Washington Law Review*, 34: 593–614.

——— (1967a). "Ideologies and Attitudes, Academic and Judicial." *Journal of Politics*, 29: 3–40.

——— (1967b). "Academic Ideology and the Study of Adjudication." *American Political Science Review*, 61: 106–29.

——— (1967c). "The Rhetoric of Constitutional Change." *Journal of Public Law*, 16: 16–50.

——— (1968a). "Behavioral Jurisprudence." *Law and Society Review*, 2: 407–28.

——— (1968b). "Political Ideology on the High Court." *Politics*, 3: 21–40.

——— (1969a). "Judicial Attitudes and Policy Making in the Dixon Court." *Osgoode Hall Law Journal*, 7: 1–29.

——— (1969b). "Justice and Reasoning: A Political Science Perspective." *Revista Internazionale de Filosofia del Diritto*, 46: 474–96.

——— (1969c). "Two Causal Models of Decision-Making by the High Court of Australia." In G. Schubert and D. J. Danelski, eds., *Comparative Judicial Behavior*, 335–66. New York: Oxford University Press.

——— (1970). *The Constitutional Polity.* Boston: Boston University Press.

——— (1972). *The Future of the Nixon Court.* Honolulu: University of Hawaii Foundation.

——— (1973). "Biopolitical Behavior: The Nature of the Political Animal." *Polity*, 6: 240–75.

——— (1974). *The Judicial Mind Revisited: Psychometric Analysis of Supreme Court Ideology.* New York: Oxford University Press.

——— (1975a). "Biopolitical Behavioral Theory." *Political Science Reviewer*, 5: 403–28.

——— (1975b). *Human Jurisprudence.* Honolulu: University Press of Hawaii.

——— (1976). "Politics as a Life Science: How and Why the Impact of Modern Biology Will Revolutionize the Study of Political Behavior." In Albert Somit, ed., *Biology and Politics*, 155–95. The Hague: Mouton.

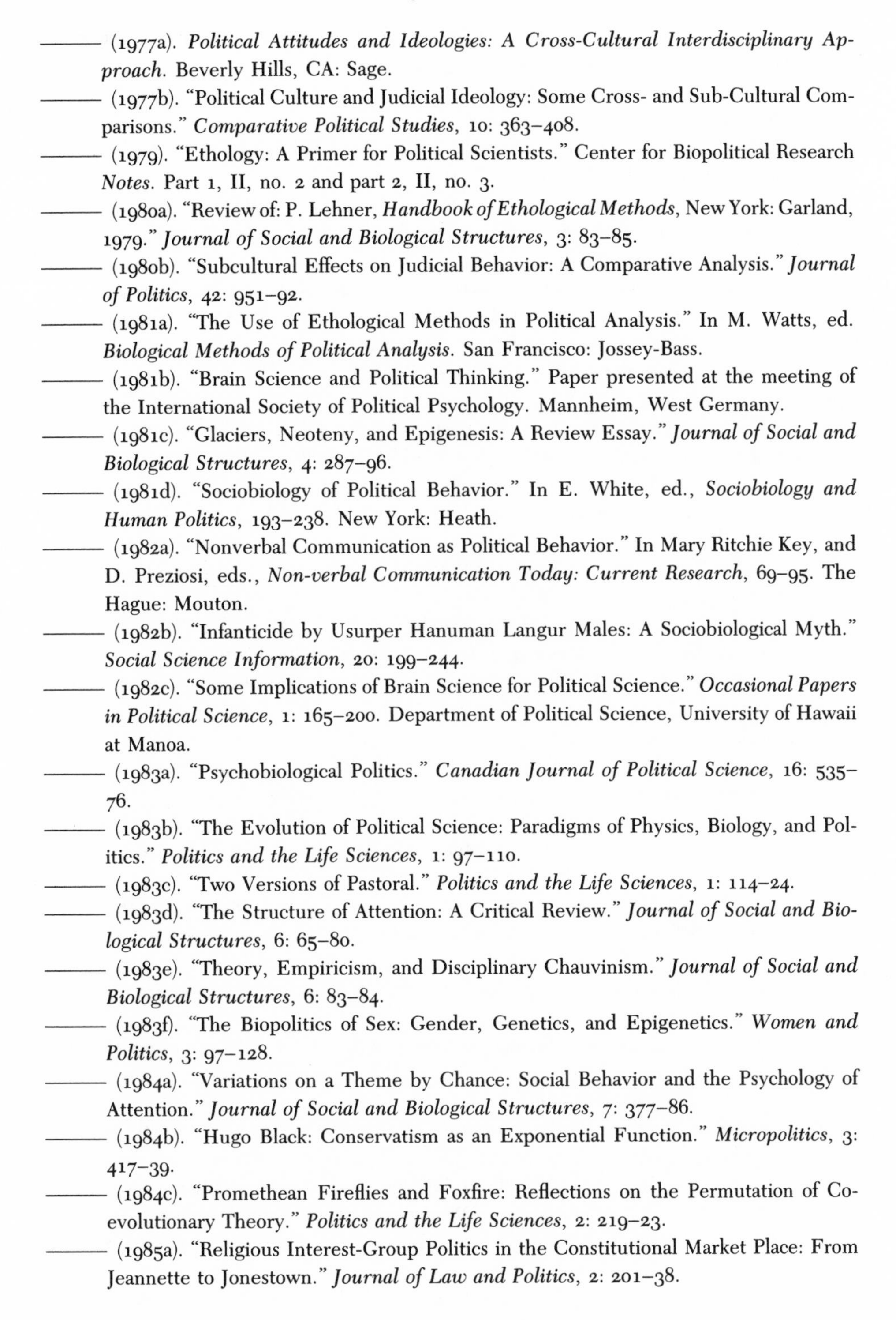

——— (1977a). *Political Attitudes and Ideologies: A Cross-Cultural Interdisciplinary Approach.* Beverly Hills, CA: Sage.

——— (1977b). "Political Culture and Judicial Ideology: Some Cross- and Sub-Cultural Comparisons." *Comparative Political Studies*, 10: 363–408.

——— (1979). "Ethology: A Primer for Political Scientists." Center for Biopolitical Research *Notes.* Part 1, II, no. 2 and part 2, II, no. 3.

——— (1980a). "Review of: P. Lehner, *Handbook of Ethological Methods*, New York: Garland, 1979." *Journal of Social and Biological Structures*, 3: 83–85.

——— (1980b). "Subcultural Effects on Judicial Behavior: A Comparative Analysis." *Journal of Politics*, 42: 951–92.

——— (1981a). "The Use of Ethological Methods in Political Analysis." In M. Watts, ed. *Biological Methods of Political Analysis.* San Francisco: Jossey-Bass.

——— (1981b). "Brain Science and Political Thinking." Paper presented at the meeting of the International Society of Political Psychology. Mannheim, West Germany.

——— (1981c). "Glaciers, Neoteny, and Epigenesis: A Review Essay." *Journal of Social and Biological Structures*, 4: 287–96.

——— (1981d). "Sociobiology of Political Behavior." In E. White, ed., *Sociobiology and Human Politics*, 193–238. New York: Heath.

——— (1982a). "Nonverbal Communication as Political Behavior." In Mary Ritchie Key, and D. Preziosi, eds., *Non-verbal Communication Today: Current Research*, 69–95. The Hague: Mouton.

——— (1982b). "Infanticide by Usurper Hanuman Langur Males: A Sociobiological Myth." *Social Science Information*, 20: 199–244.

——— (1982c). "Some Implications of Brain Science for Political Science." *Occasional Papers in Political Science*, 1: 165–200. Department of Political Science, University of Hawaii at Manoa.

——— (1983a). "Psychobiological Politics." *Canadian Journal of Political Science*, 16: 535–76.

——— (1983b). "The Evolution of Political Science: Paradigms of Physics, Biology, and Politics." *Politics and the Life Sciences*, 1: 97–110.

——— (1983c). "Two Versions of Pastoral." *Politics and the Life Sciences*, 1: 114–24.

——— (1983d). "The Structure of Attention: A Critical Review." *Journal of Social and Biological Structures*, 6: 65–80.

——— (1983e). "Theory, Empiricism, and Disciplinary Chauvinism." *Journal of Social and Biological Structures*, 6: 83–84.

——— (1983f). "The Biopolitics of Sex: Gender, Genetics, and Epigenetics." *Women and Politics*, 3: 97–128.

——— (1984a). "Variations on a Theme by Chance: Social Behavior and the Psychology of Attention." *Journal of Social and Biological Structures*, 7: 377–86.

——— (1984b). "Hugo Black: Conservatism as an Exponential Function." *Micropolitics*, 3: 417–39.

——— (1984c). "Promethean Fireflies and Foxfire: Reflections on the Permutation of Coevolutionary Theory." *Politics and the Life Sciences*, 2: 219–23.

——— (1985a). "Religious Interest-Group Politics in the Constitutional Market Place: From Jeannette to Jonestown." *Journal of Law and Politics*, 2: 201–38.

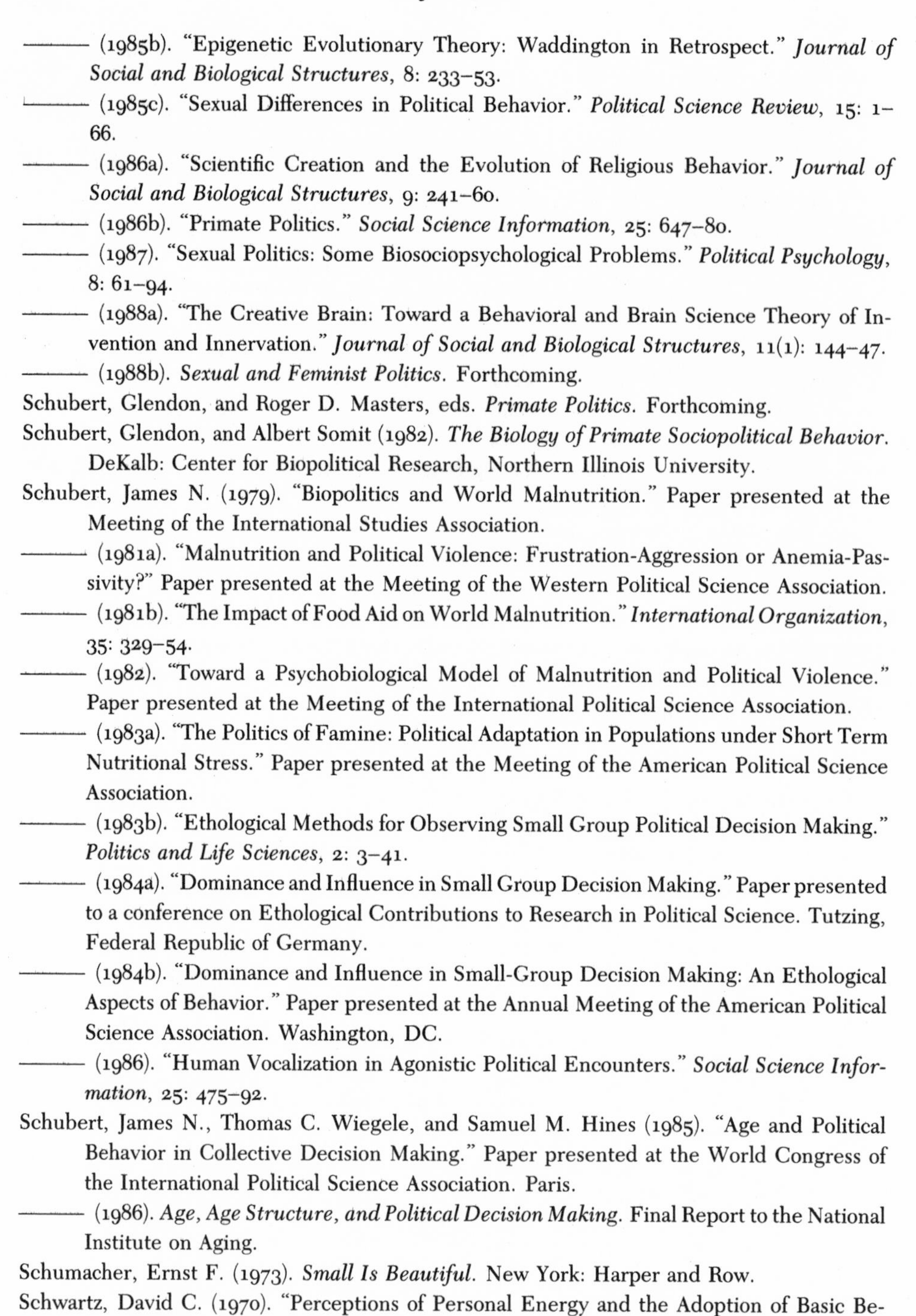

——— (1985b). "Epigenetic Evolutionary Theory: Waddington in Retrospect." *Journal of Social and Biological Structures*, 8: 233–53.

——— (1985c). "Sexual Differences in Political Behavior." *Political Science Review*, 15: 1–66.

——— (1986a). "Scientific Creation and the Evolution of Religious Behavior." *Journal of Social and Biological Structures*, 9: 241–60.

——— (1986b). "Primate Politics." *Social Science Information*, 25: 647–80.

——— (1987). "Sexual Politics: Some Biosociopsychological Problems." *Political Psychology*, 8: 61–94.

——— (1988a). "The Creative Brain: Toward a Behavioral and Brain Science Theory of Invention and Innervation." *Journal of Social and Biological Structures*, 11(1): 144–47.

——— (1988b). *Sexual and Feminist Politics*. Forthcoming.

Schubert, Glendon, and Roger D. Masters, eds. *Primate Politics*. Forthcoming.

Schubert, Glendon, and Albert Somit (1982). *The Biology of Primate Sociopolitical Behavior*. DeKalb: Center for Biopolitical Research, Northern Illinois University.

Schubert, James N. (1979). "Biopolitics and World Malnutrition." Paper presented at the Meeting of the International Studies Association.

——— (1981a). "Malnutrition and Political Violence: Frustration-Aggression or Anemia-Passivity?" Paper presented at the Meeting of the Western Political Science Association.

——— (1981b). "The Impact of Food Aid on World Malnutrition." *International Organization*, 35: 329–54.

——— (1982). "Toward a Psychobiological Model of Malnutrition and Political Violence." Paper presented at the Meeting of the International Political Science Association.

——— (1983a). "The Politics of Famine: Political Adaptation in Populations under Short Term Nutritional Stress." Paper presented at the Meeting of the American Political Science Association.

——— (1983b). "Ethological Methods for Observing Small Group Political Decision Making." *Politics and Life Sciences*, 2: 3–41.

——— (1984a). "Dominance and Influence in Small Group Decision Making." Paper presented to a conference on Ethological Contributions to Research in Political Science. Tutzing, Federal Republic of Germany.

——— (1984b). "Dominance and Influence in Small-Group Decision Making: An Ethological Aspects of Behavior." Paper presented at the Annual Meeting of the American Political Science Association. Washington, DC.

——— (1986). "Human Vocalization in Agonistic Political Encounters." *Social Science Information*, 25: 475–92.

Schubert, James N., Thomas C. Wiegele, and Samuel M. Hines (1985). "Age and Political Behavior in Collective Decision Making." Paper presented at the World Congress of the International Political Science Association. Paris.

——— (1986). *Age, Age Structure, and Political Decision Making*. Final Report to the National Institute on Aging.

Schumacher, Ernst F. (1973). *Small Is Beautiful*. New York: Harper and Row.

Schwartz, David C. (1970). "Perceptions of Personal Energy and the Adoption of Basic Behavioral Orientation to Politics." Paper presented at the International Political Science Association.

——— (1973). "Health Processes and Body Image as Predictors of Political Attitudes and

Behaviors: A Study in Political Socialization." Paper presented to the Ninth World Congress of the International Political Science Association. Montreal.

Schwartz, David C., and Sandra K. Schwartz, eds. (1975). *New Directions in Political Socialization*. New York: The Free Press.

Schwartz, David C., and Nicolas Zill (1972). "Psychophysiological Arousal as a Predictor of Political Participation." Mimeographed paper.

Schwartz, Gary, and David Shapiro, eds. (1976). *Consciousness and Self-Regulation*. Vol. 1. New York: Plenum.

——— (1978). *Consciousness and Self-Regulation*, Vol. 2. New York: Plenum.

Scott, J. P. (1978). "What Are the Expectations Regarding the Scope and Limits of Exploring the Biological Aspects of Political Behavior? Considerations of Methodology." Paper presented at the Meeting of the American Political Science Association. New York.

Sebeok, Thomas A, ed. (1977a). *How Animals Communicate*. Bloomington: Indiana University Press.

——— (1977b). "Zoosemiotic Components of Human Communication." In T. A. Sebeok, ed., *How Animals Communicate*, 1055–77. Bloomington: Indiana University Press.

Sebeok, Thomas A., and Jean Umiker-Sebeok, eds. (1980). *Speaking of Apes: A Critical Anthology of Two-Way Communication with Man*. New York: Plenum.

Secord, Paul F. (1982). *Explaining Human Behavior: Consciousness, Human Action and Social Structure*. Beverly Hills, CA: Sage.

Service, Elman R. (1962). *Primitive Social Organizations: An Evolutionary Perspective*. New York: Random House.

"The Sexual Brain" (1979). *Newsweek*, Nov. 26: 100, 103–5.

Seyfarth, Robert M. (1976). "Review of Michael R. A. Chance and Ray Larsen, eds., *The Social Structure of Attention*. New York: Wiley. *Animal Behavior*, 25: 789–90.

Shapiro, Michael (1981). *Language and Political Understanding*. New Haven, CT: Yale University Press.

Shepard, Paul (1978). *Thinking Animals: Animals and the Development of Human Intelligence*. New York: Viking.

Shneour, Elie (1974). *The Malnourished Mind*. Garden City, NY: Doubleday.

Shubs, Peter (1973). "Political Correlates of Self Body-Image." Paper presented to the Ninth World Congress of the International Political Science Association. Montreal.

Sibatani, Atuhiro (1980a). "The Japanese Brain: The Difference between East and West may be the Difference between Left and Right." *Science*, 80: 22–26.

——— (1980b). "Inscrutable Epigenetics of the Japanese Brain." *Behavioral and Brain Sciences*, 3: 255–66.

Sillen, Andrew (1981). "Strontium and Diet at Hayonim Cave." *American Journal of Physical Anthropology*, 56: 131–37.

Silverman, Paul S. (1983). "Attributing Mind to Animals: The Role of Intuition." *Journal of Social and Biological Structures*, 6: 231–47.

Simon, Herbert (1983). *Reason in Human Affairs*. Stanford, CA: Stanford University Press.

Simonton, Dean K. (1984). *Genius, Creativity and Leadership: Historiometric Inquiries*. Cambridge: Harvard University Press.

Simpson, George G. (1944). *Tempo and Mode in Evolution*. New York: Columbia University Press.

——— (1953). *The Major Features of Evolution*. New York: Columbia University Press.

Slobodkin, L. B. (1982). "A Bully Pulpit." *Behavioral and Brain Sciences*, 5: 26–27.

Smith, Annette (1979). "Playing with Play: A Test Case of 'Ethocriticism.'" *Journal of Social and Biological Structures*, 2: 197–209.

Smith, Peter K. (1982). "Does Play Matter? Functional and Evolutionary Aspects of Animal and Human Play." *Behavioral and Brain Sciences*, 5: 139–84.

Somit, Albert (1968). "Toward a More Biologically Oriented Political Science: Ethology and Psychopharmacy." *Midwest Journal of Political Science*, 12: 550–67.

——— (1972). "Review Article: Biopolitics." *British Journal of Political Science*, 2: 209–38.

——— , ed. (1976). *Biology and Politics: Recent Explorations*. The Hague: Mouton.

——— (1984). "Review of Chimpanzee Politics." *Politics and the Life Sciences*, 2: 211–13.

Somit, Albert, Steven A. Peterson, William D. Richardson, and David S. Goldfischer (1980). *The Literature of Biopolitics*. DeKalb, IL: Center for Biopolitical Research, Northern Illinois University.

Somit, Albert, and Joseph Tanenhaus (1967). *The Development of Political Science*. Boston: Allyn and Bacon.

Sommer, V. and Mohnot, S. M. (1985). "New Observations on Infanticides among Hanuman Langurs. (*Presbytis entellus*) near Jodhpur (Rajasthan/India)." *Behavioral Ecology and Sociobiology*, 16: 245–48.

Sorokin, Pitrim (1964). *Sociocultural Causality, Space, Time*. New York: Russell and Russell.

Spiro, Melford E. (1975). *Kibbutz: Venture in Utopia*. Cambridge, MA: Harvard University Press.

Springer, Sally P., and Georg Deutsch (1981). *Left Brain, Right Brain*. San Francisco: W. H. Freeman.

Sproul, Barbara C. (1979). *Primal Myths: Creating the World*. New York: Harper and Row.

Stanley, Steven M. (1979). *Macroevolution: Pattern and Process*. San Francisco: W. H. Freeman.

——— (1981). *The New Evolutionary Timetable: Fossils, Genes, and the Origin of Species*. New York: Basic Books.

——— (1986). "Is Human Evolution Punctuational?" In B. J. Williams (ed.), *On Evolutionary Anthropology*, pp. 77–89. Malibu, CA: Undena Publications.

Stapp, Henry (1972). "The Copenhagen Interpretation and the Nature of Space-Time." *American Journal of Physics*, 40: 1098–1116.

Stauffer, Robert (1969). "The Biopolitics of Underdevelopment." *Comparative Political Studies*, 2: 361–87.

——— (1971). *The Role of Drug in Political Change*. Morristown, NJ: General Learning Press.

Steiglitz, Robert R. (1982). "Numerical Structuralism and Cosmology in the Ancient Near East." *Journal of Social and Biological Structures*, 5: 255–66.

Stent, Gunther (1969). *The Coming of the Golden Age*. Garden City, NY: Natural History Press.

Stern, Elizabeth (1981). "Frontiers of Neurosciences." Lectures at National Science Foundation/Chautauqua Course, Oregon Graduate Center. Beaverton, OR.

Storing, Herbert, ed. (1962). *Essays on the Scientific Study of Politics*. New York: Holt, Rinehart and Winston.

Strayer, Fred F. (1981). "The Organization and Coordination of Asymmetrical Relations among Young Children: A Biological View of Social Power." In Meredith Watts, ed., *Biopolitics: Ethological and Physiological Approaches*, 33–49. San Francisco: Jossey-Bass.

Strehler, Bernard L. (1962). *Time, Cells,and Aging.* New York: Basic Books.

Struever, Stuart, ed. (1971). *Prehistoric Agriculture.* Garden City, NY: Natural History Press.

Strum, Shirley C. (1975). "Life with the 'Pumphouse Gang.'" *National Geographic*, 147: 673–91.

——— (1981). "Processes and Products of Change: Baboon Predatory Behavior at Gilgil, Kenya." In Robert S. O. Harding and Geza Teleki, eds., *Omnivorous Primates: Gathering and Hunting in Human Evolution*, 255–302. New York: Columbia University Press.

——— (1982a). "The Evolution of Political Behavior and Sexual Differences among Primates." Paper presented at the Annual Meeting of the Western Political Science Association. San Diego.

——— (1982b). "Agonistic Dominance in Male Baboons: An Alternative View." *International Journal of Primatology*, 3: 175–202.

——— (1983). "Use of Females by Male Olive Baboons.(*Papio Anubis*)." *American Journal of Primatology*, 5: 93–109.

——— (1987). *Almost Human: A Journey into the World of Baboons.* New York: Random House.

Strum, Shirley, and Bruno Latour (1987). "Redefining the Social Link: From Baboons to Humans." *Social Science Information*, 26: 783–802.

Sulloway, Frank J. (1979). *Freud: Biologist of the Mind.* New York: Basic Books.

Surkin, Marvin, and Alan Wolfe, eds. (1970). *An End to Political Science: The Caucus Papers.* New York: Basic Books.

Suzuki, Akira (1975). "The Origin of Hominid Hunting: A Primatological Perspective." In R. H. Tuttle, ed., *Socioecology and Psychology of Primates*, 259–79. The Hague: Mouton.

Swindler, William F. (1969). *Court and Constitution in the Twentieth Century, Volume 1: The Old Legality, 1889–1932.* Indianapolis: Bobbs-Merrill.

Symons, Donald (1977). "Effect and Function in Play." Paper presented at the Meeting of the Animal Behavior Society. Pennsylvania State University.

——— (1978). *Play and Aggression.* New York: Columbia University Press.

——— (1979). *The Evolution of Human Sexuality.* New York: Oxford University Press.

Szasz, Thomas (1965). *Madness and Civilization.* New York: Pantheon.

——— (1970). *The Manufacture of Madness.* New York: Harper and Row.

——— (1973). *The Age of Madness.* Garden City, NY: Anchor Press.

Tanenhaus, Joseph (1977). "Experimental/Physiological Research on the Meaning of Political Concepts." Paper presented at the Conference on Biology and Politics. International Political Science Association, Bellagio, Italy.

Tanenhaus, Joseph, and Mary Ann Foley (1982). "'The Words of Things Entangle and Confuse': The Ambiguous Political Concept." *International Political Science Review*, 3: 107–30.

Taylor, David (1978). "The Biases of Sex and Maturation in Lateralization: 'Isomeric' and Compensatory Left-Handedness." *Behavioral and Brain Sciences*, 1: 318–20.

Teleki, Geza (1975). "Primate Subsistence Patterns: Collector Predators and Gatherer-Hunters." *Journal of Human Evolution*, 4: 125–84.

Templeton, Alan R. (1984). "The Evolution of Man." Seminar sponsored by the Department of Genetics and Hawaiian Evolutionary Biology Program. University of Hawaii at Manoa.

TenHouten, Warren D. (1980). "Social Dominance and Cerebral Hemisphericity: Discriminating Race, Socioeconomic status, and Sex Groups by Performance on Two Lateralized Tests." *International Journal of Neuroscience*, 10: 223–32.

TenHouten, Warren D., and Charles Kaplan (1973). *Science and Its Mirror Image*. New York: Harper and Row.

TenHouten, Warren D., Andrea L. Thompson, and Donald O. Walter (1976). "Discriminating Social Groups by Performance on Two Lateralized Tests." *Bulletin of the Los Angeles Neurological Societies*, 41: 99–108.

Terrace, Herbert S. (1979). *Nim: A Chimpanzee Who Learned Sign Language*. New York: Knopf.

Thomas, Tracy Y. (1930). "On the Unified Field Theory, I." *Proceedings of the National Academy of Sciences*, 16: 761–76.

Thompson, Philip R. (1975). "A Cross-species Analysis of Carnivore, Primate, and Hominid Behavior." *Journal of Human Evolution*, 4: 113–24.

——— (1978). "The Evolution of Territoriality and Society in Top Carnivores." *Social Science Information*, 17: 949–92.

Thorbeck, William J. (1965). *A New Dimension in Political Thinking*. New York: Oceana.

Thorson, Thomas L. (1970). *Biopolitics*. New York: Holt, Rinehart and Winston.

Tiger, Lionel (1979a). "Biological Aspects of a Belief in Progress." Paper presented at the Eleventh World Congress of the International Political Science Association. Moscow.

——— (1979b). *Optimism: The Biology of Hope*. New York: Simon and Schuster.

——— (1981). "Are Humans Optimistic?" *Journal of Social and Biological Structures*, 4: 97–98.

Tiger, Lionel, and Robin Fox (1971). *The Imperial Animal*. New York: Holt, Rinehart and Winston.

Tiger, Lionel, and Joseph Shepfer (1975). *Women in the Kibbutz*. New York: Harcourt Brace Jovanovich.

Tinbergen, Niko (1976). "Ethology in a Changing World." In P. P. G. Bateson and R. A. Hinde., eds., *Growing Points in Ethology*. New York: Cambridge University Press.

Tomkins, Sylvan S. (1962). *Affect, Imagery, Consciousness: The Positive Affects*. Vol. 1. New York: Springer.

——— (1963). *Affect, Imagery, Consciousness: The Negative Affects*, Vol. 2. New York: Springer.

Travis, Cheryl (1977). "Human Ethology Abstracts II." *ManEnvironment System*, 7: 227–73.

——— (1978). "The Ethology and Ecology of Human Relations." *Human Ethology Newsletter*, 23: 8–9.

Travis, Cheryl, et al. (1977). "Human Ethology Abstracts." *Man-Environment System*, 7: 3–34.

Trivers, Robert L. (1971). "The Evolution of Reciprocal Altruism." *Quarterly Review of Biology*, 46: 35–57.

——— (1974). "Parent-Offspring Conflict." *American Zoologist*, 14: 249–64.

——— (1978). "Remarks." Presented at the sociobiology panel at the Annual Meeting of the American Behavior Society. Seattle.

Truman, David (1951). *The Governmental Process*. New York: Knopf.

——— (1955). "The Impact on Political Science of the Revolution in the Behavioral Sciences." In Stephen K. Bailey, et al., eds., *Research Frontiers in Politics and Government*, 202–32. Washington, DC: Brookings Institution.

Trump, David H. (1980). *The Prehistory of the Mediterranean*. New Haven, CT: Yale University Press.

Tullock, Gordon (1971). "Biological Externalities." *Journal of Theoretical Biology*, 33: 565–76.

Turnbull, C. M. (1973). *The Mountain People*. London: Jonathan Cape.

Tuttle, Russell H. (ed.) (1975). *Socioecology and Psychology of Primates*. The Hague: Mouton.

Ucko, Peter J., and G. W. Dimbleby, eds. (1969). *The Domestication and Exploitation of Plants and Animals*. Chicago: Aldine.

Underwood, Jane H. (1975). *Biocultural Interactions and Human Variation*. Dubuque, IA: Wm. C. Brown.

——— (1979). *Human Variation and Human Microevolution*. Englewood Cliffs, NJ: Prentice-Hall.

Ushenko, Andrew (1937). *The Philosophy of Relativity*. London: George Allen and Unwin.

——— (1946). *Power and Events*. Princeton: Princeton University.

Ussher, James (1650–1654). *Annales Veteris et Novi Testament*.

Uttal, William (1978). *Psychobiology of Mind*. Hillsdale, NJ: Lawrence Erlbaum Associates.

van den Berghe, Pierre L. (1974). "Bringing Beasts Back in: Toward a Biosocial Theory of Aggression." *American Sociological Review*, 39: 777–88.

——— (1978). *Man in Society: A Biological View*. New York: Elsevier.

——— (1979). *Human Family Systems: An Evolutionary View*. New York: Elsevier.

deWaal, Frans B. M. (1978). "Exploitative and Familiarity-Dependent Support Strategies in a Colony of Semi-Free Living Chimpanzee." *Behaviour*, 66: 268–312.

——— (1982). *Chimpanzee Politics: Power and Sex among Apes*. London: Jonathan Cape.

deWaal, Frans B. M., and Jan A. R. A. M. van Hooff (1981). "Side Directed Communication and Agonistic Interaction in Chimpanzee." *Behaviour*, 77: 164–98.

van Dijk, J. J. M. (1977). *Dominantiegedrag en Geweld: Een Multidisciplinaire Visie op de Veroorzaking van Geweldmisdrijven*. Nijmegen, Netherlands: Dekker and van de Vegt.

van Dyke, Vernon (1971). "Review of: Marvin Surkin and Alan Wolf, eds., *An End to Political Science: The Caucus Papers*, New York: Basic Books, 1970." *American Political Science Review*, 65: 793–94.

van Hooff, Jan A. R. A. M. (1962). "Facial Expressions in Higher Primates." *Symposia of the Zoological Society of London*, 8: 97–125.

——— (1972). "A Comparative Approach to the Phylogeny of Laughter and Smiling." In R. A. Hinde, ed., *Non-verbal Communication*. New York: Cambridge University Press.

——— (1973a). "The Arnhem Zoo Chimpanzee Consortium." *International Zoo Yearbook*, 13: 195–205.

——— (1973b). "A Structural Analysis of the Social Behavior of a Semi-captive Group of Chimpanzees." In Mario van Cranach and Ian Vine, eds., *Social Communication and Movement*. New York: Academic.

——— (1982). "Coalitions and Positions of Influence in a Chimpanzee Community." In G. Schubert and A. Somit, eds., *The Biology of Primate Sociopolitical Behavior*, 2–15. DeKalb, IL: Center for Biopolitical Research, Northern Illinois University.

Van Over, Raymond (1980). *Sun Songs: Creation Myths from around the World*. New York: New American Library.

Vaughn, Brian E., and Everett Waters (1980). "Social Organization among Preschool Peers: Dominance, Attention, and Sociometric Correlates." In D. Omark, F. Strayer, and D. Freedman, eds., *Dominance Relations: An Ethological View of Human Conflict and Social Interaction*, 359–79. New York: Garland.

——— (1981). "Attention Structure, Sociometric Status, and Dominance: Interrelationships to Social Competence." *Developmental Psychology*, 17: 275–88.

Vogel, Christian (1979). "Der Hanuman-Langur (*Presbytis entellus*), ein Parade-Exempel für die theoretischen Konzepte der 'Soziobiologie'?" *Verhandlungen Deutsche Zoologische Gesellschaft*. Stuttgart: Gustav Fischer.

Vogel, Christian, and H. Loch (1984). "Reproductive Parameters, Adult-Male Replacements, and Infanticide among Free-Ranging Langurs (*Presbytis entellus*) at Jodhpur (Rajasthan), India." In Glenn Hausfater and Sarah Blaffer Hrdy, eds., *Infanticide: Comparative and Evolutionary Perspective*, 237–55. New York: Aldine.

von Cranach, Mario, et al., eds. (1979). *Human Ethology: Claims and Limits of a New Discipline*. Cambridge: Cambridge University Press.

von Eckardt, Barbara (1978). "What Is the Biology of Language?" In Edward Walker, ed., *Explorations in the Biology of Language*. Montgomery, VM: Bradford.

von Glasersfeld, Ernst (1977). "Linguistic Communication: Theory and Definition." In Duane M. Rumbaugh,ed., *Language Learning by a Chimpanzee: The LANA Project*. New York: Academic Press.

Waber, Deborah (1980). "What is the Significance of Sex Differences in Performance Asymmetry?" *Behavioral and Brain Sciences*, 3: 249–50.

Waddington, Conrad Hal (1940). *Organisers and Genes*. Cambridge: Cambridge University Press.

——— (1948). *The Scientific Attitude*. 2d ed. West Drayton, Middlesex: Penguin Books.

——— (1957). *The Strategy of the Genes*. London: George Allen and Unwin.

——— (1960). *The Ethical Animal*. London: George Allen and Unwin.

——— (1962). *The Nature of Life*. New York: Atheneum.

——— (1968). *The Scientific Attitude*. London: Hutchinson.

——— (1969). *Behind Appearance: A Study of the Relations between Painting and the Natural Sciences in This Century*. Cambridge: MIT Press.

——— (1975). *The Evolution of an Evolutionist*. Ithaca, NY: Cornell University Press.

——— (1978). *The Man-Made Future*. New York: St. Martin's Press.

Wahlke, John (1976). "Observations of Biopolitical Study." In Albert Somit, ed., *Biology and Politics*, 253–59. Paris: Mouton.

——— (1979). "Pre-behavioralism in Political Science." Presidential Address, American Political Science Association, 1978. Published in *American Political Science Review*, 73: 9–31.

Walker, Alan (1984). "Extinction in Hominid Evolution." In Matthew H. Nitecki, ed., *Extinctions*, 119–52. Chicago: University of Chicago Press.

Wallace B., and A. M. Srb (1961). *Adaptation*. Englewood Cliffs, NJ: Prentice-Hall.

Wallas, Graham (1908). *Human Nature in Politics*. 3d ed. Boston: Houghton.

Wasby, Stephen (1976). *Continuity and Change from the Warren Court to the Burger Court*. Pacific Palisades, CA: Goodyear.

Washburn, Sherwood L. (1978). "The Evolution of Man." *Scientific American*, 239: 146–54.

Watson, James D. (1968). *The Double Helix*. New York: Atheneum.

Watts, Meredith W. (1976). "Desensitization of Children to Violence?: Another Look at Television's Effects." *Experimental Study of Politics*, 5: 1–24.

——— , ed. (1981). *Biopolitics: Ethological and Physiological Approaches*. San Francisco: Jossey-Bass.

——— (1982). "Biopolitics and Sex Differences." Paper presented at the Meeting of the Western Political Science Association. San Diego.

——— (1983). "Introduction: Biopolitics and Gender." *Women And Politics*, 3: 1–27.

Weber, Max (1964). *The Theory of Social and Economic Organization*. Translated and edited by Talcott Parsons. New York: The Free Press.

Webster, G., and B. C. Goodwin (1982). "The Origin of Species: A Structuralist Approach." *Journal of Social and Biological Structures*, 5: 15–47.

Weg, Ruth B. (1975). "Changing Physiology of Aging: Normal and Pathological." In Diana S. Woodruff and James E. Birren, eds., *Aging: Scientific Perspectives and Social Issues*, 229–58. New York: D. van Nostrand.

Weiss, Gerald (1973). "A Scientific Concept of Culture." *American Anthopologist*, 75: 1376–1413.

Welker, W. I. (1956). "Variability of Play and Exploratory Behaviour in Chimpanzees." *Journal of Comparative Physiological Psychology*, 49: 181–85.

West Eberhard, Mary Jane (1975). "The Evolution of Social Behavior by Kin Selection." *Quarterly Review of Biology*, 50: 1–33.

——— (1976). "Born: Sociobiology." *Quarterly Review of Biology*, 51: 89–92.

Western, Jonah D. and Shirley C. Strum (1983). "Sex, Kinship, and the Evolution of Social Manipulation." *Ethology and Sociobiology*, 4: 19–28.

Wheatley, Bruce P. (1982). "Adult Male Replacement in *Macaca Fasiculais* of East Klimantan, Indonesia." *International Journal of Primatology*, 3: 203–19.

Wheeler, Harvey (1975). "Constitutionalism." In Fred Greenstein and Nelson Polsby, eds., *Handbook of Political Science*. Vol. 5. pp. 1–91. Reading, MA: Addison-Wesley.

——— (1978). "Human Sociobiology: An Exploratory Essay." *Journal of Social and Biological Structures*, 1: 307–18.

Wheeler, Harvey, and James Danielli (1982). "Constructional Biology." *Journal of Social and Biological Structures*, 5: 11–14.

Wheeler, John A. (1962). "Curved Empty Space-Time as the Building Material of the Physical World: An Assessment." In Ernest Nagel, Patrick Suppes, and Alfred Tarski, eds., *Logic and Philosophy of Science*, 361–74. Stanford, CA: Stanford University Press.

Whitaker, Harry (1978). "Is the Right Left Over?" *Behavioral and Brain Sciences*, 1: 323–24.

Whitaker, Harry, and G. A. Ojemann (1977). "Lateralization of Higher Cortical Functions: A Critique." *Annals of the New York Academy of Sciences*, 299: 459–73.

White, Elliott (1972). "Genetic Diversity and Political Life: Toward a Population-Interaction Paradigm." *Journal of Politics*, 34: 1203–42.

——— (1975). "Genetic Diversity and Democratic Theory." Paper presented at the Annual Meeting of the American Political Science Association.

——— (1980a). "Clouds, Clocks, Brains and Political Learning." Paper presented at the Third Scientific Meeting of the International Society for Political Psychology. Boston.

——— (1980b). "The End of the Empty Organism: Human Neurobiology Classical Social Science, and Political Learning." Paper presented at the Meeting of the Midwest Conference of Political Scientists. Chicago.

——— (1981a). "The Neurobiological Basis of Human Action." Paper presented at the Annual Meeting of the Western Political Science Association. Denver.

——— (1981b). "Sociobiology, Neurobiology, and Political Socialization," *Micropolitics*, 1: 113–44.

——— , ed. (1981c). *Sociobiology and Human Politics.* New York: Heath/Lexington.

——— (1981d). "Political Socialization from the Perspective of Generational and Evolutionary Change." In E. White, ed., *Sociobiology and Politics.* New York: Heath/Lexington.

——— (1982a). "Clouds, Clocks, Brains and Political Learning." *Micropolitics,* 2: 279–309.

——— (1982b). "Self-Direction and Political Action: Conscious Purpose, Emic Analysis, and Ongoing Political Biography." Paper presented at the Meeting of the Northeastern Political Science Association.

——— (1982c). "Brain Science and the Emergence of Neuropolitics." *Politics and the Life Sciences* 1: 23–25.

White, Elliott, and Joseph Losco, eds. (1986). *Biology and Bureaucracy: Public Administration and Public Policy from the Perspective of Evolutionary, Genetic and Neurobiological Theory.* Lanham, MD: University Press of America.

White, Lynn, Jr. (1974). "Animals and Man in Western Civilization." In Joseph Klaits and Barrie Klaits, eds., *Animals and Man in Historical Perspective.* New York: Harper and Row.

Wiegele, Thomas C. (1971). "Toward a Psychophysiological Variable in Conflict Theory." *Experimental Study of Politics,* 1: 51–81.

——— (1973). "Decision Making in an International Crisis: Some Biological Factors." *International Studies Quarterly,* 17: 295–335.

——— (1977). "Models of Stress and Disturbances in Elite Political Behavior: Psychological Variables and Political Decision Making." In R. S. Robins, ed., *Psychopathology and Political Leadership.* New Orleans: Tulane Studies in Political Science.

——— (1978a). "Physiologically Based Content Analysis: An Appication in Political Communication." In B. D. Ruben, ed., *Communication Yearbook.* New Brunswick, NJ: Transaction Books.

——— (1978b). "The Psychophysiology of Elite Stress in Five International Crises: A Preliminary Test of a Voice Measurement Technique." *International Studies Quarterly,* 22: 467–511.

——— (1979a). "Signal Leakage and the Remote Psychological Assessment of Foreign Policy Elites." In Lawrence S. Falkowski, ed., *Psychological Models in International Politics.* Boulder, CO: Westview.

——— (1979b). *Biopolitics: Search for a More Human Political Science.* Boulder, CO: Westview.

——— (1980). *Psycholinguistic Analysis of Physiological Stress During International Crises.* Dekalb: Center for Biopolitical Research.

——— , ed. (1982). *Biology and the Social Sciences: An Emerging Revolution.* Boulder, CO: Westview.

Wiegele, Thomas C., Gorden Hilton, Kent Oots, and Susan V. Kisiel (1985). *Leaders under Stress.* Durham, NC: Duke University Press.

Wiegele, Thomas C., Sharon Plowman, and Robert Chrey (1973). "International Crisis, Cardiorespiratory Health, and Political Attitudes: A Literature Review and a Pilot Study." Paper presented to the Ninth World Congress of the International Political Science Association. Montreal.

Wight, Martin (1946). *Power Politics.* London: Royal Institute of International Affairs.

Wilber, Ken, ed. (1982). *The Holographic Paradigm and Other Paradoxes.* Boulder, CO: Shambhala.

Willhoite, Fred H., Jr. (1976). "Primates and Political Authority: A Biobehavioral Perspective." *American Political Science Review*, 70: 1110–26.

——— (1977). "Evolution and Collective Intolerance." *Journal of Politics*, 39: 665–84.

Williams, George (1966). *Adaptation and Natural Selection: A Critique of Some Current Evolutionary Thought*. Princeton: Princeton University Press.

——— (1971). *Group Selection*. Chicago: Aldine-Atherton.

——— (1975). *Sex and Evolution*. Princeton: Princeton University Press.

Wilson, David Sloan (1980). *The Natural Selection of Populations and Communities*. Menlo Park, CA: Benjamin/Cummings.

Wilson, Edward O. (1975). *Sociobiology: The New Synthesis*. Cambridge: Harvard University's Belknap Press.

——— (1978). *On Human Nature*. Cambridge, MA: Harvard University Press.

Wilson, Glenn D., ed. (1973). *The Psychology of Conservatism*. New York: Academic Press.

Wilson, H. T. (1977). "Attitudes Toward Science: Canadian and American Scientists." In Robert Presthus, ed., *Cross-National Perspective: United States and Canada*, 154–75. Leiden: Brill.

Wilson, James R. (1985). "Jensen's Support for Spearman's Hypothesis Is Support for a Circular Argument." *Behavioral and Brain Sciences*, 8: 246.

Wingerson, Lois (1982). "Nice Guys Finish First: A Study of Facial Politics Concludes that a Candidate Should Not Lead with His Chin." *Discover*, 3/4: 66–67.

Winick, Myron, and Pedro Rosso (1975). "Malnutrition and Central Nervous System Development." In James W. Prescott, Merrill S. Reed, and David B. Coursin, eds., *Brain Function and Malnutrition*. New York: Wiley.

Witelson, Sandra (1980). "Neuroanatomical Asymmetry in Left-Handers: A Review and Implications for Functional Symmetry." In Jeannie Herron, ed., *Neuropsychology of Left-Handedness*. New York: Academic Press.

Wittig, Michele, and Anne Petersen, eds. (1979). *Sex-Related Differences in Cognitive Functioning*. New York: Academic Press.

Wolf, Fred Alan (1981). *Taking the Quantum Leap*. San Francisco: Harper and Row.

Wood, Charles C. (1985). "Pardon, Your Dualism Is Showing." *Behavioral and Brain Sciences*, 8: 557–58.

Wood, James E., Jr. (1982). "'Scientific Creationism' in the Public Schools." *Journal of Church and State*, 24: 233.

Woodward, Bob, and Scott Armstrong (1979). *The Brethren: Inside the Supreme Court*. New York: Simon and Schuster.

Wright, Quincy (1955). *The Study of International Relations*. New York: Appleton-Century-Crofts.

Wright, Sewall (1940). "The Statistical Consequences of Mendelian Heredity in Relation and Speciation." In J. S. Huxley, ed., *The New Systematics*, 161–83. Oxford: Clarendon Press.

——— (1968). *Genetic and Biometric Foundations*. Vol. 1 of *Evolution and the Genetics of Populations*. Chicago: University of Chicago Press.

——— (1977). *Experimental Results and Evolutionary Deductions*. Vol. 3 of: *Evolution and the Genetics of Populations*. Chicago: University of Chicago Press.

——— (1978). *Variability Within and Among Natural Populations*. Vol. 4 of: *Evolution and the Genetics of Populations*. Chicago: University of Chicago Press.

Wynne-Edwards, Vero C. (1962). *Animal Dispersion in Relation to Social Behavior*. Edinburgh: Oliver and Boyd.

Yakimov, V. P. (1975). "Discussion." In Russell H, Tuttle, ed., *Socioecology and Psychology of Primates*. The Hague: Mouton.

Yanarella, Ernest J. (1984). "Slouching toward the Apocalypse: Visions of Nuclear Holocaust and Eco-Catastrophe in Contemporary Science Fiction." Paper presented at annual meeting of the American Political Science Association. Washington, DC.

Young, Michael F. D. (1961) *The Rise of Meritocracy*. Harmondsworth, Middlesex: Penguin.

Yunis, Jorge J., and Om Prakash (1982). "The Origins of Man: A Chromosomal Pictorial Legacy." *Science*, 215: 1525–29.

Yunis, Jorge J., Jeffrey R. Sawyer, and Kelly Dunham (1980). "The Striking Resemblance of High-Resolution G-Banded Chromosomes of Man and Chimpanzees." *Science*, 208: 1145–48.

Zeuner, Frederick E. (1963). *A History of Domesticated Animals*. London: Hutchison.

Zinnes, Dina (1970). "Coalition Theories and the Balance of Power." In S. Groennings, E. Kelley, and M. Leierson, eds., *The Study of Coalitions*. New York: Holt, Rinehart, and Winston.

Zivin, Gail (1977). "Facial Encounters Predict Preschoolers' Encounter Outcomes." *Social Science Information*, 16: 715–29.

Zukav, Gary (1979). *The Dancing Wu Li Masters: An Overview of the New Physics*. New York: William Morrow.

Name Index

Abeles, Ronald P., 226
Abercrombie, Nicholas, 212
Abramovitch, Rona, 75
Acquino, Corazon, 281
Adorno, Theodore, 227
Adrian, Charles, 4
Albin, Rochelle, 80
Alcock, John, 36, 77, 260
Alexander, Richard, 46, 48, 99, 128–29, 137, 140, 203
Allison, A. C., 15
Alloway, Thomas, 58, 223
Almond, Gabriel, 307, 321
Andrew, R. J., 55, 102, 219
Annette, Marian, 279
Archimedes, 299
Ardrey, Robert, 10, 34, 70, 118
Argyle, Michael, 257
Aristophanes, 70
Aristotle, 26, 29, 83, 99, 113, 216, 296–97, 299; *Politica,* 99
Armstrong, Scott, 229
Aubert, Vilhelm, 39, 320
Auel, Jean, 105
Axelrod, Robert, 101, 264

Bacon, Francis, 294
Baer, Darius, 104, 115
Baer, Denise, 282
Baerends, Gerhard, 97, 257
Baldwin, Janice, 59, 94, 163, 268, 274
Baldwin, John, 59, 94, 163, 268, 274
Ball, Donald, 4
Barash, David, 105, 163
Barkow, Jerome, 34, 38, 45
Barlow, George, 53, 128
Barner-Barry, Carol, 41, 75, 80, 85, 101
Barnes, S. B., 54
Barsley, Michael, 277
Barwick, Judith, 16
Bateson, Gregory, 212, 221
Bateson, William, 201, 212
Bay, Christian, 25
Beale, Ivan, 261
Beccaria, Cesare, 224
Beck, Henry, 56, 58, 75, 303
Beckoff, Marc, 59
Beckwith, Jonathan, 163
Beer, Colin, 36–37
Beethoven, Ludwig van, 292
Bellamy, Edward, 103
Bengtson, Vern, 14, 237, 239, 241
Bentley, Arthur Fisher, 150, 300–301
Bernhard, Prince (The Netherlands), 150
Berns, Walter, 6
Bernstein, Irwin, 70, 75, 115, 224
Binford, Lewis, 106
Birdsell, Joseph, 44, 105, 231
Birdwhistell, R. L., 219
Bischof, Norbert, 152
Black, Hugo, Jr., 229
Black, Hugo Lafayette, 104, 228–29, 235
Blank, Robert, 5, 81, 325, 327
Blaustein, Albert, 237
Bleier, Ruth, 328

Blurton-Jones, Nicholas, 34, 36–37, 46, 48, 50, 55–56, 58–60, 145
Boas, Franz, 109
Bock, Kenneth, 176
Boggess, Jane, 75, 92
Bohm, David, 113, 120, 202, 207–8, 311–12, 315, 317–19, 322
Bohr, Niels, 113, 165, 269, 306, 312
Boklage, Charles, 261, 277
Boll, Thomas, 273
Bonham, G. Mathew, 272
Bonner, John, 48, 78, 92
Boorman, Scott, 151
Borgia, Gerald, 100
Boulding, Elise, 282
Boulding, Kenneth, 4
Bourne, Geoffrey, 53
Bowlby, John, 37
Boyd, Robert, 105, 118, 129, 293
Boysen, Sally, 248
Brace, C. Loring, 99
Bradshaw, John, 166, 276, 279–80
Brady, Ivan, 231
Braidwood, Robert, 105
Brandeis, Louis D., 230, 233, 240
Brannigan, Christopher, 55
Breines, Wini, 98
Brennan, William, 104
Bromley, Dennis, 226, 232
Brothwell, Don, 224
Brown, Barbara, 97
Brown, Jethro, 225
Brown, Lester, 11
Bruce, Virginia, 275
Bruner, Jerome, 59, 269
Bryan, William Jennings, 103
Burger, Warren Earl, 229–31
Burgess, John, 296
Burhoe, Ralph, 120
Burke, Kenneth, 38, 250
Butzer, Karl, 107, 110
Bygott, J. D., 75

Caldwell, Lynton, 4, 42, 61
Callan, Hilary, 3
Calvin, William, 112, 275
Campbell, Bernard, 73, 78, 102, 105, 148
Campbell,Donald, 12, 134–35, 145, 224
Caplan, Arthur, 34, 105, 128, 141, 163
Capra, Fritjof, 317
Cardozo, Benjamin, 225
Carey, George, 301
Carneiro, Robert, 62, 70, 99
Cartwright, Dorwin, 308
Cattani, Richard, 87
Cavallaro, Sahli, 75
Cavalli-Sforza, Luigi, 105, 161, 221
Cerullo, Margaret, 98
Chagnon, Napoleon, 62, 70, 115, 224, 281
Chamberlain, Neville, 216
Chance, Michael, 47, 75, 84–87, 94, 101, 114
Chapple, Elliot, 3, 10, 116–17, 251
Chardin, Teilhard de, 4
Charles, Prince of Wales (U.K.), 150
Charlesworth, William, 50–51, 163, 166
Chase, Allan, 13
Chase, H. Peter, 286
Cherfes, Jeremy, 99
Chevalier-Skolnikoff, Suzanne, 57, 80, 223, 252–53
Chomsky, Noam, 276
Cigler, Allan, 294
Clapham, Wentworth, 18
Clarke, John H., 104, 240
Clinton, Richard, 14
Cloak, F. T., Jr., 47
Cocconi, G., 305, 322
Cohen, Felix, 225
Cohen, L. Jonathan, 271
Cohen, Mark N., 100, 105, 198, 231
Coleman, James, 307
Comfort, Alex, 113, 226, 231–32, 315–18
Comrey, Andrew, 227
Copernicus, 113, 299
Corballis, Michael, 166, 183–84, 261, 275, 277, 283
Corning, Peter, XIII, 4, 5, 25, 38, 41, 47, 75, 83, 99–100, 141, 221, 246, 281, 294, 296
Correns, Karl Erich, 292, 299
Corson, Elizabeth, 112
Corson, Samuel, 112
Cory, Gerald, 303
Crapo, Lawrence, 11, 14, 178, 226
Crick, Francis, 269
Crombie, Alistair, 297–98
Crook, John H., 46, 54–55, 71, 105, 119, 235, 273, 281, 290, 303
Curio, Eberhard, 60, 76
Curtin, Richard, 75, 92
Cutler, Neal, 14, 151, 226, 228, 233–34, 237, 239, 241

Dahl, Robert, 150, 301
Daly, Martin, 34, 38
Danelski, David, 234
Danielli, James, 113, 314
d'Aquili, Eugene, 112, 115–16
Darlington, Cyril, 48, 100, 105
Darrow, Clarence, 103
Darwin, Charles, 11, 45, 83, 113, 120, 128, 130, 141, 165, 169, 204, 292, 297, 300, 325
Davidson, Julian, 159, 264, 290

Davidson, Richard, 159, 264, 290
Davies, James C., 4, 13, 25, 27, 38, 81, 112, 223, 259, 261, 263, 285, 288, 301, 303, 330; *Human Nature in Politics*, 25
Davies, N. B., 53
Davies, Paul, 113, 263
da Vinci, Leonardo, 221
Davis, Patrick, 107
Dawkins, Richard, 36, 38, 55, 128, 137, 139, 202, 327
Dawson, John, 261
Dawson, M. E., 293
Day, Clarence, 37, 45, 62, 115, 147, 158, 164
Dean, R. F. A., 287
Dearden, John, 16
de Broglie, Louis Victor, 301
de Luce, Judith, 245
Demarest, William, 152
Denenberg, Victor, 256, 270, 276
de Nicholas, Antonio, 112
Denny, Raoul, 154
Dent, A. A., 107
Deutsch, Georg, 281, 292
Deutsch, Karl, 15, 307–8
Deutsch, Morton, 308
DeVore, Irven, 55, 84, 101, 115, 118, 145, 148
de Vries, Hugo, 292, 299
de Waal, Frans, 73, 84–85, 88–90, 94, 101
Dimbleby, G. W., 106, 122, 169
Dimock, Marshall E., 301
Dimond, Stuart, 280, 282
Dirac, Paul, 301
Dobzhansky, Theodore, 137, 171, 201, 203, 212
Dolhinow, Phyllis, 75, 92
Douglas, William O., 104, 228–29, 240
Dreyfus, Hubert, 293
Dreyfus, Stuart, 162, 293
Dror, Yehezkel, 170
Dunbar, M. J., 43, 146
Dunford, Christopher, 161, 221
Dunham, Kelly, 99
Dunne, Gerald, 229, 235
Durham, William, 48–49, 70, 94, 100, 153, 203

Easton, David, 9, 11, 150, 301–3, 307
Ebbesson, Sven, 167
Eccles, John, 264, 276, 323
Eckelmann, Herman, 110
Edelman, Gerald, 162, 166, 292, 314
Edelman, M. S., 59, 264
Edsall, Thomas, 15
Ehrman, Lee, 294
Eibl-Eibesfeldt, Irenaus, 23–24, 32–40, 44, 56, 75, 92
Einstein, Albert, 113, 177, 226, 269, 299, 304–9, 315
Eisenberg, Leon, 36
Ekman, Paul, 6, 55
Eldredge, Niles, 171–74, 176–77, 204–5, 326–27
Elliott, David, 173
Engels, Friedrich, 32–33, 147
Eshel, Ilan, 49, 118, 135, 148–49, 154
Etheredge, Lloyd, 264
Euclid, 299, 309
Eulau, Heinz, 301, 304
Eysenck, Hans, 3, 226, 273

Fagen, Robert, 163
Fairman, Charles, 228
Falger, Victor, 57
Feldman, M. W., 105, 161, 221
Fentress, John, 76
Festinger, Leon, 226
Field, Stephen J., 229
Fields, W. C., 320
Filskov, Susan, 273
Fincher, Jack, 277
Findlay, C. Scott, 246, 289–95
Firestone, Harvey, 292
Firestone, Shulamith, 282
Fisher, Elizabeth, 71, 99, 105–6, 111–12, 282
Fisher, Helen, 74
Fisher, Ronald A., 102, 142–43, 171, 175, 201–3, 222; *Genetical Theory of Natural Selection*, 222
Flannery, Kent, 99
Flash, William, 14
Foley, Mary Ann, 271
Fortas, Abe, 104, 240
Fossey, Dian, xv
Foucault, Michel, 86, 112
Fox, Michael, 52
Fox, Robin, 6, 38, 42, 45, 62, 71, 99, 115, 140, 152, 281
Frank, Jerome, 225
Frank, Pat, 150
Frankel, Mark, 15
Frankfurter, Felix, 228, 233
Franklin, Rosalind, 269
Frankovich, Kathleen, 282
Freedman, Daniel, 101, 119
Freeman, L. G., 115
Freud, Sigmund, 25, 225
Fried, Morton, 6, 70, 99
Friedman, Milton, 217
Fries, James, 11, 14, 178, 226
Friesen, W. V., 6
Fuller, Melville, 229
Futuyama, Douglas, 104, 118, 299

Galileo, Galilei, 297, 299
Gamow, George, 299, 312
Gandhi, Indira, 281
Gandhi, Mahatma, 140
Gardner, B. T., 254
Gardner, Martin, 261, 322
Gardner, R. A., 254
Gazzaniga, Michael, 267, 276
Geber, Marcelle, 287
Geertz, Clifford, 109, 116
Geigle, Ray, 303
Geist, Valerius, 47, 49, 54, 71, 79, 99, 101, 105, 109, 159, 162, 172, 185, 188–98, 203–4, 225, 293, 329
Genco, Stephen, 321
Gerstenberg, Arnold, 208
Ghiselin, Michael T., 163
Gianos, Phillip, 5, 96, 215, 264
Gibson, Kathleen, 57, 253, 275
Gilligan, Carol, 282
Gilman, Charlotte, 282
Gilman, Stuart, 224, 274
Ginsburg, Benson, 327
Glamser, Francis, 228
Glazer, Nathan, 154
Glenn, Norval, 226
Globus, Charles, 264
Gluckman, Max, 39
Godwin, R. Kenneth, 14
Goethe, Johann Wolfgang von, 104; *Faust,* 104
Goff, John S., 228
Goffman, Ervina, 85, 112
Goldberg, Arthur J., 240
Goldfischer, David, 97
Golding, William, 105
Goldman, Sheldon, 229–30
Goldman, Stanford, 314
Goldsmith, Donald, 173, 179
Goodall, Jane, xv, 72, 75, 84
Goodman, Ellen, 282
Goodwin, B. C., 106, 113, 118, 202, 206, 207–8
Gould, Stephen Jay, 76, 108, 170–73, 175, 177–78, 204–6, 222, 313, 323, 325–26
Gouldner, Alvin, 134
Gow, David, 226
Graber, Doris, 96, 272
Grady, K. E., 280
Graham, George, 301
Grant, Ewan, 55
Grant, J. A. C., 233
Grant, Lawrence, 97
Grant, Ulysses S., 235
Graubard, Mark, 165
Gray, Bennison, 120, 208
Gray, Jeffrey, 113, 162, 166, 268
Green, Halcott, 11, 17, 63, 106, 146–47, 173, 176, 182, 195, 328
Greenberg, L. M., 146
Greenwood, Davydd, 3, 42, 49, 70, 164
Gregory, Michael, 34
Gribbin, John, 99
Griffin, Donald, 71, 119, 145, 169, 255–57, 270
Grings, M. W., 293
Groos, Karl, 217
Gruber, Howard E., 163
Grunbaum, Adolf, 113, 215, 305–6, 316
Gruter, Margaret, xiii
Gualtieri, Thomas, 184, 281
Guillemin, Victor, 312
Gur, Rubin, 276–77
Gustafsson, Jan-Eric, 287
Guthrie, Stewart, 114

Haas, Michael, 4, 13, 302
Haddon, Alfred, 109
Haddow, Anna, 296
Hager, Joseph, 55
Haldane, John, 171, 201–3
Hall, Edward, 219
Hall, Roberta, 46, 81, 95, 102, 107, 251
Halliday, R. J., 224
Hallpike, C. R., 163
Hamburg, David, 115
Hamilton, William J., III, 10, 38, 115, 128–30, 137, 143, 149, 152–54, 157, 175, 202, 264
Handberg, Roger, 81, 237
Handler, Philip, 9, 11, 231, 299
Hannon, Bruce, 295
Harding, Robert, 74, 102
Harlan, John Marshall, 228–29
Harlow, Harry, 59
Harlow, Margaret, 59
Harnad, Steven, 284
Harrington, F. H., 256
Harris, Lauren J., 278, 280
Harris, Marvin, 34, 44
Hartl, Daniel L., 163, 166
Hartsock, Nancy, 322
Hass, Hans, 6, 33, 39, 56
Hawthorne, Nathaniel, 134; *Scarlet Letter, The,* 134
Head, Kendra G., 227
Hediger, Heini, 52
Heelan, Patrick, 317
Heisenberg, Werner, 177, 269, 301, 304–6, 308, 311–15, 319
Henley, Nancy, 275
Herodatus, 100, 134
Herring, Pendleton, 298

Herrnstein, Richard, 15, 34
Herron, Jeannine, 279
Hershey, Marjorie, 282
Heymer, Armin, 51
Hicks, Robert, 184, 281
Hildebrand, James, 44
Hill, Jack, 47, 49
Hill, Sandra, 281
Hinde, Robert, xv, 10, 36–37, 52, 55, 128, 164, 219, 257
Hines, Samuel, 14, 81, 224, 300
Hippocrates, 297
Hirsch, Harry, 233
Hitler, Adolf, 87, 140
Hobbes, Thomas, 6, 70
Hofer, Myron, 6, 292
Hofstadter, Richard, 216, 220, 224
Holmes, Oliver Wendell, Jr., 228, 233
Hopple, Gerald, 272
Horowitz, Irving L., 264
Hrdy, Sarah Blaffer, 73, 75, 84–85, 90–92, 94, 101–2
Hubbard, Ruth, 139, 163, 168
Hughes, Charles Evans, 229, 240
Hull, David, 54
Hummel, Ralph, 86
Humphries, David, 55
Hutt, Corinne, 217
Huxley, Julian, 171

Ingber, Lester, 314
Inhelder, Barbara, 267
Irons, William, 62, 70, 115, 224, 281
Isaak, Robert, 86
Itani, Junichiro, 73, 252
Iverson, Leslie, 120
Izard, Carroll, 267, 275

Jahn, Robert, 113, 263
Jantsch, Erich, 217
Jaros, Dean, 13–14, 97, 224, 226, 233
Jarvik, Lissy, 232
Jastrow, Robert, 104
Jay, Antony, 45, 80
Jaynes, Julian, 80, 111, 265
Jefferson, Thomas, 220
Jensen, Arthur, 287
Jerison, Harry, 280
John, E. Roy, 265–66
Johnson, Ronald, 287
Johnson, Samuel, 133
Johnson, Timothy D., 163, 256
Jolly, Clifford, 114
Jones, Alexander, 110
Jones, Diane, 80, 101
Juhasz, Victor, 318

Kamin, Leon, 293–94
Kant, Immanuel, 318
Kaplan, Abraham, 98, 301
Kaplan, Charles, 322
Karl, Barry, 298
Katchadourian, Herant, 280
Katz, Neil, 282
Kaufman, Herbert, 176
Kavanagh, Dennis, 96
Kawanaka, K., 75
Keith, Arthur, 105, 110, 118
Kelly, Petra, 281
Kennedy, John F., 221
Kennedy, Robert F., 221
Kent, Ernest, 162, 166
Key, Mary, 55, 112, 257
Khomeini, Ayatollah, 112
Kimble, Daniel, 3
Kimura, M., 141
King, Glen, 46, 72, 74, 77–78
King, William, 229, 237
Kinsbourne, Marcel, 275, 277
Kirkpatrick, Jeane, 281
Kirkpatrick, Samuel, 234
Kitchin, William, 303, 322
Klein, R. G., 146, 170
Klinghammer, Erich, 52
Kocel, Katherine, 184, 276, 279–80
Koestler, Arthur, 212; *Case of the Midwife Toad, The*, 212
Konner, Melvin, 55, 101
Kort, Fred, 105, 176
Kortlandt, Adriaan, 72, 95, 114
Kovack, Joseph K., 163
Krames, Lester, 223
Krebs, J. R., 53, 55
Krech, David, 25
Kuhn, Thomas, 98, 141, 179, 208, 220, 222, 298
Kurten, Bjorn, 105
Kurth, G., 99, 102, 105

Lamarak, Jean de, 325
Lamb, Charles, 230–31, 240
Lancaster, Jane, 47, 105, 115, 249, 257
Landau, Martin, 37, 84, 89, 139, 298, 307, 319, 322
Laponce, Jean, 5, 7, 15–16, 259, 262–63, 278, 303, 315, 330
Larven, Ray, 47, 75, 86–87
Lasswell, Harold, 13, 85, 98, 170, 264, 272, 302
Latour, Bruno, 58
Laughlin, Charles, 112, 116, 231
Lawson, Robert, 96, 271

Layzer, David, 131
Leavitt, Donald, 237
LeDoux, Joseph, 267, 276
Lee, Richard, 45, 55, 84, 101, 118, 145, 148
Lee, Robert E., 235
Lehman, Harvey, 232
Lehner, Philip, 56–57
LeMay, M., 277
Lenin, Nikolai, 217
Lenneberg, Eric, 256, 276
Lenski, Gerhard, 4
Lerner, Gerda, 282
Lerner, I. Michael, 157, 185
Leventhal, Howard, 6
Levins, R., 151
Levinson, Boris, 108
Levitt, Paul R., 151
Levy, Jerre, 277, 280–81
Levy, J. M., 280
Lewin, Kurt, 308–10
Lewin, Roger, xv
Lewontin, Richard, 34, 43, 132, 137, 141, 168, 170, 203, 293–94, 327
Lex, Barbara, 112, 116–17
Libet, Benjamin, 289–90
Lieber, Francis, 296
Lilly, John C., 52
Lippitt, Ronald, 308
List, David, 282
Livingston, Robert, 261–62, 273, 286, 315
Loch, H., 91
Lockard, Joan, 48, 55, 71, 115
Lockwood, Randall, 76
Lodge, Milton, 271, 274
Loftus, Geoffrey R., 163
Loomis, Burdett, 294
Lorenz, Konrad, 10, 23, 33–34, 36–37, 51, 70, 148
Losco, Joseph, xiii, 4–5
Lovett-Doust, Jon, 16
Lovett-Doust, Leslie, 16
Lowell, James Russell, 235
Lowther, Gordon, 60
Loy, James, 73, 84, 99
Lucy, xv, 98
Lumsden, Charles, 92, 105, 156, 170, 221, 246, 289–96, 327
Lyons, John, 55
Lysenko, Trofim Denisovich, 212

MacArthur, Robert H., 53, 162; *Geographical Ecology*, 162
Maccoby, Eleanor, 16
Machiavelli, Nicccolo, 95, 98
MacLean, Paul, 162, 166, 225, 257
Macphail, Eulan, 287
Maddox, William, 81
Madsen, Douglas, 26, 102
Mainardi, Danilo, 47–48, 78, 92, 251
Manheim, Jarol, 5, 96, 273, 294, 303
Manicas, Peter, 92, 96
Mann, Thomas, 157
Mao, Tse-tung, 140, 160; "Little Red Book," 160
Marais, Eugene, 37, 101, 115, 158
March, James G., 301
Margenau, Henry, 299, 301, 303–7, 312, 320
Markl, Hubert, 163, 166
Marler, Peter, 10
Marshack, Alexander, 71, 109
Marshall, John, 229
Marshall, Thurgood, 104
Martin, Paul, 146, 170
Maruyama, Magorah, 283
Marx, Karl, 32, 62, 120, 127, 142
Maslow, Abraham, 25, 27, 38, 220
Mason, William, 217
Masters, Roger D., xiii, xv, 5, 10–11, 34, 41, 45, 47–48, 56, 61, 71, 75, 78, 83, 85, 97, 133, 158, 163, 178, 220, 275, 296
Mateer, Catherine, 275
Maynard-Smith, John, 79, 129–30, 163, 202–3
Mayr, Ernst, 172, 204
McCarthy, Joseph R., 179
McClain, Ernest, 110–11, 113
McClelland, Charles, 309
McClosky, Herbert, 227
McCoy, Charles, 301
McEacheron, Donald, 104, 115, 118
McGlone, Jeannette, 280
McGown, Elizabeth, 115
McGrew, William, 36, 249–50
McGuigan, F. J., 267
McGuinness, Diane, 42, 102, 265, 279–81
McGuire, Michael, 73, 81, 91, 102, 330
McKelvey, Richard, 127
McKenna, Joseph, 228
McReynolds, James C., 240
Mead, Margaret, 221
Means, R., 4
Mecacci, Luciano, 283
Mech, L. D., 75–77, 256
Medawar, Peter, 44, 46, 168
Meier, Golda, 281
Mendel, Gregor, 32, 113, 292, 297, 325
Menzel, Emil, 252, 254, 256–57
Merriam, Charles E., 258, 298, 301–2
Mersky, Roy, 237
Merton, Robert, 307
Messinger, Harley, 4
Metcalf, David, 286
Mey, Harold, 308–9
Miller, Eugene, 9

Miller, Samuel F., 237–38
Minkowski, Hermann, 304
Mirabile, C. S., 262
Mittwoch, Ursula, 184
Monad, Jacques, 164, 169, 279; *Chance and Necessity*, 164
Monckeberg, Fernando, 285
Montagu, Ashley, 34, 36
Moody, William, 228
Moran, Greg, 76
Morgan, Charles, 48
Morgan, Elaine, 174
Morgan, Michael, 166, 183–84, 277
Morgan, Thomas Hunt, 201
Morowitz, Harold, 318, 322–23
Morris, Desmond, 10, 39, 56, 70, 88, 114–15, 148, 257
Mortenson, F. Joseph, 36–37
Moses, 112
Mountcastle, Vernon, 162, 166, 264, 314
Munro, William Bennett, 304
Murphy, Bruce Allen, 104, 230
Murphy, Frank, 104

Nadel, Lynn, 268
Nagoshi, Craig, 287
Nash, John, 27
Nettleton, N. C., 166, 276
Newman, Oscar, 80
Newman, Robert, 110
Newmeyer, John, 227
Newton, Isaac, 37, 113, 246, 297, 299–300
Nishida, Toshisada, 72, 75
Nishimura, A., 252
Nitecki, Matthew, 146, 170
Noe, Ronald, 73
Northrop, F. S. C., 311
Nye, Russel B., 238

O'Connor, Sandra, 240, 281
Odum, Eugene, 17, 120
Odum, Howard T., 140
Ohta, T., 141
Ojemann, G. A., 280
O'Keefe, John, 268
Olton, David, 268
O'Malley, J., 120
Omark, Donald, 59, 101, 114–15
O'Neil, Robert, 104
Ordeshook, Peter, 89, 101, 127
Ornstein, Robert, 113, 117
Orwell, George, 112
Overton, William, 104

Pagels, Heinz, 113, 263
Paige, Jeffery, 110
Paige, Karen, 110
Panksepp, Jaak, 162, 166, 261, 268
Parker, Sue Taylor, 57, 275
Parrington, Vernon, 134; *Main Currents in American Thought*, 134
Parsons, Jacquelynne, 280
Parsons, P. A., 294
Parsons, Talcott, 87
Pascal, Blaise, 289
Paxson, Frederic, 238
Pelletier, Kenneth, 263–64, 266, 276, 314
Peters, Roger, 71, 75–77, 108
Petersen, Anne, 280
Peterson, Steven, 57–58, 73, 76, 96–97, 165, 220, 223–24, 271–74, 303, 314, 323
Pettersson, Max, 43, 62, 80, 98, 105, 152, 328
Pettman, Ralph, xiii, 5, 42, 90, 96, 281, 329
Pfeiffer, John, 249
Piaget, Jean, 208–9, 267
Pierce, Charles S., 322
Pilbeam, David, 6, 10
Pitcairn, Thomas, 35
Planck, Max, 269, 299, 306
Plato, 98, 113, 246
Playford, John, 301
Pliner, Patricia, 223
Plutchik, Robert, 163, 166
Poirier, Frank, 53, 57, 59, 80, 223, 253
Poortinga, Ype, 287
Pope, Kenneth, 266–67
Popper, Karl, 208, 320–23
Porter, Joshua, 108
Prakash, Om, 72, 99
Pranaer, Robert, 4
Premack, David, 248, 257
Presthus, Robert, 36
Preziosi, David, 112
Pribram, Karl, 42, 102, 159, 161, 166, 261, 263, 265–66, 273, 281, 314
Pritchett, C. Herman, 104, 237
Pryor, Gordon, 284, 286
Pulliam, H. Ronald, 161, 221
Pyke, Graham, 53
Pythagoras, 113

Ra, Chong Phil, 292
Rabbitt, P. M. A., 287
Rajecki, D. W., 37
Rainier, H. S. H., III, 53
Rao, B. S. Madhava, 301, 305, 312
Rashevsky, Nicholas, 307
Raskin, Allen, 232
Ray, James Lee, 310
Ray, R. M., 43
Reagan, Ronald, 97
Reed, Charles, 107

Reed, Thomas, 302
Rehnquist, William H., 238
Reichenbach, Hans, 113, 215, 304–6, 311
Restivo, Sal, 113
Reynolds, Peter, 235
Reynolds, Vernon, 71, 81, 93, 128
Ricardo, David, 127
Richardson, William, 97
Richerson, Peter, 105, 129, 293
Riecken, Henry W., 226
Riemann, Georg Friedrich, 305
Riesman, David, 154
Riker, William, 89, 101, 127
Riley, Matilda, 226
Roberts, Morley, 4, 208
Roberts, Owen J., 240
Robertson, Alec, 199–200, 203, 207–8, 210, 214–15
Robinson, John P., 227
Rodman, John, 263–64, 302
Roederer, Juan, 159, 161, 263, 268–70, 273, 278, 292, 314
Roell, A., 37
Roelofs, Joan, 233, 240
Rogers, Edward S., 4
Rogers, Linda, 283, 322
Roitblat, H. L., 267
Rokeach, Milton, 227, 273
Roosevelt, Franklin Delano, 229
Rose, S., 293–94
Rosen, Robert, 314
Rosenbaum, Walter A., 97–98
Rosenberg, Alexander, 163
Ross, Barry, 152
Rossi, Alice, 98
Rosso, Pedro, 285
Rousseau, Jean Jacques, 6, 62, 70, 83
Rubenstein, Robert, 85
Rumbaugh, Duane, 245, 248
Rummel, Rudolph, 309–11
Ruse, Michael, 299
Rushton, J. Phillipe, 287
Rusk, Jerrold G., 227
Russell, Bertrand, 104, 207, 215, 226; "A Free Man's Worship," 104
Russell, William R. S., 180
Rutledge, Wiley, 104
Ruyle, Eugene, 203
Ryan, John Paul, 237
Ryder, Norman, 234

Sacher, George, 232
Sachs, Mendel, 306, 308, 312, 319
Sackheim, Harold, 276
Sacksteder, Richard, 113
Sagan, Carl, 105, 144, 162, 235
Sahlins, Marshall, 34, 128
Salmon, Wesley, 113, 215, 305, 308
Sambursky, Samuel, 299
Sapir, Edward, 283
Sasanuma, Sumiko, 283
Savage-Rumbaugh, E. Sue, 245, 248
Sawyer, Jeffrey, 99
Schacter, Stanley, 135, 226
Schaller, George, 46, 60, 188
Scher, Jordan, 264
Schleidt, Margaret, 35
Schmalhausen, I. L., 212
Schmidhauser, John, 226, 229, 234
Schopf, Thomas, 177
Schrodinger, Erwin, 301, 313, 320
Schubert, Glendon, xv, 5, 7, 16, 28–30, 39, 41, 49, 51, 56, 73, 75, 81, 83, 85–86, 91–92, 97, 99, 103–4, 112, 114–15, 128, 151, 159, 161–63, 165–66, 172, 179, 193, 202–4, 208, 220, 223, 227, 229, 232, 235–37, 240, 254, 257, 272–73, 275, 278, 280–82, 296–98, 300–303, 307, 309, 314, 320, 322–23, 330
Schubert, James, 5, 14, 56, 81, 284, 288
Schumaker, Ernst, 263
Schwartz, David C., 13–14, 223
Schwartz, Gary, 261, 264
Schwartz, Sandra, 223
Scott, John Paul, 61
Sebeok, Thomas, 53, 71, 81, 245
Secord, Paul, 96
Service, Elman, 99
Seyfarth, Robert, 75
Shakespeare, William, 75, 95, 98, 134, 144, 166; *Titus Andronicas*, 75
Shapiro, David, 261, 264
Shapiro, Michael, 271–72, 322
Sharp, Elizabeth, 6
Sharp, Henry, 46, 95, 102, 107, 251
Shepard, Paul, 71, 108–9, 112, 114, 119, 169, 235, 273
Shepfer, Joseph, 98, 163
Shneour, Elie, 284
Shubs, Peter, 13
Sibatani, Atuhiro, 283
Sibly, R., 55, 145
Silverberg, James, 53, 128
Silverman, Paul, 119
Simon, Herbert, 246
Simon, Robert L., 224
Simonton, Dean K., 290, 293–94
Simpson, George, 148, 171
Singer, J. E., 135
Singer, Jerome, 266–67
Slagter, Robert, 97
Slobodkin, L. B., 163
Smith, Adam, 127

Smith, Annette, 268
Smith, Euclid, 70, 115, 224
Smith, Peter, 269
Socrates, 6, 62, 133
Somit, Albert, 3–5, 7, 13, 25, 34, 58, 73, 83–85, 97, 115, 164, 220, 259, 274, 296, 298, 300–302, 304
Sorokin, Pitrim, 309
Sperry, Roger, 264
Spiro, Melford, 98
Springer, Sally, 281, 292
Sproul, Barbara, 110
Srb, A. M., 46
Stacey, Judith, 98
Stanley, Steven, 105, 172–73, 178, 204, 275
Stapp, Henry, 315
Stauffer, Robert, 4, 13, 288
Steiglitz, Robert, 107, 113
Steklis, Horst, 284
Stent, Gunther, 314
Stern, Elizabeth, 286
Stevens, John Paul, 237
Stini, William, 3, 42, 49, 70, 164
Stone, Harlan Fiske, 104, 228–29, 240
Storing, Herbert, 6, 301
Strauss, Leo, 70
Strayer, Fred, 75, 101
Strehler, Bernard, 231
Struever, Stuart, 100
Strum, Shirley, xv, 58, 74–75, 85, 93–94, 99, 102
Sulloway, Frank, 225, 264
Surkin, Marvin, 25
Suzuki, Akira, 72
Swindler, William, 237
Symons, Donald, 59, 74, 80–81, 268
Szasz, Thomas, 86, 112

Tanenhaus, Joseph, 57, 271, 296, 298, 300–302, 304
Taney, Roger B., 229
Tate, C. Neal, 237
Tattersall, Ian, 171, 173–74, 176–77, 326–27
Taylor, David, 184, 280
Teitelbaum, Michael S., 226
Teleki, Geza, 73, 78, 102, 115
Templeton, Alan, 99
Ten Houten, Warren, 284, 322
Terrace, Herbert, 53, 245
Thatcher, Margaret, 281
Thomas, Tracy, 306
Thompson, Malcolm, 104
Thompson, Philip, 46, 60, 73, 75, 77, 102
Thorbeck, William, 4
Thorson, Thomas, xiii, 4
Thouless, R. H., 208
Tiger, Lionel, 6, 38, 42, 45, 62, 71, 74, 98–99, 102, 109, 113, 115, 117–20, 140, 152, 235, 273, 281; *Optimism, The Biology of Hope*, 235
Tinbergen, Niko, 23, 37, 45–46, 50–51, 97, 257
Tobach, Ethel, 36
Tomkins, Sylvan, 6, 37, 267–68
Travis, Cheryl, 34–36, 38, 55
Tresolini, Rocco J., 103
Trivers, Robert, 34, 38–39, 91, 128, 132–37, 145–46, 148–49, 203
Truman, David, 301
Trump, David H., 105
Tschermak von Seysenegg, Erich, 292, 299
Tsunoda, Tadanobu, 283
Tulloch, Gordon, 4
Turnbull, Colin, 44
Tuttle, Russell, 115

Ucko, Peter, 106, 122
Umiker-Sebeok, Jean, 53, 71, 245
Underwood, Jane, 45, 47, 49, 185
Ushenko, Andrew, 304–5
Ussher, James (Archbishop), 113
Uttal, William, 261, 264, 283

van den Berghe, Pierre, 73, 79–81, 115, 163, 224
van Dijk, J. J. M., 60, 78, 153
van Dyke, Vernon, 25
van Gulick, Robert, 163
van Hooff, Jan, 55, 73, 88, 102, 219
Van Over, Raymond, 110–11
Vaughn, Brian, 75
Veblen, Thorstein, 186
Vogel, Christian, 75, 91–92
von Cranach, Maria, 257
von Eckardt, Barbara, 276
von Glasersfeld, Ernst, 249
von Neumann, John, 301

Waber, Deborah, 279, 283
Waddington, Conrad Hal, xvi, 49, 54, 62, 105, 157, 159, 161–62, 164, 166–67, 172, 186, 191, 199–222, 293, 329; *Behind Appearance*, 221; *Ethical Animal, The*, 161, 215; "Evolution in the Sub-Human World" 217; *Evolution of an Evolutionist, The*, 161; *Organizers and Genes*, 201; *Scientific Attitude, The*, 216–17; *Strategy of the Genes, The*, 161
Wahlke, John, 271, 274, 302–3
Walker, Alan, 72
Wallace, B., 46
Wallas, Graham, 25, 225, 258, 300–301; *Human Nature in Politics*, 225
Warren, Earl, 104, 237
Wasby, Stephen, 237

Washburn, Sherwood, 43, 105, 275
Waters, Everett, 75
Watson, James, 269, 314
Watts, Meredith, 5, 60, 74, 102, 281
Weber, Max, 86–87, 149
Webster, G., 106, 113, 118, 206–8
Weg, Ruth B., 226, 232
Weiss, Gerald, 251
Welker, W. I., 217
West Eberhard, Mary Jane, 129, 131, 145
Western, Jonah David, 94
Wheatley, Bruce, 92
Wheeler, Harvey, 48, 53, 111, 113, 314
Wheeler, John A., 306
Whitaker, Harry, 280
White, Edward D., 229
White, Elliott, xiii, 5, 11, 15, 34, 42, 97, 163, 223–24, 259, 274, 298, 303, 321–23
White, Lynn, 108
Whitehead, Alfred North, 207–8, 215, 221; *Principia Mathematica*, 207
Wholwill, Joachim F., 163
Whorf, Benjamin Lee, 283
Wiegele, Thomas C., 4, 5, 13–14, 56–57, 81, 97, 223, 257, 302
Wight, Martin, 89–90
Wilber, Ken, 120, 263
Wilder, Hugh T., 245
Willhoite, Fred, 45, 52, 99, 140
Williams, George, 83, 104, 127–28, 130, 132, 137–38, 140, 142, 146–49, 151, 163, 170, 175, 179, 202, 231, 281, 299
Wilson, David Sloan, 118
Wilson, Edward O., xx, 34, 36, 38, 92, 105, 128, 131, 135, 137, 140, 146, 151, 156–57, 162, 167, 170, 203, 221, 246, 249, 289, 293, 296, 327; *Genes, Mind, and Culture: The Coevolutionary Process*, 161–63, 167; *On Human Nature*, 161; *Promethean Fire: Reflections on the Origins of Mind*, 161–64, 167, 170; *Sociobiology: The New Synthesis*, 161
Wilson, Glenn, 226–28, 273
Wilson, H. T., 36
Wilson, James R., 287
Wilson, Margo, 34, 38
Wingerson, Lois, 80, 85
Winick, Myron, 285
Witelson, Sandra, 277
Wittgenstein, Ludwig, 200, 208; *Philosophical Investigations*, 208
Wittig, Michele, 280
Wolf, Fred Alan, 113, 263
Wolfe, Alan, 25
Wood, Charles, 289–90
Wood, James, 104
Woodruff, Guy, 248, 257
Woodward, Bob, 229
Wright, Quincy, 90, 309–10; *The Study of International Relations*, 309
Wright, Sewall, 49, 90, 119, 129, 131–32, 136–37, 142–43, 148–51, 155, 157, 172, 179, 201–3, 328; *Evolution and the Genetics of Populations*, 203
Wynne-Edwards, Vero, 53, 61, 129–30, 142–43, 202

Yakimov, V. P., 73
Yanarella, Ernest, 170, 179
Young, Michael, 15
Yunis, Jorge, 72, 99

Zegura, Stephen L., 224
Zeuner, Frederick, 107, 169
Zill, Nicholas, 13
Zinnes, Dina, 89–90
Zivin, Gail, 55
Zukav, Gary, 263, 305, 312–13, 316–17

Subject Index

Adaptation, 171, 189; animal, 173, 175; to artificial environments, 279; of chimpanzees, to human laboratories, 250; endogenous, 20; exogenous, 210, 212, 327; human, 17–19, 26, 34, 39, 43–46, 48, 50, 61, 79, 87, 105, 121, 128, 130, 147, 149, 152–54, 159, 169, 189, 226, 273, 327; plant, 173
Africa, 34, 55, 60, 105–6, 118, 174, 196; East, 98; North, 189; South, 174, 218
Aggression theory, 42, 60, 77–79, 81, 86–88, 91, 93, 152–53, 190, 194; and sexual behavior, 330
Agriculture, 100, 105, 115, 121, 136, 147, 149, 152, 154–55, 170, 185, 195
American Civil War, 104, 221, 233, 238
American Political Science Association, 4, 301–2
American Political Science Review, 15
American Society of Primatologists, 97, 250
American Sociological Review, 33, 134
Analysis, levels of, 89, 94, 96–97, 130, 292
Animal behavior, 194, 248–57, 260, 299, 324–25; dominance, 86–91, 93, 100, 114–15; experimental studies of, 217; feral, 249–50, 252, 325; of pets, 108, 139; responses to environmental information, 194; social, 50–52, 129; social learning, 48, 83
Animal Behavior Society, 97, 132, 140, 149
Animal communication, 54, 256–57
Animal culture, 217, 251
Animal thinking, 108–9, 119, 158, 248–57; and memory, 269
Ann Arbor Science for the People Editorial Collective, 38, 128
Anthropology: cultural, 251, 291, 301; physical, 52, 67, 71, 80, 151, 260; social, 151
Anthropomorphism, 54, 68, 85, 88–89, 91, 95–98, 115, 138–39, 249
Appetitive behavior, 29, 31
Archeology, 157, 195
Arnhem Zoo, 73, 84, 88, 97
Art, 200, 203, 214, 221; cave wall paintings, pretransition, 109
Asia, 105–6, 111, 118, 185, 195, 234, 259, 328; Southeast, 154
Association for Politics and the Life Sciences, 4, 302
Astronomy, 104, 175, 179, 263; the Universe, 104, 207, 292, 311, 314
Athens, Greece, 6, 70, 80, 141
Attention "structure," 68, 75, 84–87
Australopithecids, 70, 78, 98, 148, 174; gracile, 174; robust, 174. *See also* Hominid(s)

Behavior(al): attachment, 37, 133; coalitional, 88–90; conservatism, 186–87; ecology, 17–19, 49, 53, 128, 193–94, 217, 290, 324; epigenetics, 202–6, 209, 294, 326–27, 329; "fixed action patterns," 40, 51, 56; genetics, 128, 134, 158, 299, 327; plasticity, 144–45, 326–27; psychophysiological, 13–14; scavenging, 74, 78–79, 94, 99
Behavioral and Brain Sciences, 5, 33, 161, 163, 167–68, 245
Biobehavioralism, 121, 245–46, 258, 302; biocultural epigenetics, 204; bioenergetics, 193;

bioethics, 169; biogenetic engineering, 325, 327; biogenetic structuralism, 116
Biofeedback, 30, 49, 261
Biological competitive exclusion, 179
Biological determinism, 225, 260, 262
Biological organization, 203, 259–64
Biological "wisdom," 218–19
Biology, 221–22, 290, 303, 319; antiquarian (Aristotle), 246, 299, 307; biometrics, 203; classical (Darwinian), 299–300, 307; distinguished from culture, 290; feminist fears of, 177; modern biology and modern physics: unified theory of, 314; modern (Mendelian synthesis), 9–12, 164, 169–70, 200, 222, 258, 299, 303, 309; pre-Darwinian epistemology, 321; social, 128, 224, 275, 293–94; *un*biological approaches, 248, 326
Biophysics, 245–46, 303, 307; of the brain, 313–19, 329. *See also* Psychobiology
Biopolitical behavior, 12–16, 28–31, 98, 258
Biopolitical theory, 6, 8–9, 95–102, 273–74
Biopolitics, 4–5, 97, 260, 288, 302–4
Biosocial behavior, 259–60
Biosocial science, 3–4
Biosphere, 130, 146, 189
Biospheric niche, human, 179–80
Birds, 221
Birth control, 153
Birth order, 198
Brain anatomy, 259–64, 323; resistance to malnutrition, 316; versus mind metaphysics, 289–95
Brain development, 183–87, 192, 259; and accelerated prenatal growth, 285; cultural effects on, 259, 282–84; decrements in, 259; environmental modification of, 284; epigenetic, 284, 292–95; "Japanese" brain, the, 283; laterality, direction, and timing of, 183; models of, 260–64; "neuroplasticity" in, 284; nutritional effects on, 284; plasticity in, 274–75; and political equality, 274–88; postnatal, 293; postnatal language bilaterality, 276; RH decline with age, 277
Brain evolution; and speech, 275; and simian preadaptation for speech, 276
Brain functions, higher, 73, 292, 328
Brain science, 113, 120, 122, 128, 136, 156, 159, 162, 164, 166–67, 245–46, 257–58, 303, 307; joint arousal of aggression and sex, 194, 330; sexual differences in brain structure specified, 280–81
Brain structure, three-axis, 260–61; and limbic system, 267

Cambridge University, 200–201, 203, 207–8, 215
Canada, 150
Canadian Rockies, 185, 188, 203
Cannibalism, 134, 152–53, 195
Catastrophe theory, 24, 126, 152, 175, 204–6, 327–28; ecocatastrophes, 118, 126, 152, 328; nuclear holocaust, 176, 179, 218–19, 282; water pollution, 19, 328
Caucus for a New Political Science, 25, 302
Center for Advanced Study of the Behavioral Sciences, 70, 84
Chance, 313, 319–20, 322, 327; and natural randomness of environmental exploration, 273
Chauvinism, 217; of age, 217; ideological, 240; racist, 240; religious, 240; sexist, 217, 220, 240; of social class, 216–17
Chimpanzee(s), 72–74, 77–78, 84, 88–89, 115, 158, 325; Austin, 249–54; cognition, 248–55; *Pan troglodytes*, 249, 252; Sarah, 253–55; Sherman, 249–54; socialization of, 252
China, 33, 111, 149, 154, 329
Cognition, human, 113, 165–66, 179–80, 235, 245, 254, 264–70, 315; consciousness, 245, 258, 261, 268, 272, 303–4, 316–18, 321–23; distinguished from nonhuman, 248–49; emotional effects on, 267; engram, 273; explicate, 318; implicate, 318; intentionality, 317–18; language processing, 267, 270; manifestation (of the present), 316–18; memory, 245, 259, 265–70, 274, 303, 314; neocortex, 261; neurobiological models, 259–62; of the past and future, 297, 316; processing nonverbal communication, 267; reification, 160; schemata, 271–72; *self*-consciousness, 264–66, 329; and sensory perception, 266, 268–70, 311, 315, 330; subjective relevance, importance of, 273, 312–13, 317–18; templates, 269; of verticality, 260
Cognitive development, human, 108–9, 144, 225; epigenetic, 330; Piagetian, 57, 292; and rigidity 228; senescence in, 227
Cognitive ethology, 57, 94, 102, 108–9, 119, 245, 256; evolutionary continuity of mental experience, 257; and theory of chimpanzees' computer minds, 250
Cognitive mapping, 76, 272
Cognitive psychology, 120, 166, 271, 291; heuristics of, 271–72
Colonizing populations, 185, 189, 197, 205, 329; adaptive radiation, 173; demographic mobility, 148; dispersal phenotype, the, 185, 191–92, 197, 221; neoteny of, 190; in space, 190–91
Comparative primate phylogenetic evolution, 95, 99, 101–2
Conservatism, 219, 226–31; attitudinal, 226; behavioral, 226; correlates of, 228; defined, 227; judicial, 224; political, 176, 187, 216, 226–28,

263; psychological, 226; theories of, 224, 227–28
Consociational democracy, 150
Constitutional policy (U.S.), 104; historical, 228; interpretations of, 220
Constitution (U.S.), 62, 70, 287, 319
Creation myths, 69, 109–111, 113, 121–22; Genesis, 6, 69, 110–11, 113; Judaic-Christian, 68, 106–7, 110–12, 115
Creativity, 149, 165–66, 186, 197, 199, 208, 222, 232, 246, 289–95; diversity in, 291–92; and environmental entropy, 295; genetic potential for, 294; genius, 293; individual, 294–95; innervation versus innovation, 289–90, 292, 294; in the Middle Ages, 295; invention versus discovery, 289–90; in vivo, 292; and political fragmentation, 294–95; and population group diversity, 290; schizoid, 119–20; and serendipity, 290; Simonton's statistical law of social recognition, 290
Cultural determinism, 224, 324
Cultural evolution, 43, 47–49, 54, 62, 92, 95, 105–6, 109, 112, 121, 125, 135, 145, 157, 162, 176, 196, 219, 246, 293–95, 325; biologically transmitted in vivo, 326; Lamarckian, 325–26; and written language, 256
Cultural selection, 129, 134, 137, 147, 154; production versus reproduction in, 326
Culture: defined, 325; Eastern, 160; national, 103; political, 104, 151, 179, 259; religious, 104; stimulation by, 186; Western, 160, 165

Design, environmental, 194–95; artificial, 189, 263–64, 273–74; healthful, 189
Development, phylogenetic, 108, 117–18, 162, 165, 183, 214, 296, 329–30; and emotion, 267–68; and "the phylogeny fallacy," 189
Developmental neurobiology, 158, 167, 186, 299, 328
Dinosaurs, 105, 175
DNA, 150, 209, 328; defined, 325
Domestication, 69, 100–101, 106–8, 111, 121–22, 126, 150–51, 169–70, 263; and human slavery, 121–22; plants, 107, 170; self-domestication of humans, 106, 108, 151, 170, 326
Dualism, mentalistic, 247, 264, 289–95, 323; unself-consciousness, 289

East African Rift, 85
Economic rationalism, 127; econometrics, 127, 246, 264, 294; Economic Man, 127; game theory, 127, 273
Egalitarianism, of age and sex, 150
Egypt, 107, 111, 167
Emotion, 33, 79, 81, 85–87, 96–97, 102, 112, 117, 121, 133–34, 162, 165–66, 198, 225, 227, 245, 253, 255–57, 259, 261, 273, 303, 329; and activation/arousal, 265, 267–68; and consciousness, 268–69; and limbic system, 267; and memory, 267, 271; and motivation, 267–68; and neonate consciousness, 267
Empirical evidence, 158–59, 165, 167, 173, 206, 294
Empiricism, 130, 165, 300
England, 107, 139, 199, 216, 218
English language, 138, 159, 249
Entropy (Second Law of Thermodynamics), 246, 316
Environmental change, 202, 221, 274
Environmental degradation, 91, 176, 195
Environmental determinism, 264
Environmental exploration, 273
Environmentalism, 52, 147, 264, 282
Environmental resources: abundant, 188–89, 191–92, 195, 197; minimal, 191–93
Environmental stases, 202, 295
Environmental stress, 86, 91, 94, 99, 106, 115, 126, 130, 170, 172, 178, 191–92, 204, 210, 212, 214, 220, 324; and behavior, 192, 210–11; physiological responses to, 213; selects against creativity, 192; stimulates psychopathic behavior, 192
Environments: artificial, 189, 263–64, 273–74; natural, 263
Epigenesis, 24, 49, 62, 106–8, 119, 130, 146, 157–59, 161–62, 164–68, 170, 178, 186, 191, 198–222, 246, 277, 292–93, 326–27, 329; defined, 211; exogenous, 214
Epigenetic consequences, 204, 292–93
Epigenetic genotypic reorganization, 204, 292
Epigenetic imprinting, 37, 51, 58
Epigenetic "rules," 165–66, 246, 292–93, 295
Epistemology, 186, 200, 207–8
Eras of primary socialization, 236–38; classified as decision eras, 238–39; defined, 236
Ethnocentrism, 195, 329
Ethnography, 39–40, 44, 53–54, 129, 136
Ethology, 51, 80, 128, 148, 153, 188, 210, 217, 252, 257, 260, 324; classical, 32–33, 36–37, 51–52, 194; and ethnography, 33; and experimental psychology, 33, 36, 38; human, 32–38, 40, 42, 45, 52, 54–56, 115, 193; methodology, 57; modern, 51–52; naturalistic observation and description, 50, 102, 143; political, 9–11, 23, 41, 55, 58; and sociobiology, 33
Eurasia, 186, 196
Europe, 106–7, 109, 112, 147, 149, 153, 185, 189, 195; pre-Renaissance, 246; Western, 12, 32, 36, 174, 183, 216, 231, 259, 328
Evolutionary biology, 80, 96, 104–6, 113, 118, 128, 148, 170, 176, 205, 291; analogical comparison, 71–72, 76–77, 79, 95, 97; convergent

evolution, 72, 79, 190; homological comparison, 45–47, 67, 71–72, 95, 97, 99, 101–2, 139
Evolutionary change, 171–75, 177, 204, 246
Evolutionary ethics, 215–21
Evolutionary extinction, 43–44, 52, 108–9, 155, 326–27; background, 126, 147, 327; human, 180, 190, 282, 295, 328; mass, 126, 175, 328; random, 175
Evolutionary genetics, 132, 141, 145, 149, 164, 326
Evolutionary paradigm, the classic, 200–206; Darwinian gradualism, 24, 43, 47, 69, 125–26, 140, 171–72, 175–79, 186, 204–6, 220, 246, 326, 328–29; neo-Darwinism-Mendelism synthesis, the, 199–206, 212, 222, 225, 293, 299, 329; sociobiology, 202, 293–94
Evolutionary paradigm, the new, 172, 200–206, 222; colonizing population/genetic drift, 172, 328–29; constructional biology, 113, 200, 206, 209, 314; epigenetic/dispersal change, 172; punctuationism/catastrophism, 172–75, 220–22
Evolutionary philosophy, 200, 207–8
Evolutionary progress, 216
Evolutionary stable strategies, 57, 79
Evolutionary stasis, 174, 176
Evolutionary systematics, 171
Evolutionary teleology, 231
Evolutionary theory, 34, 42–51, 62, 71, 80, 83–84, 97–99, 102–3, 121–22, 125–26, 139, 141, 143–44, 165, 168–69, 173, 176, 186, 200–202, 204, 215, 220–22, 260, 289, 299, 321–22, 324–25, 328–29; biotic, 154; Great Ladder of Being, 216–17; macro, (interspecific), 173, 205; micro (infraspecies), 173–74; parallel, 95, 101; parental investment, 53, 133, 138, 150; parent-offspring conflict, 150; phyletic, 173–74, 325; progressive, 216; punctuated equilibria, 126, 173–79, 205, 329; synthetic (defined), 205
Existentialism, 116
Experimental embryology, 201, 205, 221
Experimental psychology, 166

Family groups, 98–100, 113–14, 118, 127, 129, 149–50; 196
Female leadership, 88–89
Female social structures, 68, 73–74, 88–89, 102
Fertile Crescent, 69, 98, 107
Field theory, 306, 308–11
Fitness: cultural, 293; genetic, 118, 125, 129, 146, 150, 153; inclusive, 53, 128–29, 131, 133, 135, 139–40, 144–45, 149, 153, 155, 157, 202, 294; political, 220
Flood(s): Biblical, 107, 111, 118; glacial, 206
Food: politics, 18, 259; sharing, 118, 134; surpluses, 196
Fossil record, 174, 177–78, 205
Freudianism, 225, 264
Freudian slips, 165, 235

Galapagos, 45–46, 169
Garden of Eden, the, 68, 107, 110
Gatherer-hunters, contemporary, 136, 147, 149, 151–53, 162, 256; Kalahari San, 100–101, 136; ! Kung San, 55; Lapps, 152; New Guinea highlanders, 136; Philippine mountain Pygmies, 136; Zhun/twa San, 60
Gathering/hunting-band niche, the, 43–45, 60, 62, 69, 79, 94, 101, 107–8, 111, 115, 118–19, 121, 134, 136, 143, 147–49, 151, 154, 170, 195–96, 273, 279, 325, 329; group size, 196; mammoth hunters, 153; Natufian aboriginals, 69, 106–7, 110–11; Paleolithic, 101, 186, 231
Generation cohorts, 151, 178–79, 224–26, 233–34; transfer of information between, 251
Gene(s), 165, 201, 325–28; expression, 126, 172, 186, 212–14, 329; pool, 54, 118–19, 131, 139, 143–46, 170, 176, 189, 201, 210, 220
Genetic assimilation, 157, 201, 209, 212–14, 221; defined, 213
Genetic determinism, 128, 134, 145, 147–48, 158, 166, 177, 293, 326; of sex differences in brain structure, 280
Genetic(s), 201, 245, 292; "bean bag," 294; biotechnology, 126, 325, 327; canalization, 157, 201, 210–12; change, 172, 178, 221; competition, 130–31, 139–40, 147, 175–76; consequences, 157; diversity, 108, 137; drift, 126, 173, 329; engineering, 150, 170, 325–27; environment, 132; evolution, 105–8, 125, 135–36, 153, 157, 169–70, 325; eugenics, 4, 169–70, 195; molecular, 299, 314; mutation, 142, 171, 173, 176, 201–2, 205, 211–14, 327; plasticity, 140, 144, 191; pleiotropy, 54, 119, 158, 201, 213, 294; polygenic inheritance, 54, 158, 294; population, 128, 130, 139, 142, 148, 151–52, 154, 158, 171; random drift, 142–43, 155; recombination, 173, 326; research regulation of, 150; variation, 171, 211–12, 214
Genetic selection, 131; of neuronal groups, 292
Genotypes, 54, 62, 186, 201–2, 210, 260, 274, 293, 327; assimilated, 212; genetic reserves, 327; genomic repertoire, 327; genotypic organization and reorganization, 212, 327–28
Geology, 200, 206
Germany, 32, 218, 308
Gerontology, 232; political, 151
God(s), 110–11, 114–17, 139–40; Buddha, 140; deistic meaphors, 69, 259, 297; Jesus Christ, 114, 140; Jove, 140

Group(s): autogenous, 217; face-to-face, 115, 148, 154; hunting, social behavior of, 46, 196–97; learning processes, territories, 138

Habitat(s): chosen by animal(s) ("exploitive system"), 209, 211–12; depletion, 147; novel, 185; variation of, in space and time, 179
Harvard Kalahari Research Group, 55
Harvard University, 142, 163, 168, 207
Health, 189, 197–98; animal, 197; behavior, 185; disease, 213
Himalayas, 188
Hominid(s), 67–68, 70, 76–79, 84, 105, 108, 114, 118, 121, 136, 140, 151, 153, 169, 179, 196; *Australopithecus afarensis*, 174; *A. africanus*, 174; *A. boisei*, 174; *A. robustus*, 174; culture, preverbal, 251, 275; evolution, 173–74, 249, 273, 325; family *Hominidae*, 174, 327; genus *Homo*, 136, 143, 158, 174, 190, 192, 196, 273, 328; *Homo archaic sapiens*, 174; *H. erectus*, 78, 105, 174, 189; *H. habilis*, 78, 174; *H. sapiens*, 101, 105, 117, 136, 140, 174, 196, 325, 327; *H. sapiens Cro-Magnon*, 105, 109, 185; *H. sapiens neanderthalis*, 61, 105,109, 140, 146, 174, 188; *H. sapiens sapiens*, 105, 107, 121, 154, 169, 327; speciation, 174
Human aging, 223–40; behavioral differences, caused by, 76, 81, 88, 91; effects of, on decision making, 229; effects of, on political alienation, 237; judicial, 224, 228–35
Human development, 183, 196–97, 245–46; biocultural, 187, 236; biology of, 231–33, 236; of brain, 259, 262, 292–93; of children, 133; language, 197, 256; life-span, 231; neonatal, 263; 292–93; prenatal, 183–84, 210, 259, 262, 292–93; retarded rate for males (y chromosome), 184, 280; senescent, 186–87, 198, 224, 226, 232, 235–36; senility, 186, 231; sex differences in conversation, 227
Human Ethology Newsletter, 34–35
Human evolution, models of, 104–13, 188; biocultural, 113, 189–91; biological, 111, 113, 121; the brain, 265–68; and entomology, 157; Laetoli footprints, the, xv, 48, 98–99; pulses of glaciation, the driving force in, 189
Humanism, 251, 322, 327; softcore, 290, 310
Humanities, 300, 304, 324
Human needs, 25, 37–38, 63, 80, 290; psychological, 27, 108–9, 116, 272; psychophysiological, 29–30, 112, 116, 120; sensory stimulation, 26–27, 33, 107, 116, 120, 328; sexual interaction, 27, 116
Human(s): aggression, 190; biology, 258; culture (defined), 251; demography, 188; diagnostic features, 197; feral, 151; genetics, 191–93; nature, 258, 311
Human thinking, 108–9, 112, 116–19, 122, 158, 162, 166, 177, 179, 258–88, 323; artificial, 264, 273–74; modal female, 281; modal male, 281, 330; natural, 274; physiological sex differences in, 280–81
Human thinking, models of, 264–70; hemispheric dominance (Eccles), 265; holographic image-processing (Pribram), 265–67, 270; synergistic multipotentiality (John), 265

Ice ages, 174, 188, 206; Bering landbridge, 185; interglacial, latest, 188; periglacial environment, 185, 188–89, 196, 293, 329; Wurm-Wisconsin Glacial, the, 100, 109, 111, 185, 188
Ideological change: causes, 223; psychobiological (senescence), 223, 225, 239–40; sociocultural (socialization), 224, 233, 237–40; sociopsychological (learning), 159, 234, 241
Ideology, 97–98, 118, 125, 127, 134, 137–38, 176–77, 179, 186, 216–17, 226, 237, 241, 275, 328; academic, 6, 296, 302; authoritarian, 61, 218, 220; capitalism, 216, 219; conservatism, 221, 234, 240; judicial, 236; lateralization of, 277, 330; left-right dimension of, 278, 330; liberal, 234; Libertarian, 219; Marxism, 118, 142, 160, 264, 322; Reaganomic, 219; socialism, 216
Incest, 157, 163, 167
Indeterminacy, 201, 206, 213, 246, 312–13, 319–20, 322, 330. *See also* Chance
India, 84, 90, 111, 218; Jodhour, 90; Mount Abu, 84, 90; Southern, 218
Individualism, 149, 151, 177, 189, 199, 220, 274
Industrialism, 79, 106, 108, 115, 121, 136, 146–47, 149, 155, 176, 183, 195, 259, 328–29; post-industrialism, 106, 108, 326
Infanticide, 68, 75, 90–91, 134, 153
Insects, 127–28, 150, 158
Intelligence, 76–77, 94, 102, 105, 108–9, 131, 143, 152, 154, 175, 189, 259, 269; animal, 252–53; artificial, 162, 264, 273–74, 290, 314; and conservatism, 228; defined, 270; differences in, 253; genius, 226, 293; Imo, 217; malnutrition and IQ deficits, 286–88; measured by Piagetian criteria, 286
International Political Science Association, 4
International Primatological Society, 83, 94, 97
International Society for Human Ethology, 35
Israel, tribe of, 107, 110

Japan, 111, 136, 147, 149, 234, 283; Hiroshima, 219; history, 234; Supreme Court of, 234; Taisho democracy, 234
Jericho, 107, 111
Journal of Social and Biological Structures, 5
Judges, 112; Chief Justice of the United States, the, 80, 104; decision-making behavior of, 223;

District of Columbia circuit, 230; federal courts of appeals, 230; radical, 104, 240; of United States Supreme Court, 225–31, 233, 235–40

Kalahari Desert, 101, 145
Kenya, 83, 85, 94

Language behavior, 54, 95–96, 98, 102, 105, 144, 157, 165, 178, 193, 197, 267, 271; of apes, 245; artificial languages, 250; between humans and chimpanzees, 250; biolinguistics, 275, 322; of chimpanzees, 249; computer languages, 250, 314; development, 208, 253, 274–75; emotion and speaking, 276, 329; ethnolinguistic study of Hopi, 283; female verbal superiority, 280–81; grammatical models of, 38, 96, 248; learning, 276; linguistic phenomenology, 322; linguistic-relativity hypothesis, 283; mathematical, 315; "natural" languages, 249, 315; psychology of, 322; scientific, 297; speech, 249, 256–57; voice stress analysis of, 257; writing, 112–13, 256, 275, 314. *See also* Nonverbal communication
Language evolution, 136, 249; and the brain, 275; deep structure, 256; speech, 256, 275
Laterality: boys more sinistral than girls, 279; distinguished from lateralization, 277; girls more dextral than boys, 279; ideological, 262, 277–79; left-handedness correlated with left-wing politics, 278; majority are dextrals, 277; sinistrals a more heterogeneous group, 277
Lateralization, hemispheric, 105, 113, 115–17, 121, 162, 183–84, 246, 260–61, 275–84, 329; and aging, 184; cultural effects on, 282–84; females more symmetrical than males, 279, 281; greater variation within than between sexes, 280–81; left (functions), 261, 265, 270, 276, 292; left hemisphere slower to develop for males, 184, 280–81; left-hemispheric dominance, 166, 183–84, 265; left-hemispheric English language, 283–84; right (functions), 261, 292; right-hemispheric decline with aging, 184–85; right-hemispheric dominance, 117, 121; right-hemispheric Hopi language, 283–84; right-hemispheric learning of *kanji* ideographs, 283; sex differences, 162, 279–82, 330
Life science, 260, 323
Lifetime: developmental, 213–14; evolutionary, 140, 205, 214; physiological, 214

Mammalian biology, 128, 190; brains, 268; expansion and radiation, 328; sexual reproduction, 326
Mammals, 105, 108, 174–75, 183, 221, 256, 260, 324, 329; bighorn sheep, 185, 188, 205; canids, 256, 325; carnivore social behavior, 71–72, 76–77, 188, 217, 325; cetacea, 146, 148; herbivores, 188; mice, 190; moose, 188; mountain goats, 185, 188; rats, 190; superfelids, 158; ungulates, 194
Mathematics, 207, 289, 292–93, 307, 310, 326; Euclidean geometry (linear), 305; Riemannian geometry (curved), 305
Mediterranean Sea, 69, 111, 185
Mendelian genetics, 479; law of segregation and reassortment of genes, 292; quantum theory of biotic change, 299; rediscovery of, 292, 299
Mesolithic age, 195
Mesopotamia, 69, 98–99, 105, 107, 110, 112
Methodology, 236–37, 254, 297, 300, 324; data, 237–39, 310, 324; modal judicial age (defined), 237; research design, 236–37, 252, 300; research findings, 239–40; statistics, 300, 310
Moralism, 133–35, 138, 186, 235
Moralistic aggression, 134, 240
Music, 215
Myth(s): internal dynamics of, 116; psychiatric, 112; scientific, 122, 218; Victorian, 176

Natural science, 3, 31, 67, 120–21, 140, 297, 304; history of, 297–99, 322; macroparadigms, 298–300; paradigm change in, 298; revolutionary change in, for biology and physics, 300
Natural selection, 125, 127, 130–31, 134, 141, 146, 148–49, 153, 170–71, 173, 175, 178, 191, 203, 210–11, 213, 231, 292; distinguished from evolution, 189
Nature, 138, 146
Near East, 100, 106, 111, 121
Neolithic age, 110, 148, 153–54
Neologisms, 211
Neoteny, 101, 185, 190–91, 195, 197–98, 203; human males more than females, 280–81
Nepotism, 149–50
Netherlands, the, 84, 89, 97, 153
Neurophysiology, 128, 156, 162, 166–67, 223, 245, 258, 260, 273–74, 291, 313, 323
Neuropsychology, 245, 260
New generations as colonists, 146, 151, 179. *See also* Speciation: allopatric
Noah's Ark, 107, 111, 118
Nonverbal communication, 33, 35, 39–40, 54–57, 77, 81, 97, 102, 105, 112, 117, 219, 249, 251, 257, 267; and activation/arousal/emotion, 267, 275; Clever Hans, 253; memory of, 267; and sensory perception, 267; submissive gestures, 197
North America, 107, 147, 149, 183, 185–96, 203, 259, 328

Nutrition, 274–275, 284–88; effects of artificial nursing in urban ghettos, 285–86; effects on intelligence, 286–88; effects of malnutrition pre-, peri-, and postnatally, 285–88; malnutrition, 284–86; modal racial differences in 288

Objectivity, 165, 319
Ontogenetic development, 108, 183, 186, 214, 220, 330; and emotion, 268
Optimism, biology of, 117–20, 273; and endorphins, 117, 120; and manic diabetics, 119; and schizophrenia, 119

Paleontology, 171, 173, 200, 206, 208, 300
Palestine, 107, 110, 179; Jordan River Valley, 106; Mount Ararat, 107; Sea of Galilee, 106
Pastoralism, 100, 105, 121, 153–54, 170, 185
Perception, 113, 158–59, 165–66, 180, 246, 255, 259–64, 303, 315, 323, 330; and art, 197; extrasensory, 165; holistic, 289; misperception, 273; and music, 197; observer bias, 246, 262, 266, 268, 315, 322; observer paradox, 165, 259, 313, 315, 317–19, 322–23; observer projection, 259, 266, 272, 315, 317–19, 323; prenatal, 292–93; quantum physicist observer of quantum phenomena, 315; sensory, 107, 116–17, 165, 292, 304, 321; structure of, 262–63, 323; of verticality, 262–63
Phenotypes, 191, 200–202, 260, 274, 293, 327; wild type, the, 210
Phenotypic diversity, 206
Phenotypic selection, 202–4, 213
Phenotypic uniformity/genotypic heterogeneity, 210
Philosophy, 221–22; epistemology, 319, 322; ethics, 215, 217–18, 327; jurisprudence, 224–25; metaphysics, 322; methodological individualism, 50, 177; political, 6, 83, 99, 149, 200, 215, 297; of science, 122, 130–31, 177, 200, 208, 298, 306, 316, 322; teleology, 138
Physics, antiquarian, (500 BC–AD 1500), 262, 300, 304; Arabic, 297; Euclidean, 269; Greek (Aristotelian), 296–98, 308; medieval, 296–98
Physics, classical, (1500 BC–AD 1917), 269, 299, 304, 307–8, 320–21; directionless time, 305; Galilean, 269; gravity, 260, 262; mechanics, 301, 305–6; Newtonian, 269, 299, 311, 317, 320; relativity, 213–15, 246, 292, 301, 304–7
Physics, dominant macroparadigms: (modern physics) synthesis of quantum and relativity, 203, 303–4, 309, 311–20, 330; (neoclassical, relativity theory) deterministic, objective, 299; (quantum) indeterminate, subjective, probabilistic, 299, 312–13
Physics, quantum (AD 1900 to present), 113, 177, 207, 214, 246–47, 299, 304, 306, 308, 310–13, 319–22; Bell's theorem, 317, 319 (defined, 316); complementarity, 312; discontinuity, 311–12, 317; duality of particles and waves, 312; Hilbert space, 317; manifest phenomena, 317; measurables, 317; particles, 246, 317–18; quantized fields, 317; quantum mechanics, 301, 319; statics, 312; statistical, 313; time irrelevant, 311–12, 316. *See also* Field theory; Transactionalism
Physiology, 67, 128, 134–35, 156, 159, 191–92, 204, 215, 231, 258, 260, 268, 275, 297, 318, 320, 324
Play, 217; aggressive, 59; randomness, 273
Pleistocene epoch, 105, 136, 138, 155, 174; Upper, 188, 196
Pliocene epoch, 76
Poets, 221
Politica, 70, 99
Political attitudes, 150, 258–59; change in, 240; effects of aging, 186–87, 226–28, 231, 234, 239; females more liberal than males, 282; subjective age, 234; tough/tender-mindedness, 226
Political behavior, 5–9, 28–31, 38, 41, 53–54, 68, 71, 84–85, 94–96, 99, 101–2, 125–28, 151, 179, 183, 215, 223, 226, 246, 259, 296, 310, 324–25; aggression, 42, 46, 330; alienation, 237; defined, 98, 101; and the gender gap, 282; post-behavioralism, 264
Political change, 24, 42–43, 61–63, 90, 92, 126, 151, 175–80, 220–21, 328; impact of malnutrition on, 288; lag in, 240; revolutionary, 179, 220–21, 328
Political communication, 56, 81, 99
Political decentralization, 246, 294–95
Political decision-making, 57, 246, 254, 259, 272–74, 303–4, 319–23, 330
Political "development," 183, 219, 259
Political ecology, 9–11
Political elite(s), 125, 149, 151–52, 163, 167, 223, 226, 254, 270–71, 293; and differences among, 11–16
Political equality, 12, 81, 101, 177, 219–20, 246, 271, 275, 282, 302
Political evolution, 9–10, 63, 70, 99–102
Political integration, 150
Political leadership, 81, 86–87, 90–92, 102, 112, 118, 149, 273; charismatic, 86–87, 119; female, 281–82; and followership, 149, 282
Political organization, 98–101
Political participation, 150, 259; malnutrition effects on, 4, 288; sexual discrimination in, 271, 282
Political Psychology, 245
Political science, 7, 34, 41–42, 58, 67, 71, 80, 82, 84, 87, 96, 127, 130, 149–50, 154, 159, 163,

167, 170, 176, 183, 186, 215, 220–22, 245–46, 258–60, 262, 271, 294, 297–98, 311, 327–28, 330; antiquarian, 298, 300, 319; Aristotelian framework of, 298, 300, 319; and classical physics, 307–9, 321; field theory in, 308–11; life-science approach, 324–30; methodology in, 309–10; and modern physics, 311, 319, 321; paradigm change in, 264, 288, 296–304, 324; positivism, 264, 310; preclassical, 298; and preclassical biology, 307; Rashevsky models, the, 307
Political science, subdominant paradigms, 300–304; behavioral, 4–5, 300–304; conventional 5, 300, 303, 324; policy science 321; prebehavioralism, 303; traditional, 4–5, 300, 302, 324
Political socialization, 24, 42, 58, 67, 80, 102, 104, 150–51, 159, 215, 233–34, 259, 274; adult, 241; biocultural model of 223, 241; desocialization, 14, 81, 223; eras of 187, 236; primacy principle, the, 186, 234, 236, 238, 241
Political stasis, 62, 125, 221
Political theory, 83, 85, 89, 92, 94, 97–98, 179, 186, 200, 215, 221–22, 245, 259, 272, 297–98, 300, 304, 324; and quantum physics, 319–23; revolutionary, 24, 62, 176, 179; systems, 307
Political thinking, 245–47, 258, 264–74, 323, 330; female 281–82; language differences in, 284; quantizing, 246
Politics, 221–22, 258, 260, 319, 324–25; androgynous, 281, 330; international, 89–90, 154, 172, 272, 309–10, 330; interest-group, 150, 246, 294; life-science, 324–30; radical, 163, 168; violent, 288
Politics and the Life Sciences, 5, 170
Polities, 178, 221; American, 240; Australian, 235; British, 217; Western societies, 319
Polynesians, 163, 167
Population biology, 53, 128, 299; growth, global, 27, 146, 155; growth, spatial, 151; isolated populations, 174, 177–78, 329; population neoteny, 185
Predation, 47, 59–60, 74, 80, 91, 94, 99, 133, 138, 178–79, 190, 193, 213, 329; disease as, 178
Prescience, 120
President (U.S.), 95, 226, 229–30
Primate(s), 138, 190, 194, 260, 324–25; baboons, 74–75, 84–86, 93–94; evolution, 108; gorillas, 115, 188; langurs (Hanuman), 75, 84, 90–92; macaques, 48, 252; orangutans, 115; *Pan* (genus), 192; play, 59–60, 268, 273; politics, 68, 78, 83–94; pongids, 115; *Ramapithecus*, 78, socialization, 57, 223, 245; training, 249–54. *See also* Chimpanzee(s)
Primatology, 39, 42, 45–47, 51–53, 55, 59, 67–68, 71, 80, 83–87, 93–95, 97–99, 101–2, 114–15, 118, 128, 217, 248–57, 260
Progress, 153, 170, 176–78, 216, 219, 306
Prophets, 120–21
Psychiatry, 85–86, 95
Psychobiological politics, 258–88, 303–4, 307, 320–22
Psychobiology, 87, 96, 102, 105, 113, 121, 160, 162, 166–67, 177, 245–46, 259–71, 284, 288, 303, 319–23, 329–30
Psychology, 245, 258, 271, 297, 308, 318; biological, 302; comparative, 53, 128, 252, 256; developmental, 158, 170; Freudian, 272; Gestalt, 309; holistic, 308; nineteenth-century, 290; political, 259, 273; social, 245, 291, 301, 308
Psychometrics, 309; factor analysis, 309–10
Psychopharmacology, 13–14
Public: attitudes, 103; law, 236; policy, 150–51, 154–55, 169–70, 179, 259, 275, 302, 328

Quantification, 297, 301, 324; data, 297; dimensions of, 291; explanation, 297; prediction, 297

Rationalism: occupational psychosis, 203; Reaganomics, 264, 282; the "thousand-year rule," 158, 162, 170; trained incapacity, 186
Rationality, 157, 165, 170, 207–8, 231, 247, 256–57, 259, 264, 270, 272, 297, 306, 316, 319–20, 323, 329; logic, 165; optimality theory, 53
Reality, 164–65, 320; cognitive, 165; construction of, 165, 316, 319–20. *See also* Perception: observer projection
Recent epoch (geological), 105, 188
Relativity theory, 141, 308–9, 321, 330; four-dimensional (space-time) manifold, 305, 308, 313; simultaneity, 305, 308; time measures change, 305
Relativity theory, epigenetic, 213–15, 292; four-dimensional (space-time) organismic structure, 213–15; the organismic "present," 215
Religion, 119–22, 140, 227; Bible, the, 110, 134; Christmas legend, the, 114; Islam, 115, 179; Koran, the, 128; Old Testament, the, 68, 110, 133–34; Pentateuch(al) Hebrews, 111–12
Religious behavior, 68–69, 103, 113–21, 153; charisma, 119–20; leadership, 112, 118; meditation, 116; priests, 112, 120; neurobiological effects of ritual, 116–17
Renaissance, the, 70, 95
Rhythms, biological and social, 117

Science: Hellenic, 113; modern, 113, 121
Scientific criticism, 161, 163, 167, 187, 241
Scientific evidence, 216, 254, 258
Scopes trial, 6, 103

Scotland, 199
Selection: artificial, 170; diffusion of, 142; epigenetic, 191; group, 49, 53, 118, 129–30, 132–33, 136–37, 142–43, 151–52, 155, 172, 175, 194, 329; K, 125, 132, 185, 192–93, 196, 198; kin, 53, 127–29, 133, 136, 138, 142–43, 149–50, 153, 163, 175, 194; Malthusian, 201; neo-Lamarckian, 212; non-Darwinian, 141; r, 192, 196; sexual, 147; species, 173
Sex differences in behavior, 74, 76, 88, 90–91, 93–94, 102, 217, 259; dimorphism, human, 186, 195–96, 281; male composition, 68, 73, 81, 88–89. *See also* Female social structures
Siberia, 185
Simianizing hominid/human behavior, 94, 115, 121
Social classes, 215–17
Social Darwinism, 176–77, 216
Social hierarchy, 78, 86
Social science, 128, 134, 137, 139–44, 147, 154, 157–60, 163–64, 167, 179, 186, 189, 199, 215, 221, 224, 251, 260, 274, 291, 293–94, 297, 300, 304, 309, 313, 324, 327; cannibalization of, 160; field theory in, 309–10; and natural science, 3, 140, 289–90, 309
Societies, Anglo-American, 216
Sociobiology, 34–35, 38, 42, 53, 90–92, 119, 127–28, 136, 139–40, 144, 149, 160–63, 176–77, 246–47, 290, 293, 299, 326; altruistic behavior, 39, 53, 128, 131–38, 143, 145–46, 148–50, 153–55; hardcore, 127–33, 137–39, 141, 143, 145–47, 149–52, 154; softcore, 127, 133–36
Sociology, 128, 130, 153, 176, 226, 228, 233, 291, 297, 300, 309
Space: Cartesian, 261; Euclidean, 261–62, 310, 330; factorial, 310; quantized, 305
speciation: allopatric, 205; directed, 173–74; divergent, 173; interspecific, 146; quantum, 24, 126, 173–74, 327, 329; sympatric, 205
Speech physiology: Broca's area (articulation), 276; conscious control over behavior generally, 276; Wernicke's area (comprehension), 276
Supreme Court (U.S.), 103–4, 187, 208, 223, 225–31, 233, 236–38, 240; Burger Court, the, 237; decisions of, 236–41; "Four Horsemen," the, 230; liberal majority of 1960s, 240; "Nine Old Men," the, 230; Rehnquist Court, the, 103; Stone Court, the, 104; Taft Court, the, 104; tenure of justices, 229; token representation of minority groups on, 240; Warren Court, the, 103; White Court, the, 237
Switzerland, 56
Synergy, 176, 294

Technology, 216, 218, 273, 326, 328
Teenage gangs, 153
Territoriality, 46, 73, 77–78, 90–92, 101, 137, 142, 152; tribal competition, 100, 143, 152–53
Theology, 120–21, 164, 297; Christian, 216; Scientific Creationism, 68, 82, 103–4, 122, 297; Second Coming of Christ, the, 226
Third World, 147, 149, 259, 287, 329
Time, 213–15, 236; artificial measurement of change, 320; developmental, 213–14; epigenetic construction of human biology and culture, 316, 318; evolutionary, 140, 214, 256; geological, 154, 171, 221, 269, 325; historical, 209, 215, 256, 305, 325; phenotypically cognitively unidirectional, 316; physiological, 214, 269; quantum, 246–47, 305; relativity, 246
Transactionalism, 12, 27–28, 41, 49, 58, 63, 105–6, 136, 148, 156, 166, 172, 183, 186, 221, 223, 241, 260, 263, 275, 292, 320, 323, 329–30. *See also* Epigenesis
Transcendentalism, 164–5, 254. *See also* Dualism, mentalistic
Transition, from gathering-hunting to cultivation-pastoralism, 98–100, 105–8, 113, 121–22, 135, 151–52, 154, 169, 185, 329
Turkey, 107

United Kingdom, 89, 150
United States of America, 36–37, 89, 107, 120, 150–52, 154, 179, 195, 216, 308, 329; chronic malnutrition in, 287
Urbanism, 105, 112, 130, 136, 146, 152, 154
U.S.S.R., 16, 33, 149, 154, 216, 329

Voice stress analysis, 56

Warfare, 60–61, 118, 134, 138, 151–54; modern, 154; primitive, 152–54
Whiteheadian concept of process ("becoming"), 201, 206, 221
White House, the, 31, 81
Wolf and Man: Evolution in Parallel, 46
Wolf behavior, 52, 67, 76–77; social structure, 76
World War I, 142, 233–34
World War II, 146, 149, 179, 201, 233, 246, 313, 324

Zambia, 39
Zipf's Law, 193
Zoology, 128

Glendon Schubert grew up in upstate New York and received his degrees from Syracuse University. His forty-year postdoctoral teaching career began at UCLA, and includes appointments at Howard University, Syracuse University, Rutgers University, Franklin and Marshall College, the University of Minnesota, the University of Oslo, and Michigan State University; and also academic chairs as the William Rand Kenan, Jr., Professor at the University of North Carolina at Chapel Hill, and University Professor at York University in Toronto.

During the past twenty years he has been a biobehavioral political scientist with a principal interest in a life-science approach to politics, emphasizing primatology, feminism, and psychology. This approach is exemplified by his work as associate or advisory editor to such interdisciplinary and international journals as *The Behavior and Brain Sciences*, the *Journal of Social and Biological Structures*, and *Politics and the Life Sciences*. The author of more than one hundred articles published in professional journals, and the author or editor of twenty-five books, his books include (in press) *Primate Politics* and *Sexual and Feminist Politics: An Androgynous Biosocial Approach.*

Presently he is University Professor at the University of Hawaii, Manoa, and Research Professor at Southern Illinois University at Carbondale.